PLANTS INVADE THE LAND

Critical Moments and Perspectives in Earth History and Paleobiology

PLANTS INVADE THE LAND

Evolutionary and Environmental Perspectives

Editors

PATRICIA G. GENSEL
DIANNE EDWARDS

Columbia University Press
New York

Columbia University Press
Publishers Since 1893
New York   Chichester, West Sussex

Copyright © 2001 Columbia University Press
All rights reserved

Library of Congress Cataloging-in-Publication Data

Plants invade the land : evolutionary and environmental perspectives / Patricia G. Gensel
and Dianne Edwards, editors.
p. cm. — (Critical moments and perspectives in earth history and paleobiology series)
Includes bibliographical references (p.).
ISBN 978-0-231-11161-4 (pbk. alk. paper)
1. Paleobotany.   2. Plants–Evolution.   I. Gensel, Patricia G., 1944-   II. Edwards, D.
(Dianne)   III. International Organization of Paleobotany Conference (6th : 1996)   IV.
Series

QE905 .P55 2000
561–dc21   00-057021

Casebound editions of Columbia University Press books are printed on permanent
and durable acid-free paper.
Printed in the United States of America

To THE MEMORY OF Professor Winfried Remy, Paläontologisches Institüt, West-fälisches-Welhelms Universität, Germany, a distinguished and stimulating paleo-botanist who died in December, 1995. Professor Remy made numerous significant contributions to our knowledge about Carboniferous and Permian plants during his long career, and, more recently, to our knowledge of early land plants, based mainly on careful studies of the permineralized Rhynie Chert remains. His documentation, prepared in collaboration with his wife Renate Remy and colleagues Hagen Hass, Stepan Schultka, Hans Kerp, and Tom Taylor, of convincing gameto-phytes; of the detailed anatomy of some early land vascular plants, including several stages of their development; and of fungi and algae preserved in the Rhynie Chert represent major advances in elucidating aspects of early terrestrial ecosystems. He maintained a strong interest in the ecology of ancient plants, and the material he published on these aspects was based largely on the Rhynie Chert studies. Thus, the theme of this book, interaction of early land plants with their environment and other organisms in an evolutionary context, is especially appropriate for commemorating Professor Remy and his contributions.

CONTENTS

CONTRIBUTORS

(senior authors are indicated by an asterisk)

THOMAS J. ALGEO*. H.N. Fisk Laboratory of Sedimentology, Department of Geology, University of Cincinnati, Cincinnati, Ohio 45221-0013

JAMES F. BASINGER. Department of Geolgical Sciences, University of Saskatchewan, Saskatoon, Saskatchewan, Canada, S7N5E2

ROBERT BERNER*. Department of Geology, Yale University, New Haven, Connecticut, 06520-8109

CHRISTOPHER M. BERRY*. Department of Earth Sciences, University of Wales, College at Cardiff, Cardiff, CF1 3YE, United Kingdom

J. S. BRIDGE. Department of Geological Sciences, Binghamton University, Binghamton, New York 13902-6000

GILLIAN A. COOPER-DRIVER*. Department of Biology, Boston University, Boston, Massachusetts 02215

STEVEN G. DRIESE*. Department of Geological Sciences, University of Tennessee, Knoxville, Tennessee 37996

DIANNE EDWARDS*. Department of Earth Sciences, Cardiff University, Cardiff, CF10 3YE, United Kingdom

MURIEL FAIRON-DEMARET. Services Associés de Paléontologie de l'Université de Liège, Place du Vingt-Août 7, B-4000, Liège, Belgium

PATRICIA G. GENSEL*. Department of Biology, University of North Carolina, Chapel Hill, North Carolina 27599-3280

LINDA GRAHAM*. Department of Botany, University of Wisconsin, Madison, Wisconsin 53706

JANE GRAY. Department of Biology, University of Oregon, Eugene, Oregon 97405

D. H. GRIFFING. Department of Geography and Earth Sciences, University of North Carolina at Charlotte, Charlotte, North Carolina 28223-0001

SHOU-GANG HAO*. Department of Geology, Peking University, Beijing 100871, China

HAGEN HASS. Abteilung Paläobotanik am Geologisch-Palaeontologisch Institut und Museum, Hindenburgplatz 57-59, D-48143, Münster, Germany

CAROL L. HOTTON*. Department of Paleobiology, NHB MRC 121, National Museum of Natural History, Washington, D.C. 20560

FRANCIS M. HUEBER. Department of Paleobiology, NHB MRC 121, National Museum of Natural History, Washington, D.C. 20560

HANS KERP*. Abteilung Paläobotanik, Westfälisches-Wilhelms Universität, Hindenburgplatz 57-59, D-48143, Münster, Germany

MICHELE E. KOTYK. Department of Biology, University of North Carolina, Chapel Hill, NC 27599-3280

J. BARRY MAYNARD. H. N. Fisk Laboratory of Sedimentology, Department of Geology, University of Cincinnati, Cincinnati, Ohio 45221-0013

CLAUDIA I. MORA. Department of Geological Sciences, University of Tennessee, Knoxville, Tennessee 37996

VOLKER MOSSBRUGGER. Department of Geology and Palaeontology, Eberhard Karls-Universität, Siwarrtstrasse 10, D-72076, Tübingen, Germany

STEPHEN E. SCHECKLER. Departments of Biology and Geological Sciences, and Museum of Natural History, Virginia Polytechnic Institute and State University, Blacksburg, Virginia 24061-0406

PAUL SELDEN. Department of Earth Sciences, University of Manchester, Manchester M12 9PL, United Kingdom

WILLIAM SHEAR*. Department of Biology, Hampden-Sydney College, Hampden-Sydney, Virginia 23943

CHARLES WELLMAN. Department of Earth Sciences, The University of Sheffield, Brook Hill, Sheffield, S37HF, United Kingdom

PLANTS INVADE THE LAND

Introduction

Patricia G. Gensel

The invasion of the land by organisms is one of the major events in the evolution of life, permanently altering the conditions on Earth and essentially resulting in the diversification of the kingdom Plantae as presently defined. As a result of recent research, remains of terrestrial plants occur as early as basal Ordovician. The record of terrestrial animals now dates from the Middle–Late Silurian, and an earlier timing is possible. It has been postulated for some time that fungi and other types of organisms also may have inhabited the land since at least 450 million years ago. That terrestrial plants strongly influenced environmental conditions on Earth, essentially initiating major global change, is well established, but details remain to be worked out. This book focuses on a current understanding of the fossil record of land plants and other organisms; recent environmental interpretations based on geology, geochemistry, and fossils; assessment of the impact of plants on geological processes; and the nature of interactions of plants with fungi, animals, and their environments.

The chapters are based on talks given at a symposium held during the Fifth International Organization of Paleobotany Conference in 1996, and a few more authors were invited to cover additional pertinent topics. Our goals for the symposium and this book are to consider the present state of knowledge and recent develop-

ments concerning the evolution of plants on the land from both an evolutionary and an environmental (geological) perspective, including the interplay of plants and their environments, or plants and other coeval organisms (animals, fungi), and to suggest future avenues of investigation. The time span is Ordovician—Upper Devonian, with emphasis on Silurian through Middle Devonian, thus covering the period of major global change that resulted from the diversification of plants and their impact on the environment. Many of the chapters are synthetic, but these also include new primary data, while others document new findings. We hope this combination of approaches and topics presents a comprehensive view of what we presently know about early terrestrial organisms and their environment and will serve both as a useful reference and as a stimulus for additional research.

The first several chapters set the stage by documenting early land inhabitants ranging from protistan to plant and animal; this is done via a survey of the record of earliest terrestrial plants (basal Ordovician to middle Devonian), the description of new taxa or newly discovered plant structures, and assessment of biogeographical patterns. Later chapters deal with data and interpretation of critical structural, biochemical, and/or physiological adaptations necessary for terrestrial habitation; environmental parameters such as atmosphere and substrate types; plant—animal—soil interactions; and impact of environmental change caused by plants inhabiting the land. Themes emerging from these chapters include the following:

1. The value of careful study of morphological and anatomical detail for inferring growth patterns and ecological adaptations, and for whole-plant reconstruction and taxonomic clarity.

2. The acquisition of new data. New data about the basal regions [i.e., rhizomes (horizontal stems) and rooting structures] in early land plants are documented more extensively than previously, both from a plant structural and from an ecological/geological viewpoint. Spores as an important source of data about presence, distribution, lineage, and evolutionary change in early land plants also are highlighted.

3. The recognition of up-to-now overlooked evolutionary events. This is especially significant for the Middle Devonian, which is more than a continuation of what happened in the Late Silurian–Early Devonian. Turnover of taxa and a radiation of trimerophyte-derived lineages and new types of lycopsids, and the beginnings of stratification of vegetation with the onset of small tree–sized plants, were initiated near the end of the Emsian and amplified during the Middle Devonian.

4. The power of integration of several types of data. Considering the dispersed spore record with the megafossil record of early land plants has altered ideas on timing of origin and may aid in interpretation of affinities of early embryophytes. Considerations of the extant closest putative relative(s) to land plants and adaptations to terrestriality may show us ways that the invasion of land was achieved. The animal record and the plant record both contribute uniquely to examining the probable role of animals in early terrestrial ecosystems vis-à-vis plants—Were there any early herbivores? What do the sediments tell us about early terrestrial environments and biota, and what are the possible factors controlling the distribution of taxa on the landscape, biogeochemical changes in early soils, and contributions of both to the early atmosphere? Attention is given to root–soil interactions, to plant–atmosphere interrelationships, and to biochemical pathways that may have been important to the success of plants on the land. How plants may have influenced events in other parts of the ecosystem (e.g., as causal factors for the Middle–Late Devonian marine anoxic events and biotic crises) is also developed.

These chapters demonstrate that while the current state of knowledge is sufficient to address broad questions and provide new information, much more primary data and more integrated studies are needed. Continuation of such studies and incorporation of new approaches should yield even more insights into the nature and dynamics of early terrestrial ecosystems.

2

Embryophytes on Land: The Ordovician to Lochkovian (Lower Devonian) Record

Dianne Edwards and Charles Wellman

This chapter is primarily concerned with the documentation of the record of land plants from their earliest occurrences in the Ordovician to the end of the Lochkovian in the Lower Devonian. It covers the initial radiation of the embryophytes in the Ordovician; the emergence of tracheophytes, including the lycophytes, in the Silurian; the proliferation of axial plants with terminal sporangia around the Silurian–Devonian boundary; and, at the end of the Lochkovian, the first major radiation of the zosterophylls. The record remains a very patchy one and makes any conclusions on phylogenetic relationships, evolutionary and migration rates, and global distribution highly conjectural. The fossils themselves are usually very fragmentary, the plants sometimes represented by spores only, so that whole-plant reconstructions are speculative and concentrate on the distal aerial parts. There is very little, if any, information on possible subterranean parts, although the earliest lycophytes are presumed to have roots. Estimates of primary productivity and trophic relationships with consumers and decomposers in the early terrestrial ecosystems have very little factual foundation (see chapter 3). Indeed, although modeled concentrations of atmospheric carbon dioxide in the Phanerozoic emphasize the dramatic impact of plants on land via increase in chemical weathering (chapter 10), quantitative data on coeval vegetation and extant analyses are sadly lacking. But reservations/inadequacies such as these should not obscure the progress in the last decade in documenting and understanding early phases of land plant vegetation.

The earliest direct evidence for land plants comprises spores and fragments (phytodebris) that were preserved in considerable numbers in a variety of Ordovician and Early Silurian environments, both continental and marine. Fossils of the producers are first recorded in the Llandovery (Early Silurian) some 35 million years after the first microfossil records, but at approximately the same time as the first unequivocal trilete monads believed to derive from tracheophytes.

THE MEGAFOSSIL DATABASE

Since Edwards (1990) compiled a database for Silurian and Gedinnian (Lochkovian) assemblages (tables 2.1, 2.2), developments have included both new and paleogeographically satisfyingly widespread localities, and additions or revisions to existing assemblages. These are summarized next; assemblages described from localities 1 to 4 are new.

1. Bathurst Island, Canadian Arctic Archipelago (Basinger et al. 1996)

Ludlow to Pragian rocks have yielded abundant, well-preserved assemblages dated by graptolites. The Ludlow examples in the lower Bathurst Island Beds contain the oldest fertile plants of vascular plant aspect in the New World. One

Table 2.1.　Silurian localities with plant megafossils

Locator*	Age	Authors	Geographic Area	Composition	Basis for Age
1.	Llandovery (?Telychian)	Schopf et al. (1966)	Maine, USA	*Eohostimella heathana*	Invertebrates
2.	Homerian	Edwards et al. (1983)	Tipperary, Ireland	*Cooksonia* sp.	Graptolites
3.	Late Wenlock/Ludlow	Toro et al. (1997)	S. Bolivia	Sterile rhyniophytes	Graptolites
4.	Gorstian	Edwards et al. (1979)	Powys, Wales	*Cooksonia pertoni, C.* sp.	Graptolites
4.	Ludfordian	Edwards and Rogerson (1979)	Powys, Wales	*Cooksonia* sp., *Steganotheca striata*	Invertebrates, acritarchs
5.	?Late Ludow	Tims and Chambers (1984)	Victoria, Australia	*Baragwanathia longifolia, Salopella australis, Hedeia* sp., zosterophylls	Graptolites
6.	Ludlow/?Ludfordian	Edwards (work in progress)	North Greenland	*Salopella* sp.	Graptolites
7.	Ludlow	Basinger et al. (1996)	Bathurst Island, Arctic Canada	Cooksonioid types	Graptolites
8.	?Ludlow/Prídolí	Morel et al. (1995)	Tarija, Bolivia	*Cooksonia* cf. *caledonica*	Field relation/ spores
9.	Prídolí (?ultimus)	Obrhel (1962) Schweitzer (1983)	Bohemia, Czechoslovakia	*Cooksonia C. bohemica*	Graptolites Graptolites
10.	Basal Prídolí		Ludford Lane, Ludlow, England	3 new taxa (rhyniophytoid)	Spores/inverte- brates
10.	Prídolí (?ultimus)	Lang (1937); Fanning, Edwards, and Richardson (1990, 1991); Fanning, Richardson, and Edwards (1991)Edwards (1979)	Hereford, England	*Cooksonia pertoni, C. cambrensis, Salopella, Pertonella dactylethra, Caia langii*	Invertebrates and spores
10.	Prídolí (early)		Dyfed, Wales	*Cooksonia pertoni, C. cale- donica, C. cambrensis, C. hemisphaerica, Tortili- caulis transwalliensis, Psilophytites* sp.	Spores
10.	Prídolí (early)	Edwards and Rogerson (1979)	Dyfed, Wales	*Cooksonia* sp, *Steganotheca striata*	Spores
10.	Prídolí (?early)	Rogerson et al. (1993)	Shropshire (Little Wallop), England	*Cooksonia pertoni*	Spores and invertebrates
11.	Prídolí (bouceki)	N. Petrosyan (pers. comm.)	Kazakhstan, USSR	*Cooksonia* sp., *Zostero- phyllum* sp.	Graptolites
12.	Prídolí (late)	Senkevitch (1975)	Balkhash area, Kazakhstan	*Cooksonella* sp., ?*Baragwa- nathia* sp., ?*Taeniocrada* sp., *Jugumella burubaensis*	Graptolites
12.	Prídolí (late)	Cai et al. (1993)	Junggar Basin, Xinjiang, China	*Cooksonella* sp., *Junggaria spinosa,* ?*Lycopodolica, Salopella xinjiangensis, Zosterophyllum* sp.	Graptolites
13.	Prídolí (late)	Ishchenko (1975)	Podolia, USSR	*Cooksonia pertoni, C. hem- isphaerica, Eorhynia (Salopella),* ?*Zosterophyllum* sp., *Lycopodolica*	
14.	Prídolí (?late)	Banks (1973)	New York State, USA	*Cooksonia* sp.	Conodonts
15.	Prídolí	Daber (1971)	Libya	?*Cooksonia* sp.	Graptolites

* Locations are noted by numbers on figures 2.5 and 2.6.

example is a smooth main axis about 2 mm in diameter with alternate fertile complexes. Each comprises a sporangium, oval in face view, with a possible narrow marginal band and subtended by a short, presumed distally curved axis or stalk that increases in diameter below the sporangium. It thus broadly resembles a *Cooksonia caledonica* sporangium in shape, while sporangial organization is closer to a *Zosterophyllum* and is more complex than coeval *Cooksonia*. In this preliminary report, plant localities are also recorded in the upper-most Lochkovian in the upper member of the Bathurst Island Beds, a time of major diversification in other parts of the Old Red Sandstone Continent (e.g., Allt Ddu, Brecon Beacons, in Wellman, Thomas, et al. 1998).

Table 2.2 Lochkovian localities with plant megafossils

Locator*	Age	Authors	Geographic Area	Composition	Facies	Basis for Age
1.	*Uniformis* zone	Obrhel (1968)	Bohemia, Czechoslovakia	*Cooksonia downtonensis* (=*C. hemisphaerica*)	Marine (deep)	Graptolites
2.	*Micrornatus-newportensis* (lower–middle)	Lang (1927), Edwards (1975)	Forfar, Scotland	*Zosterophyllum myretonianum, Cooksonia caledonica*	ORS (internal facies)	Spores, fish
2.	*Micrornatus-newportensis* (lower–middle)	Edwards (1972)	Arbilot, Scotland	*Z. fertile*	ORS (internal facies)	Spores, fish
3.	*Micrornatus-newportensis* (lower)	Edwards and Fanning (1985)	Targrove, Shropshire, England	*Tortilicaulis transwalliensis, Resilitheca, Uskiella reticulata, Tarrantia salopensis, C. hemisphaerica, C. pertoni, C. cambrensis, C. caledonica, Salopella marcensis*	ORS fluviatile (distal)	Spores, fish
3.	*Micrornatus-newportensis*	Edwards et al. (1994)	Brown Clee Hill, Shropshire, England	*Salopella* cf. *marcensis, Tortilicaulis offaeus, Resilitheca salopensis, C. pertoni, Grisellatheca salopensis, C. hemisphaerica,* cf. *Sporogonites, Pertonella* sp., *Fusitheca fanningiae, Culullitheca richardsonii,* Spherical sporangia × 5, oval sporangia × 5, *Tarrantia salopensis,* reniform sporangia × 2, various other unnamed rhyniophytoids	ORS fluviatile (distal)	Spores
3.	?	Edwards and Kenrick (unpublished)	Cwm Mill, Gwent, Wales	*Z. fertile, Cooksonia* sp.	ORS fluviatile (distal)	?
4.	*Micrornatus-newportensis* (upper)	Leclercq (1942)	Nonceveux, Belgium	*Z. fertile*	Marine (Rhenish)	Spores, fish
3.	Upper Gedinnian	Edwards and Richardson (1974) and unpublished	Newton Dingle and environs, Shropshire, England	*Z.? fertile, Salopella allenii*	ORS fluviatile (medial)	Spores
3.	*Breconensis-zavallatus*	Edwards and Kenrick (unpublished)	Allt Ddu, Brecon Beacons, Powys, Wales	*Salopella allenii, C.* cf. *caledonica, Deheubarthia splendens, Gosslingia breconensis, Z. fertile, Z.* sp., *Uskiella* sp.	ORS fluviatile (medial)	Spores
4.	*Breconensis-zavallatus*	Steemans and Gerrienne (1984)	Gileppe, la Vesdre, Belgium	*Gosslingia breconensis* and other abundant remains (work in progress)	Marine (Rhenish)	Spores
5.	Upper Gedinnian	Schweitzer (1983)	Rhineland, Germany	*Drepanophycus spinaeformis, Taeniocrada* sp.?, *Zosterophyllum rhenanum*	Marine littoral	Field relations
6.	?Basal Gedinnian (Ainasu)	Senkevitch (1978)	Balkhash, Kazakhstan	*Cooksonella sphaerica, Taeniocrada pilosa, Jugumella burubaensis*	Marine	Faunas, graptolites
7.	?Upper Gedinnian (Kokbaital)	Senkevitch (1978)	Balkhash, Kazakhstan	*Tastaephyton bulakus, Taeniocrada pilosa, Mointina quadripartita, Jugumella jugata, J. burubaensis, Balchaschella tenera*		
7.	Gedinnian undet.	Stepanov (1975)	Kuzbass, Siberia	*Zosterophyllum, C. pertoni, Stolophyton acyclicus, Juliphyton glazkini, Uksunaiphyton ananievi, Pseudosajania pimula, Salairia bicostata*	Marine	?

continued

Table 2.2 (continued)

Locator*	Age	Authors	Geographic Area	Composition	Facies	Basis for Age
8.	Gedinnian	Li and Cai (1978)	E. Yunnan, south-west China	*Zosterophyllum* sp.	?	?
9.	?Lochkovian undet.	J. Tims (pers. comm.)	Tyers, Victoria, Australia	*Baragwanathia longifolia, Zosterophyllum* n. sp., *Baragwanathia* n. sp.	Marine	Faunas (corals)
10.	Gedinnian undet.	Høeg (1942)	Spitsbergen	Sterile remains only: *Hostinella, Taeniocrada, Zosterophyllum*	ORS (internal)	Spores
11.	Pragian—pre-Pragian ? Gedinnian	Janvier et al. (1987)	Viet Nam	Undet. terminal sporangia	ORS	Plants and fish
12.	Gedinnian	Alvarez-Ramis (unpublished abstract, (1988)	Badajoz, Spain	*Sciadophyton steinmanni*	?	—
13.	?Lochkovian	Edwards et al. (work in progress)	Precordillera, Argentina	2 new rhyniophytoids, 1 plant with enations	Marine	Spores
14.	Uppermost Lochkovian	Basinger et al. (1996)	Bathurst Island, Arctic Canada	Not specified	Marine	Invertebrates including graptolites
15.	?Uppermost Silurian, Lochkovian	Mussa et al. (1996), Gerrienne (1999)	Paraná Basin, Brazil	*Cooksonia* cf. *pertoni, Sporogonites* sp. nov., cf. *C. cambrensis, Pertonella* sp., *Salopella* sp., ?leafy and spinous axes	?Marine	Spores

* Locations are noted by numbers on figures 2.7 and 2.8.

2. Southern Bolivia (Morel et al. 1995; Toro et al. 1997)

Dating is less secure, but field relationships and hence indirect invertebrate evidence points to an upper Ludlow-Prídolí age for the Kirusillas Formation at Tarija, southern Bolivia, which contains abundant plant debris at certain horizons and three well-preserved specimens of *Cooksonia* with marginal thickening characteristic of *Cooksonia caledonica* (compare figures 2.1B and 2.3A) (Morel et al. 1995). Poorly preserved spores place a maximum Gorstian age on the plant horizons. Toro et al. (1997) have described plant debris from the same formation further north at Cochabamba, comprising unbranched axes and *Hostinella* associated with *Rhynia* sp., *Cooksonia* sp., *Zosterophyllum* sp., and *Drepanophycus* sp., although examination of their illustrations suggests that at least some of their identifications fall into the "optimistic category" of Edwards (1990). The fossils are highly fragmented, partially coalified, and associated with graptolites indicative of a Late Wenlock—Early Ludlow age.

3. Ludlow, Shropshire, England (Jeram et al. 1990)

Historically, this is an important locality in that it contains the Ludlow Bone Bed, once thought to mark the Silurian—Devonian boundary, but now approximately equivalent to the Ludlow—Prídolí boundary. Excavations to make the bank safe yielded fresh rock just above the bone bed, which on maceration produced at least three new rhyniophytoid genera (e.g., figure 2.2A) and smooth sterile axes with stomata, the earliest yet demonstrated (Jeram et al. 1990), and numerous discoidal and fusiform spore masses (figure 2.4E,J,L), containing the dispersed hilate monads *Laevolancis divellomedia* s.l. (Wellman, Edwards, and Axe 1998b). The locality has also yielded the earliest body fossil evidence for trigonotarbids and various myriapods (see chapter 3).

4. Parana Basin, Brazil (Mussa et al. 1996)

Assemblages from two localities in the upper part of the Furnas Formation, dated by spores either as latest Silurian or Early Lochkovian, are

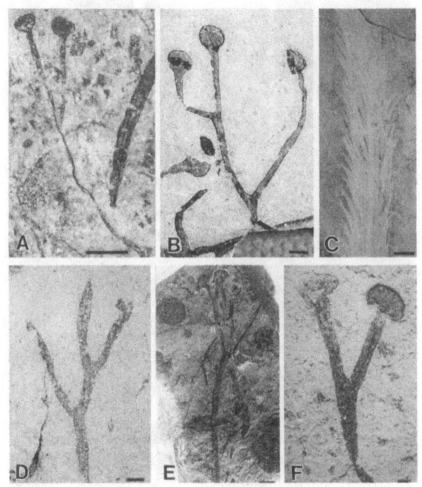

Figure 2.1.
Coalified Silurian fossils (except for the impression of *Baragwanathia*).
A: *Cooksonia cambrensis.* Freshwater East, Wales. Prídolí. NMW77.6G.21. Scale bar
= 1 mm. **B:** *Cooksonia* cf. *caledonica.* Tarija, Bolivia. LPPB 12744. ?Ludlow/Prídolí.
Scale bar = 1 mm. **C:** *Baragwanathia longifolia.* Victoria, Australia. Ludlow. Scale
bar = 10 mm. **D:** *Salopella* sp. Wulff Land, Greenland. Ludlow. MGUH. 17464
from GGU 319207. Scale bar = 1 mm. **E:** *Cooksonella sphaerica/Junggaria spinosa.*
North Xinjiang, China. Prídolí. D4. Scale bar = 10 mm. **F:** *Cooksonia pertoni.* Hereford, England. Prídolí. NMW90.41G.1. Scale bar = 1 mm.

currently being reinvestigated by Philippe Gerrienne (1999). Abundant coalified remains are
dominated by naked, sometimes isotomously
branching axes terminated by sporangia of
Cooksonia pertoni morphology. Also recorded are
a new species of *Sporogonites*, in which longitudinally striated sporangia are covered by minute
spines; cf. *Cooksonia cambrensis*; *Pertonella* sp.;
Salopella sp.; and axes with possibly microphyllous emergences or spines. The sporangia lack
any spores that are considered critical for the
identification of these taxa to specific or even
generic level.

5. Xinjiang Province, northwest China
(Cai et al. 1993)

The Wutubulake Formation has been dated as Prídolí, based on graptolites, and was probably
deposited on one of the Kazakhstan plates, thus
explaining similarities in the plants with those
from the Balkhash area in Kazakhstan (Senkevich
1975, 1986). Cai et al. (1993) presented detailed
descriptions of *Junggaria spinosa* Senkevich. The
most complete specimens show pseudomonopodial branching with sporangia characterized
by a broad margin terminating lateral dichotomously branching structures (figure 2.1E).

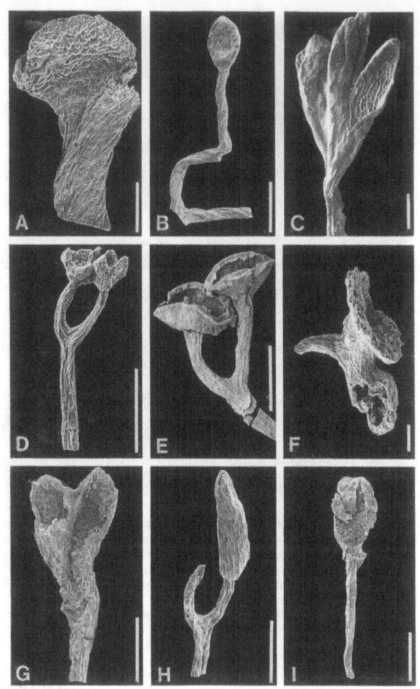

Figure 2.2.

Scanning electron microscopy of mesofossils. **A:** New taxon, Ludford Lane, England. Prídolí. NMW96.11G.7. Scale bar = 500 µm. (B) through (I) from north Brown Clee Hill, England. Mid-Lochkovian. **B:** *Tortilicaulis offaeus.* NMW99.11G.1. Scale bar = 500 µm. **C:** *Salopella* cf *marcensis.* NMW93.98G.8. Scale bar = 100 µm. **D:** New taxon. NMW96.5G.6. Scale bar = 500 µm. **E:** *Cooksonia pertoni.* NMW94.60G.14. Scale bar = 1 mm. **F:** Sporangia with ornamented *Velatitetras.* NMW96.11G.3. Scale bar = 100 µm. **G:** *Grisellatheca salopensis* with smooth *Velatitetras.* NMW94.76G.1. Scale bar = 500 µm. **H:** *Fusitheca fanningiae* containing permanent dyads. NMW97.42G.4. Scale bar = 500 µm. **I:** *Culullitheca richardsonii* with permanent dyads. NMW96.11G.6. Scale bar = 500 µm.

Variation in the degree of lobing of the margin may indicate the presence of two species. Sterile axes at the locality show isotomous, pseudo-monopodial, and more complex H-type, K-type, or clustered branching. Many bear a longitudinal central line, hairs, and small spines. Of particular interest is a single axis covered by crowded linear emergences up to 3 mm long with slightly swollen bases. It resembles a microphyllous lycopod, but there is no anatomical evidence to support such an affinity. Similar remains of approximately the same age have been assigned to *Protosawdonia* (Ishchenko 1975, in Podolia), *Lycopodolica* (Ishchenko 1975), or even *Baragwanathia* (Senkevich 1975; Cuerda et al. 1987). They demonstrate vegetative complexity in Silurian plants but in the absence of anatomical and reproductive information it is unwise to assume this is at a lycophyte grade. The use of *Baragwanathia*, a reasonably well-defined taxon for this kind of fossil, is particularly misleading.

6. Wallop Hall Quarry, Long Mountain, Shropshire (Rogerson et al. 1993)

Although concentrating on *in situ* spores, Rogerson et al. (1993) illustrate quite well-preserved, newly collected, coalified specimens of *Cook-sonia pertoni* in a fine-grained gray matrix in the Prídolí Temeside Formation. The presence of a row of similarly oriented sporangia, some isolated, others attached to axes of varying length, infers a much-branched specimen with a minimum height of 25 mm. Because of its low diversity, the locality is atypical for one containing Prídolí plants, although other very fragmentary specimens with terminal sporangia may derive from additional species of *Cooksonia*. This may demonstrate taphonomic bias in that sediments at localities such as Perton Lane and Ludford Lane were deposited on the shelf, and those at Long Mountain in the basin (Richardson and Rasul 1990). Mesofossils have not been isolated.

7. Perton Lane, Woolhope Inlier, Hereford (Fanning, Richardson, and Edwards 1990, 1991).

New collecting from the type locality of *Cooksonia* (figure 2.1F) at Perton Lane has revealed two new genera of rhyniophytoids with *in situ* spores (*Pertonella* and *Caia*) and diversity in *Salopella* based on *in situ* spores (Fanning, Richardson, and Edwards 1991).

8. Targrove, Ludlow, Shropshire (Fanning et al. 1992)

Exhaustive sampling at a very small exposure revealed a diverse assemblage dated as mid-Gedinnian (Lochkovian) on spores. Historically, this was the exposure where Lang (1937) found tracheids in sterile axes and concluded that *Cooksonia hemisphaerica* was a vascular plant, this then being the only higher plant taxon recorded at the locality. The fossils are fragmentary and coalified, the large number of isolated sporangia allowing circumscription of morphological variation and the demonstration of *in situ* spores. The nine fertile taxa present all had terminal sporangia and lacked evidence of their conducting tissues. They were therefore described as rhyniophytoid. Isolated strands of G-type tracheids were also recorded.

9. Stream section on the north side of Brown Clee Hill, Shropshire (Edwards et al. 1992; Edwards 1996, and references therein; Wellman, Edwards, and Axe 1998a)

To date, at least 26 taxa have been discovered at this locality (see list in Edwards 1996). Many are known from single specimens and, while these are well preserved and often with *in situ* spores, we have been reluctant to name them on the basis of such limited data. However, we feel it important to illustrate such mesofossils (figure 2.2B–E), as they demonstrate far greater diversity in ground-hugging vegetation of diverse affinity (see parts 3 and 4 of General Observations on Gross Morphology, Growth Habit, and Reproductive Biology, Based on the Limited Meso- and Megafossil Record, later in this chapter). Exceptions to our naming protocols are the sporangia containing permanent dyads and tetrads, the first evidence for the morphology of the plants that produced spores similar to those found isolated in Ordovician and Silurian rocks (Wellman, Edwards, and Axe 1998a; Edwards,

Wellman, and Axe 1999). Representatives are shown in figure 2.2F–I.

10. Viet Nam (Janvier et al. 1987)

The Dô Son Formation containing fragmentary plants, some attributed to *Cooksonia*, was originally considered Lower Devonian. However, on the basis of more recently collected placoderms, large chasmatispid arthropods, and herbaceous lycophytes (named *Colpodexylon* cf. *deatsii*), it is now, at least in part, thought to be Givetian or Frasnian (Janvier et al. 1989).

11. Western Argentina (Cuerda et al. 1987)

The report of *Baragwanathia* from the Villavicencio Formation of the Precordillera of western Argentina was based on a coalified unbranched axis covered with crowded enations, approximately 2 mm long with probably truncated tips. Recollecting at the San Isidro locality in Mendoza has failed to produce further examples but has yielded a relatively diverse assemblage (eight taxa) of axial fossils, some of which are fertile (Morel et al., in press). None can be assigned with confidence to existing taxa but some sporangia resemble those of *Salopella, Tortilicaulis*, and *Sporogonites*, all of which require anatomical or spore characters for verification. Particularly noteworthy is a much-branched specimen comprising longitudinally and irregularly furrowed isotomous axes with terminal sporangia. The age of the locality based on field relationship and palynology (Rubinstein 1993a,b) is probably Lochkovian, but the spores are poorly preserved, and most range between mid-Lochkovian and end of the Pragian (Rubinstein 1993b). Most of the fossils are axial fragments of similar dimensions to those in coeval sediments in the present northern hemisphere. Their abundance, also noted at another Lochkovian locality at Villavicencio to the north where they are associated with a possible isolated sporangium of *Salopella*, suggests abundant vegetation on land in the Early Devonian of southwestern Gondwana.

12. Libya (Streel et al. 1990)

The diverse assemblages in the Acacus (Klitzsch et al. 1973) and Tadrart-Emi Magri (Lejal-Nicol 1975; Lejal-Nicol and Massa 1980) formations were dated as Silurian and Lower Devonian on field relationships, although the organization of the plants themselves is typical of much younger assemblages elsewhere. The dating of these formations remains contentious. In the most recent review summarizing the palynological data for the area (although not the rocks with the plants) and revising the lithostratigraphical correlations, Streel et al. (1990) concluded that the plants themselves provide the most reliable evidence for dating—namely, a Pragian or younger age for the Acacus Formation and a mid-Devonian one for the Tadrart-Emi Magri.

AFFINITIES BASED ON MEGAFOSSILS

The morphological simplicity of Siluro-Devonian axial forms and absence of modern representatives prevent ready establishment of affinity. In addition, the lack of anatomical information for all fertile Silurian fossils and many Lower Devonian ones precludes confident placing, even in the tracheophytes. This applies both to the rhyniophyte complex—namely, plants with terminal sporangia and axial dichotomously branching systems (cooksonioids, rhyniophytoids, polysporangiates)—and to a lesser extent to those with enations reminiscent of the microphyllous lycophytes. Indeed, even though Lochkovian *Cooksonia pertoni* has been demonstrated to possess tracheids, it cannot be assumed that Silurian representatives (see, e.g., figure 2.1A) were vascular, because sequential acquisition of homoiohydric characters is a possibility (e.g., Raven 1993). In that it has already been demonstrated that the ornament on crassitate spores of *Cooksonia pertoni* differs in Silurian and Devonian plants, it is likely that there were also internal anatomical changes relating to increased efficiency in metabolism and water relationships, but that these were not apparent in plants of such simple morphology. It is, however, equally probable that the problem of affinity relates to poor preservation, coupled with lack of detailed scrutiny. The latter is time consuming (and expensive in scanning electron microscope time), but the recent demonstration of a new kind of vascular tissue in an axial mesofossil (Edwards, Wellman, and Axe 1998) is encouraging.

Among the "leafy" and presumed microphyllous fossils, *Baragwanathia*, with its more or less continuous record (figure 2.1C) from Ludlow into Emsian in Australia, is the least contentious (Tims and Chambers 1984), but the smaller and more fragmentary examples found in small numbers in isolated and geographically widespread localities (e.g., Argentina and Xinjiang) are more difficult to evaluate (see localities 5 and 11 in the preceding list), and in the absence of diagnostic characters, the use of relatively well circumscribed taxa such as *Baragwanathia* should be avoided. *Baragwanathia* itself is not recorded in Laurentia until the Emsian (Hueber 1983a); the cosmopolitan lycophyte *Drepanophycus spinaeformis* (Schweitzer 1983) appears in the Rhineland at the end of the Gedinnian (latest Lochkovian).

The earliest fertile zosterophylls, *Z. myretonianum* and *Z. ?fertile*, occur toward the base of the Lochkovian in Scotland (figure 2.3E) and the Welsh Borderland (figure 2.3B; lower-middle *micrornatus—newportensis* zone) (see Edwards 1990; Wellman et al., 2000) with *Z. fertile* appearing in the upper part of the zone in Belgium (Leclercq 1942).

Possible sterile H-shaped branching systems (equivocal in that it is not known whether such configurations are confined to zosterophylls) are recorded earlier. There are sporadic occurrences of *Zosterophyllum* cf. *fertile* spikes later in the Lochkovian in the Welsh Borderland, but a diversification of zosterophylls (figure 2.3D), including the earliest record of nonstrobilate and spiny forms (e.g., *Gosslingia* and *Deheubarthia*), occurs near the top of the Lochkovian (Allt Ddu, Brecon Beacons, South Wales). The latter assemblage also contains *Salopella* (figure 2.3C; Kenrick 1988). The vast majority of these fossils comprise

coalified or iron-stained impressions; identity is based on comparative morphology.

To date, those records in southern Britain represent the most complete succession of Lochkovian assemblages. They include two localities, Targrove and north Brown Clee Hill, where rhyniophytoids predominate, and which show diversification in sporangium form and dehiscence, and many taxa with *in situ* spores. Detailed scrutiny of those assemblages clearly distorts diversity curves; examination of dispersed spore assemblages suggests more widespread occurrences of the plants recorded as mega- and mesofossils, and that their absence in the record is probably a consequence of facies bias.

What is clear is that by the end of the Lochkovian, the vegetation structure that would dominate Pragian landscapes was already established. The younger stage is also marked by the appearance and diversification of the trimerophytes (Gerrienne 1997), plants that later in the Emsian show marked increase in size and height, and hence productivity, with greater impact on substrate (see chapter 5).

THE MICROFOSSIL DATABASE

Spores

The earliest dispersed spores, occurring some 40 million years prior to the oldest generally accepted plant megafossils [Llanvirn (mid-Ordovician); Strother et al. 1996] have been termed cryptospores, the name reflecting their unfamiliar appearance relative to the trilete monads (typical of vascular plants) and a lack of knowledge regarding the nature of their producers (Richardson 1996a and references therein). They comprise monads and obligate dyads and tetrads, which are either naked and usually laevigate, or enclosed within a thin, laevigate or variously ornamented envelope (e.g., figure 2.4G–I). The nature of cohesion between these permanently united spores is unclear (Gray 1991; Wellman 1996). They are termed unfused if there is a superficial line marking the junction between the spores, with cohesion

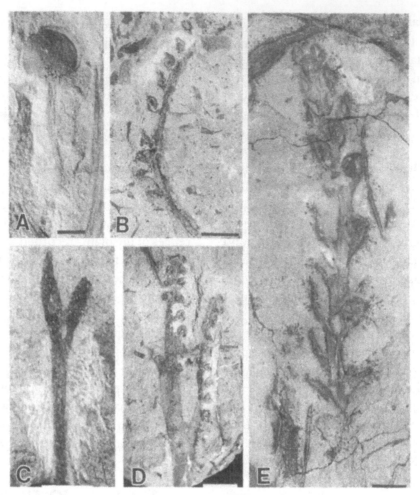

Figure 2.3.
Lochkovian coalified fossils from the United Kingdom. **A:** *Cooksonia caledonica.*
Forfar, Scotland. Lochkovian. Scale bar = 1 mm. **B:** *Zosterophyllum* cf. *fertile.*
Newport, Wales. Lochkovian. NMW99.13G.1b. Scale bar = 5 mm. **C:** *Salopella
allenii.* Brecon Beacons, Wales. Lochkovian. AD67. Scale bar = 5 mm. **D:** New
zosterophyll with spines. Brecon Beacons, Wales. Lochkovian. AD62A. Scale bar
= 10 mm. **E:** *Zosterophyllum myretonianum* Penhallow. Aberlemno Quarry, Scotland.
Lochkovian. RSM 1964.31.3. Scale bar = 5 mm.

probably resulting from localized exospore links
or bridges rather than large-scale fusion. They
are termed fused if there is no discernible line
and there is probably more extensive fusion
over most or all of the contact area. Cohesion
may also result from enclosure within a tight-
fitting envelope. The composition of cryptospores (wall and envelope) has not been chemically analyzed, but their preservation in such
ancient deposits suggests they contain sporopollenin or a sporopollenin-type macromolecule.

Space does not permit documentation of all
spore assemblage publications for the Ordovician to Lochkovian, but because they are our

only sources of information on Ordovician to
Early Silurian land vegetation, occurrences in
this earlier time interval are given in table 2.3.
For post-Llandovery assemblages characterized
by ever-increasing numbers of trilete monads,
see Richardson and McGregor (1986); a more
recent review of Late Silurian spore assemblages
is provided by Burgess and Richardson (1997).
New information on important assemblages
includes those from the Wenlock of Bohemia
(Dufka 1995), the Late Silurian of Saudi Arabia
(Steemans 1995), the Ludlow of Turkey (Steemans et al. 1996), the Ludlow of Gotland
(Hagstrom 1997), and the Ludlow of Argentina

(Rubinstein 1992) and Colombia (Grosser and Prossl 1991). Reviews of Lochkovian spore assemblages are provided by Richardson and McGregor (1986) and Steemans (1989). Since then, additional data have been published on well-known sequences of spore assemblages from Scotland (Wellman 1993a,b; Wellman and Richardson 1996), southern Britain (Barclay et al. 1994; Richardson 1996c; Wellman, Thomas, et al., 1998), and northern France (Moreau-Benoit 1994), in addition to descriptions of spore assemblages from new areas such as Argentina (Rubinstein 1993a,b; Le Herisse et al. 1996).

While there are copious examples of spore assemblages derived from continental deposits from the Late Silurian and Early Devonian, few or no such examples exist for the Ordovician—Early Silurian interval. These findings are an artefact of the stratigraphical record: The Ordovician—Early Silurian was a time of persistently high sea levels: Very few continental deposits are known, and those that do exist possess geological characteristics unsuitable for the preservation of palynomorphs (e.g., inappropriate lithologies or high thermal maturity). The earliest known spore assemblages preserved in continental deposits are from the Llandovery (Johnson 1985; Pratt et al. 1978; Strother and Traverse 1979; Wellman 1993b; Wellman and Richardson 1993). All include cryptospores and trilete spores, except that of Johnson (1985), which is the oldest (Early Llandovery) and contains only cryptospores.

There was a major change in the nature of spore assemblages in the Late Llandovery (Early Silurian) (Gray 1985, 1991; Burgess 1991; Richardson 1996a; Wellman 1996). While naked cryptospores (monads, dyads, and tetrads) continued to dominate spore assemblages, envelope-enclosed forms virtually disappeared. At this time, trilete spores and hilate cryptospores (derived from dyads that dissociate prior to dispersal) first appear. Earlier reports of trilete spores are believed to be erroneous either because the age designation is incorrect, the spores are contaminants, or the authors have described palynomorphs that resemble trilete spores but in fact represent either fortuitously folded acritarchs or spores physically broken out of cryptospore permanent tetrads (see Chaloner 1967; Schopf 1969; Wellman 1996). Both trilete spores and hilate cryptospores were initially rare, but their abundance and diversity increased throughout the Late Silurian. Both morphotypes developed sculpture in the Late Wenlock (Late Silurian), and structural/sculptural innovations ensued as both groups proliferated throughout the remainder of the Silurian and earliest Devonian. During the Early Devonian, hilate cryptospore numbers began to decline, until cryptospores became a minor component of spore assemblages, which were now dominated by a wide variety of trilete spores (e.g., Richardson and McGregor 1986).

AFFINITIES BASED ON SPORES

The use of the dispersed spore record (cryptospores and trilete spores) in determining affinities of the earliest land plants derives from three main sources: inferences based on comparison with the spores of extant land plants (occurrence and morphology), comparative morphological and anatomical studies of meso- and megafossils with *in situ* spores, and analysis of spore wall ultrastructure. There is abundant evidence suggesting that most, if not all, trilete spores represent the reproductive propagules of land plants (see Gray 1985, 1991 and references therein), but the affinities of certain megafossils, themselves with *in situ* spores, are often highly conjectural (as discussed previously). Direct evidence for the affinities of the producers for the cryptospores is even less compelling, although few would now doubt that they derived from land plants (see Gray 1985, 1991; Strother 1991; Taylor 1996; Edwards, Duckett, and Richardson 1995; Edwards, Wellman, and Axe 1998).

Based largely on analogy with the reproductive propagules of extant embryophytes, Gray (1985, 1991, and references therein) has argued persuasively that cryptospore permanent tetrads derive from land plants at a bryophyte, most likely hepatic, grade of organization. She noted

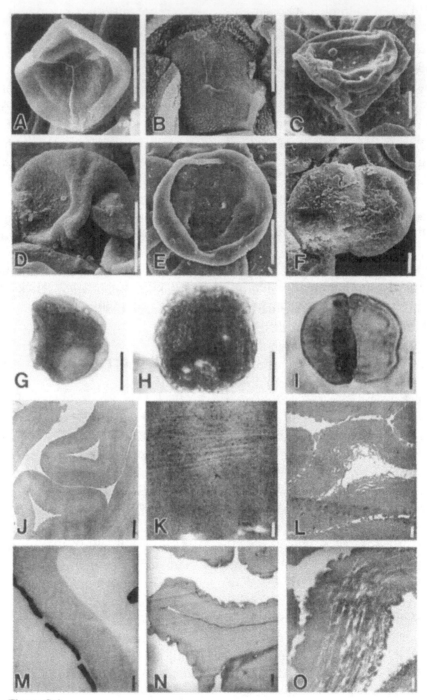

Figure 2.4.

Prídolí and Lochkovian spores. (A) through (F), scanning electron microscopy. **A:** Spore from *Tortilicaulis offaeus.* North Brown Clee Hill, England. Lochkovian. NMW99.11G.2. Scale bar = 10 μm. **B:** Spore from new rhyniophytoid taxon. (Similar to that illustrated in figure 2.2A.) Ludford Lane, England. Prídolí. NMW94.60G.9. Scale bar = 10 μm. **C:** *Velatitetras* sp. from *Grisellatheca salopensis* (see figure 2.2G). North Brown Clee Hill, England. Lochkovian. NMW94.76G.1. Scale bar = 10 μm. **D:** Ornamented *Velatitetras* sp. from terminal sporangia (see figure 2.2F). North Brown Clee Hill, England. Lochkovian. NMW96.11G.3. Scale bar = 10 μm. **E:** *Laevolancis divellomedia* type A from an irregular spore mass. Ludford Lane, England. Prídolí. NMW97.1G.3. Scale bar = 10 μm. **F:** Permanent dyad with extra-exosporal material from a fragment of sporangial cuticle with adhering dyads. North Brown Clee Hill, England. Lochkovian. NMW97.42G.2. Scale bar = 10 μm. (*continued*) that among extant free-sporing embryophytes, only hepatics regularly produce permanent tetrads, some of which are contained within an envelope similar to those enclosing certain cryptospore tetrads. The affinities of cryptospore monads and dyads are more equivocal, primarily because such morphologies do not have an obvious modern counterpart (Wellman, Edwards, and Axe 1998a). Dyads rarely occur in extant (nonangiosperm) embryophytes, and only through meiotic abnormalities (Fanning, Richardson, and Edwards 1991; Gray 1993; Richardson 1996b; Wellman, Edwards, and Axe 1998a,b). The abundance of dyads in early land plant spore assemblages indicates that they were produced by a number of taxa in which all spores within a sporangium were dyads. Their occurrence is most comfortably explained by invoking successive cytokinesis, with separation occurring following the first meiotic division and sporopollenin deposition on the products of the second division. However, an important observation is that monads, dyads, and tetrads have been reported co-occurring in identical envelopes, and it has been suggested that they are closely related, possibly even deriving from a single species (Johnson 1985; Richardson 1988, 1992; Strother 1991). Unlike many cryptospore morphotypes, trilete spores have a clear counterpart among extant embryophytes, where their production is widespread among free-sporing tracheophytes and also occurs sporadically among bryophytes (e.g., Gray 1985).

In Situ Spores

Studies of *in situ* spores, the only direct link between the palynomorph and plant megafossil records, are critical to our understanding of the affinities of dispersed spore types in diversity and paleogeographical studies. Plant megafossils are rare until the Late Silurian, probably because the vast majority of plants lacked the appropriate recalcitrant tissues suitable for preservation (e.g., Gray 1985; Edwards et al. 1999). Hence there are no *in situ* spore records for the first 65 million or so years of land plant evolution. The earliest fossils are uncommon and are usually preserved as coalified compressions, and *in situ* spores are generally absent or extremely poorly preserved in this mode of preservation (e.g., Edwards 1979b; reviews in Allen 1980, Gensel 1980). Occasionally, exceptional preservation of plant fossils, such as are found at Ludford Lane (figure 2.4B; Late Silurian—Prídolí) and north Brown Clee Hill (figure 2.4A; Early Devonian—Lochkovian) from the Welsh Borderland (Edwards and Richardson 1996), includes *in situ* spores. In both localities, the plants are preserved as coalified, but relatively uncompressed, mesofossils (Edwards 1996).

In situ cryptospore permanent tetrads (figure 2.4C,D) have been reported from five specimens from Ludford Lane and north Brown Clee Hill (Edwards et al. 1999) (see table 2.4). They differ in morphological and ultrastructural characteristics of both the mesofossil and *in situ* spores, respectively. All the fossils are too fragmentary to

(G) through (I), light microscopy of tetrads and a dyad from the Acton Scott Beds, A489 roadcut exposure in the type area for the Caradoc (Ordovician), River Onny Valley, Shropshire, England. Scale bars = 10 μm. **G:** Naked permanent tetrad of *Cheilotetras* sp. FM 812. Slide CA15/1/C. E.F. no. N55/1. **H:** Envelope-enclosed permanent tetrad of *Velatitetras rugulata* Burgess 1991. FM 850. Slide CA15/1/A. E.F. no. T36. **I:** Naked unfused dyad of *Dyadospora murusdensa* Strother and Traverse emend. Burgess and Richardson 1991. FM 824. Slide CA15/1/A. E.F. no. C61/4.

(J) through (O), transmission electron microscopy. **J:** *In situ Laevolancis divellomedia* type A spores from a discoidal spore mass. Unoxidized and unstained. Ludford Lane, England. Prídolí. NMW96.11G.1. Scale bar = 1 μm. **K:** White-line-centered lamellae in an *in situ Laevolancis divellomedia sensu lato* type C spore from a discoidal spore mass. Unoxidized but stained. North Brown Clee Hill, England. Lochkovian. NMW96.30G.5. Scale bar = 100 nm. **L:** *In situ Laevolancis divellomedia senso lato* type B spores from an elongate sporangium. Oxidized and stained. Ludford Lane, England. Prídolí. NMW96.30G.3. Scale bar = 500 nm. **M:** Permanent dyad from *Fusitheca fanningiae* (see figure 2.2H). Oxidized but unstained. North Brown Clee Hill, England. Lochkovian. NMW97.42G.4. Scale bar = 200 nm. **N:** *In situ Velatitetras* sp. from a discoidal sporangium. Unoxidized but stained. Ludford Lane, England. Prídolí. NMW98.23G.1. Scale bar = 500 nm. **O:** *Tetrahedraletes* sp. from a spore mass. Unoxidized but stained. North Brown Clee Hill, England. Lochkovian. NMW98.23G.4. Scale bar = 200 nm.

Table 2.3. Records of Ordovician and basal Silurian spore assemblages

Locator*	Age	Authors	Geographic Area
1.	Llanvirn	Vavrdova (1984, 1990), Corna (1970)	Bohemia
2.	Llanvirn	McClure (1988), Strother et al. (1996)	Saudi Arabia
3.	Caradoc	Richardson (1988), Wellman (1996)	Southern Britain
4.	Caradoc	Gray et al. (1982)	Libya
5.	Ashgill	Gray and Boucot (1972), Gray et al. (1982), Gray (1985, 1988)	Kentucky, USA
		Gray (1988)	Tennessee, USA
		Gray (1988)	Georgia, USA
		Strother (1991)	Illinois, USA
		Gray (1988)	Ontario, Canada
6.	Ashgill	Burgess (1991)	Southern Britian
7.	Ashgill	Vavrdova (1982, 1984, 1988, 1989)	Bohemia
8.	Ashgill	Reitz and Heuse (1994)	Germany
9.	Ashgill	Steemans et al. (1996)	Turkey
10.	Ashgill	Richardson (1988)	Libya
11.	Ashgill	Gray et al. (1986)	South Africa
12.	?Ashgill	Lakova et al. (1992)	Bulgaria
13.	?Ashgill	Foster and Williams (1991)	Australia
14.	Ashgill	Wang et al. (1997)	Xinjiang, China
15.	Rhuddanian	Burgess (1991)	Southern Britain
16.	Rhuddanian	Richardson (1988)	Libya
17.	Rhuddanian	Miller and Eames (1982)	New York State, USA
	Rhuddanian, ?Aeronian	Johnson (1985)	Pennsylvania, USA
	?Llandovery	Strother and Traverse (1979)	Pennsylvania, USA
18.	Llandovery	Bar and Riegel (1980)	Ghana
19.	Llandovery	Gray et al. (1985)	Brazil
20.	Late Llandovery	Gray et al. (1992)	Paraguay

provide information on the growth habit or indeed gross morphology of the original plants. The most complete demonstrate branching in an axial system with terminal sporangia containing either *Velatitetras* (figure 2.4D,H) or *Tetrahedraletes,* or in the case of *Grisellatheca* (figure 2.2G) a bifurcating apex with two deepseated, possibly interconnecting sporangial cavities containing *Velatitetras* (figure 2.4C). The latter has the most extensive sporangial walls, whose superficial cells form the only "conventional" cellular construction preserved in the fossil and cover a layer in which it is impossible to detect any organization. Although some of the characters are those of hepatics (e.g., the tetrads themselves, the presumed elater, differentially thickened tubes), there are no exactly comparable extant anatomical features. Indeed, we now believe that some of the tubes with regular internal thickenings, initially thought to represent rhizoids, are fungal pathogens (Edwards, Duckett, and Richardson 1995; Edwards and Richardson 2000). The other two specimens may have been cuticularized and, at least morpho-

logically, seem closer to the contemporary rhyniophytoids.

In situ cryptospore permanent dyads have been reported from two specimens from north Brown Clee Hill (Wellman, Edwards, and Axe 1998a). Two consist of a single sporangium terminating an axis. *Culullitheca* is cup shaped and attached to an unbranched axis, and it contains naked, laevigate, unfused dyads (figure 2.2I). *Fusitheca* is elongate (similar in shape to *Salopella*) and attached to an isotomously branching axis (figure 2.2H), and it contains dyads that appear to be enclosed within an envelope (figure 2.4M). Two further specimens are fragments of (presumably sporangial) cuticle with adhering dyads (figure 2.4F). One is of irregular shape with naked, unfused, laevigate dyads that are associated with abundant extraexosporal material. The other is elliptical with dyads that are possibly envelope enclosed. Sporangial cuticle of this type is found in the dispersed spore record from the Wenlock onwards. The production of dyads in a branching sporophyte would seem to provide evidence against

Table 2.4. Summary of ultrastructure in Ordovician to Devonian spores

Taxon / Morphotype	Occurrence / Age	Exospore Ultrastructure	Reference
Pseudodyadospora sp. (naked pseudodyad)	Dispersed (Ashgill/ Llandovery)	Homogeneous	Taylor (1996)
Dyadospora murusdensa (naked true dyad)	Dispersed (Ashgill/ Llandovery)	Bilayered with lamellate inner layer and homogeneous outer layer	Taylor (1996)
Dyadospora murusattenuata type 2 (naked true dyad)	Dispersed (Ashgill/ Llandovery)	Exospore spongy, exhibiting varying degrees of sponginess	Taylor (1996)
Dyadospora murusattenuata type 1 (true dyad enclosed in a thin envelope)	Dispersed (Ashgill/ Llandovery)	Exine trilayered: inner layer spongy, middle layer lamellate, outer layer homogeneous	Taylor (1996)
Segestrespora membranifera (envelope-enclosed dyad)	Dispersed (Ashgill)	Homogeneous to somewhat spongy with irregular globular units and lamellae at the exterior; envelope stains darker than wall	Taylor (1996)
Tetrahedraletes medinensis (naked permanent tetrad)	Dispersed (Ashgill)	Homogeneous, possibly with subtle lamellation	Taylor (1995b)
Laevolancis divellomedia type 1 (laevigate hilate cryptospore)	Discoidal sporangia/ spore masses (Prídolí)	Bilayered: outer layer homogeneous and electron dense, inner layer homogeneous and less electron dense; layers of similar thickness	Wellman, Edwards, and Axe (1998b)
Laevolancis divellomedia type 2 (laevigate hilate cryptospore)	Elongate sporangia (Prídolí)	Bilayered: outer layer homogeneous; inner layer comprising laterally continuous, concentric lamellae lacking further structure	Wellman, Edwards, and Axe (1998b)
Laevolancis divellomedia type 3 (laevigate hilate cryptospore)	Discoidal spore mass (Lochkovian)	Bilayered: outer layer thin and homogeneous, inner layer comprising laterally discontinuous, overlapping and irregularly spaced lamellae with tripartite structure	Wellman, Edwards, and Axe (1998b)
Laevolancis divellomedia type 4 (laevigate hilate cryptospore)	Small discoidal sporangium (Lochkovian)	Exospore homogeneous	Wellman, Edwards, and Axe (1998b)
Laevolancis divellomedia type 5 (laevigate hilate cryptospore)	Large discoidal sporangia (Lochkovian)	Bilayered: thin outer homogeneous layer, and thick, grainy, and homogeneous inner layer	Wellman, Edwards, and Axe (1998b)
Permanent dyad (naked)	Adhering to sporangial cuticle (Lochkovian)	Wall homogeneous; EEM abundant but does not appear to form an envelope	Wellman, Edwards, and Axe (1998a)
Permanent dyad (naked)	*In situ* in *Culullitheca richardsonii* (Lochkovian)	Wall homogeneous; very little EEM present	Wellman, Edwards, and Axe (1998a)
Permanent dyad (naked)	Elongate spore mass (Lochkovian)	Wall homogeneous; EEM abundant but does not appear to form an envelope	Wellman, Edwards, and Axe (1998a)
Permanent dyad (envelope-enclosed)	*In situ* in *Fusitheca fanningiae* (Lochkovian)	Wall homogeneous; spores enclosed in a discrete envelope	Wellman, Edwards, and Axe (1998a)
Permanent dyad (envelope-enclosed)	Adhering to sporangial cuticle (Lochkovian)	Wall bilayered with thick, homogeneous inner layer, and narrow, homogeneous outer, sometimes sporadic, layer. Outer layer darker than inner when stained (? envelope comprising EEM)	Wellman, Edwards, and Axe (1998a)
Tetrahedraletes sp. (naked permanent tetrad)	Ovoid spore mass (Prídolí)	Homogeneous with no EEM	Edwards, Wellman, and Axe (1999)
Velatitetras sp. (envelope-enclosed permanent tetrad)	Discoidal sporangia (Lochkovian)	Homogeneous exine with envelope comprising globules	Edwards, Wellman, and Axe (1999)
Velatitetras sp.? (envelope-enclosed permanent tetrad)	Bifurcating axis (Lochkovian)	Bilayered with the electron dense envelope passing around the tetrad and between the spores	Edwards, Wellman, and Axe (1999)
Velatitetras sp. (envelope-enclosed permanent tetrad)	Bifurcating axis (Lochkovian)	Bilayered with a more electron dense outer layer	Edwards, Wellman, and Axe (1999)
Tetrahedraletes sp. (naked permanent tetrad)	Discoidal sporangium (Lochkovian)	Trilayered	Edwards, Wellman, and Axe (1999)

continued

Table 2.4. *(continued)*

Taxon / Morphotype	Occurrence / Age	Exospore Ultrastructure	Reference
Tetrahedraletes sp. (naked permanent tetrad)	Irregular spore mass (Lochkovian)	Bilayered with an inner spongy and outer homogeneous layer	Edwards, Wellman, and Axe(1999)
Tetrahedraletes sp. (naked permanent tetrad)	Irregular spore mass (Lochkovian)	Homogeneous, possibly with a narrow outer layer	Edwards, Wellman, and Axe(1999)
Ambitisporites sp. (crassitate trilete spore)	*In situ* in *Cooksonia pertoni* (Prídolí)	Bilayered with peripheral coalified layer	Rogerson et al. (1993)
Synorisporites verrucatus (crassitate trilete spore)	*In situ* in *Cooksonia pertoni* (Lochkovian)	Bilayered with peripheral coalified layer; layering contours sculpture	Edwards, Davies, et al. (1995)
Streelispora newportensis (crassitate trilete spore)	*In situ* in *Cooksonia pertoni* (Lochkovian)	Bilayered with peripheral coalified layer; layering does not contour sculpture	Edwards, Davies, et al. (1995)
Aneurospora sp. (crassitate trilete spore)	*In situ* in *Cooksonia pertoni* (Prídolí)	Bilayered with peripheral coalified layer; layering does not contour sculpture	Edwards, Davies, et al. (1995)
Synorisporites downtonensis (crassitate trilete spores in tetrads)	Elongate spore masses (Prídolí)	Homogeneous	Edwards, Davies, et al. (1996)
Scylaspora sp. (crassitate trilete spore)	Elongate sporangia (Lochkovian)	Bilayered: inner layer lamellate, outer layer homogeneous	Wellman (1999)
Retusotriletes sp. (retusoid trilete spore)	*In situ* in *Resilitheca salopensis* (Lochkovian)	Faint traces of lamellae; abundant exosporal material	Edwards, Fanning, et al. (1995)
Retusotriletes cf. *coronadus* (retusoid trilete spore)	Discoidal spore masses (Prídolí)	Narrow inner darker layer and essentially homogeneous outer layer but with faint striations	Edwards, Davies, et al. (1996)
Apiculiretusispora brandtii (retusoid trilete spore)	*Psilophyton forbesii* (Emsian)	Thick inner homogenous layer with a thin outer, often detachable, layer representing the ornament	Gensel and White (1983)

EEM, extra-exosporal material.

affinity with bryophytes based on extant characters. Unfortunately, axial cellular preservation is lacking, and the position within the embryophytes remains conjectural. However, they demonstrate variation in the morphology of the dyad producers.

Similarly, the ultrastructure of laevigate hilate cryptospores from Ludford Lane (figure 2.4J,L) and north Brown Clee Hill (figure 2.4K) combined with gross morphology of spore masses and isolated sporangia (discoidal versus elongate) suggests production by at least five different plants (Wellman, Edwards, and Axe 1998b). Those from Ludford Lane occur in either discoidal sporangia and spore masses or elongate sporangia, and those from north Brown Clee Hill all occur in discoidal sporangia and spore masses. Various taxa of ornamented hilate cryptospores have also been discovered *in situ* (unpublished research). Again, none of the mesofossils with hilate cryptospores are attached to axes with preserved anatomy, and their affinities remain conjectural. However, in any one spore mass or sporangium, all the spores are of the same type, as was the case for the permanent dyads, thus allowing the conclusion that dyads were not sporadic meiotic abnormalities.

It is also important to emphasize that although some progress has been made in discovering cryptospore producers, they are relict taxa occurring at a time when cryptospore abundance in dispersed assemblages was diminishing, and it is unclear to what extent the later cryptospore producers are related to those of Ordovician and Early Silurian examples. Branching axes in permanent dyad- and tetrad-containing plants indicate more complex sporophytes than would be expected in plants at a bryophyte grade, based on extant evidence, although this might be characteristic of stem-group embryophytes. The specimen with a branching axis (*Fusitheca*) is at least 65 million years younger than the earliest cryptospore dyads, more than ample time for the evolution of a more complex sporophyte.

Studies on *in situ* triradiate monads provide additional characters for simple axial plants (see summary in Edwards and Richardson 1996) and in some cases, where ultrastructure is preserved, contribute to assessments of relationships between coeval taxa [see, e.g., *Cooksonia pertoni* (Edwards, Davies, et al. 1995), *Pertonella* (Edwards, Davies, et al. 1996)]. However, in mid-Paleozoic examples, affinity of the producers is less secure, even though Lochkovian *Cooksonia pertoni* has been demonstrated to possess tracheids (see first paragraph of the section Affinities Based on Megafossils). Not all trilete spore morphotypes have been discovered *in situ*. We have yet to discover *in situ* patinate spores, and *in situ Emphanisporites*-type spores are extremely uncommon (although both spore types are abundant in dispersed spore assemblages, and even in coprolites, from the mesofossil-bearing horizons). These findings may be a consequence of either (1) preservational artefact (the missing spore types may derive from plants with low preservation potential), (2) regional paleogeographical effects (the missing spore types may derive from plants growing beyond the catchment area of the depositional basin), or (3) facies effects (the missing spore types may derive from larger plants not represented in the assemblages of small, highly sorted, mesofossils).

Spore Wall Ultrastructure

Spore wall ultrastructural characters can be extremely valuable when attempting to ascertain the phylogenetic relationships of extant land plants, and similar research has been extrapolated back in time and is now routinely undertaken on fossil spores (e.g., Kurmann and Doyle 1994). Such research is also of paramount importance in studies of spore wall development. Recently, there has been a surge of interest in wall ultrastructure in early land plant spores (table 2.4), with the expectation that exploitation of this potentially extremely rich data source will provide characters useful in ascertaining the affinities of these ancient plants (e.g., application of spore data in the cladistic analysis of Kenrick and Crane 1997a) and shed light on the nature of spore wall development.

Two principal lines of enquiry have been explored in studies of wall ultrastructure in early land plant spores—analysis of isolated dispersed spores from the Late Ordovician (Ashgill)—Early Silurian (Llandovery) of Ohio (Taylor

1995a,b, 1996, 1997), and of *in situ* spores exceptionally well preserved in mesofossils from the latest Silurian (Ludford Lane) and earliest Devonian (north Brown Clee Hill) localities in the Welsh Borderland (Rogerson et al. 1993; Edwards, Davies, et al. 1995, 1996; Edwards, Duckett, and Richardson 1995; Edwards, Fanning, et al. 1995; Edwards et al. 1999; Wellman, Edwards, and Axe 1998a; Wellman, Edwards, and Axe 1998b). Studies on Late Silurian—Early Devonian dispersed spores remain an unexploited but potentially extremely useful data source. Taylor has examined early examples of naked and envelope-enclosed cryptospore permanent tetrads (Taylor 1995a, 1996, 1997) and naked and envelope-enclosed cryptospore dyads (Taylor 1995b, 1996, 1997). Later cryptospore tetrads (Edwards, Davies, et al. 1996; Edwards et al. 1999), dyads (Wellman, Edwards, and Axe 1998a), and hilate cryptospores (Wellman, Edwards, and Axe 1998b) have also been examined, as have a variety of trilete spore taxa (Rogerson et al. 1993; Edwards, Davies, et al. 1995; Edwards, Fanning, et al. 1995; Wellman 1999). Table 2.4 summarizes the work undertaken and results obtained to date. The lack of studies on patinate trilete spores (discussed previously) is a major omission, which can be ascribed to their absence or paucity in the mesofossil record.

In terms of ascertaining phylogenetic relationships, findings to date (see table 2.4) are rather difficult to interpret. No clear patterns emerge regarding the relationships between the different types of mesofossil, *in situ* spore, and spore wall ultrastructure. This is most likely a consequence of the sketchiness of the database. However, interpretation of wall ultrastructure in early land plant spores is also problematic because of a number of technical and theoretical factors:

1. Different workers employ slightly different techniques, each with its potential artefacts.
2. Unlike extant plants where complete ontogenetic sequences can be studied, fossils preserve only a particular ontogenetic

state, and the exact stage of maturation is often unclear, although assumed to be mature.

3. For extant plants, developmental stages are usually far more important than mature spores in assessment of affinity.

4. Diagenetic effects may vary, rendering comparisons problematic, particularly when comparing extant with fossil plants, or fossils from different localities with different diagenetic histories.

5. It is unclear to what extent the spore wall ultrastructure characters of extant plants are representative of those seen in such distant ancestors.

6. It is difficult to detect convergence resulting from similarities in developmental processes.

Nonetheless, these problems are not insurmountable, and it is likely that in the future, spore wall ultrastructural studies will play an increasingly important part in phylogenetic analysis of early land plants, particularly if the database continues to increase at its current rate. Taylor (1997) recently summarized his findings to date and proposed a tentative hypothesis for evolutionary relationships among early cryptospore producers. He suggests that at least two separate lineages may be distinguished, but he concludes that "the phylogenetic relationships between these groups and to more recent land plants remains uncertain," although he has suggested possible hepatic affinities for some of the dyads (e.g., Taylor 1995b).

In terms of understanding spore wall development in early land plants, lamellae have been recognized in trilete spores (Wellman 1999), cryptospore dyads (Taylor 1995b), and hilate cryptospores (Wellman, Edwards, and Axe 1998b), including the presence of typical white-line-centered lamellae in the latter. Such findings provide the earliest fossil evidence for the antiquity of such structures, and they provide further evidence that sporopollenin deposition on these structures is the most primitive mode of sporopollenin deposition among land plants

(e.g., Blackmore and Barnes 1987). Information has also been provided on the nature of cryptospore envelopes (Taylor 1996, 1997; Wellman, Edwards, and Axe 1998a,b) and junctions between units in cryptospore permanent polyads (Taylor 1995a, 1996, 1997; Wellman, Edwards, and Axe 1998a; Edwards et al. 1999).

PHYTODEBRIS

Enigmatic dispersed fragments (phytodebris), believed to derive from land plants and/or fungi, have long been known from the Ordovician—Early Devonian and have provided an important contribution to our understanding of early land plants and terrestrial ecosystems (e.g., Gray 1985; Gensel et al. 1990). They consist primarily of fragments of cuticle, tubular structures (aseptate), and filaments (septate), which often occur in complex associations.

The earliest fragments of cuticle are reported from the Caradoc (Late Ordovician), and they are relatively abundant in the Llandovery (Early Silurian). From the Wenlock (Late Silurian) onward, ornamented cuticles (Edwards and Rose 1984), and forms with alignment of cells, are present. Most are characterized by having one smooth (possibly outer) surface and another that has a raised reticulum, thought to reflect the pattern of the outermost cellular components of the underlying tissue [*Nematothallus* (Lang 1937), but see Strother 1993]. Less common are unistratose sheets of cells. For reviews of the Ordovician—Early Devonian dispersed cuticles, see Gray (1985), Gensel et al. (1990), and Edwards and Wellman (1996). The earliest tubular structures are smooth forms from the Ashgill (Late Ordovician). "Ornamented" tubes (including types with external ridging and others with internal annular or spiral thickenings) appear in the Llandovery (Early Silurian). A proliferation of different types of tubular structures occurs in the Wenlock (Late Silurian), and similar diverse forms exist until at least the Early Devonian. Tubular structures vary in their internal and external ornament, presence or absence of branching, and nature of terminal structures (if present). They commonly occur in complex associations. The earliest reported filaments are from the Llandovery (Early Silurian), and they have been reported sporadically throughout the Silurian and Lower Devonian. The filaments are usually branched, sometimes with flask-shaped protuberances, and the septa may or may not be perforate. The occurrence of Ordovician—Early Devonian tubular structures and filaments is reviewed by Gray (1985), Burgess and Edwards (1991), Gensel et al. (1990), Wellman (1995), and Edwards and Wellman (1996).

AFFINITIES BASED ON PHYTODEBRIS

The affinities of much of this debris are controversial because convincing extant counterparts are lacking, although recent advances have gone some way toward clarifying their biological relationships. They undoubtedly derive from nonmarine organisms, because they occur in continental deposits, and in marine sediments they have a distribution comparable to that of fragments derived from land plants (cuticles, conducting tissues) and fungi (filaments) in similar present-day depositional environments.

In the Wenlock (Late Silurian), cuticles that can be unequivocally assigned to higher land plants (tracheophytes) first appear. They include forms showing marked alignment of epidermal cells, often with stomata (post-Prídolí), which clearly derive from axial fossils, and others with very well defined, more rectangular or lenticular outlines, occasionally with attached spores that derive from their sporangia. However, the earlier forms (*Nematothallus sensu lato*), which persist until at least the Middle Devonian, are more enigmatic. They lack stomata and have cellular patterns unlike those in extant land plants. Their resilience suggests that they were of similar composition to tracheophyte cuticle and hence represented adaptations for survival in terrestrial habitats (Edwards, Abbot, and Raven 1996). However, preliminary chemical

analyses provide evidence against similarities with coeval higher plant cuticles.

The dispersed tubular structures have less convincing analogues among structures present in extant land plants. It has long been noted that forms with internal annular or spiral thickenings resemble the tracheids of extant land plants (and they are often referred to as tracheid-like tubes in the older literature). They are frequently associated with wefts of tubes (originally described as *Nematothallus*) and have also been reported from organisms of tubular construction close to *Prototaxites*. The latter has recently been interpreted as a fungus (Hueber 1996), although Lang (1937) considered the nematophytes to be a distinct group neither algal nor higher plant. Evidence supporting Hueber's hypothesis is the recent discovery of internally thickened tubes seemingly growing on higher land plants (Edwards et al. 1996) and within tissues of others (Edwards and Richardson 2000). It is possible that these tubes were either attached to the plant while it was alive (as pathogens) or dead (as decomposers). Another interesting possibility is that some of the nematophytes represent lichens, the discovery of lichenized cyanobacteria in the Pragian Rhynie Chert confirming the antiquity of these organisms (Taylor, Hass, et al. 1995). If either fungal or lichen affinities were proven, this might explain the anomalous chemical composition of the *Nematothallus* cuticles.

The dispersed filaments bear a striking resemblance to the hyphae of certain extant fungi, and it has been suggested that they provide evidence for the earliest terrestrial fungi in the Silurian (Sherwood-Pike and Gray 1985). Additional evidence is in the form of associated spores (including multiseptate spores) of probable fungal affinity.

Recently, Kroken et al. (1996) suggested that some of the dispersed cuticles and tubular structures may represent the products of fragmentation of fossil bryophyte sporangia, based on their observations on the fragmented remains of extant bryophyte sporangia. However, we find little support for this hypothesis, principally because of constraints emerging from our observations on the size and symmetry of the fragments we have isolated.

GENERAL OBSERVATIONS ON GROSS MORPHOLOGY, GROWTH HABIT, AND REPRODUCTIVE BIOLOGY, BASED ON THE LIMITED MESO- AND MEGAFOSSIL RECORD

For interpretations of the nature of Ordovician–Lower Silurian vegetation and its impact on atmospheric composition and lithosphere weathering, see chapter 8.

1. The few records indicate an increase in height and sporangial/axial dimensions in *Cooksonia* from its first appearance in the Wenlock, such that larger-scale versions occur in the Prídolí and Lower Devonian (see Bolivia, New York, and Scotland). Such a trend is less obvious in Wales and the Welsh Borderland.

2. Larger Late Silurian rhyniophytoids from Bathurst Island and Kazakhstan/Xinjiang show a change in growth strategy—that is, a departure from isotomous branching with terminal sporangia typical of *Cooksonia*, and the appearance of pseudomonopodial axes and lateral production of sporangia. Their overall dimensions, however, are similar to those of the larger cooksonias. Change in growth pattern and hence sporangial positions has major consequences in reproductive biology in that spore output/axis can be increased and sporing can occur over a longer period of time. The presence of marginal features on sporangia may infer modifications of the sporangial wall for dehiscence. Whether its timing was controlled by environmental conditions conducive for spore dispersal remains conjectural.

3. Also present in Late Silurian assemblages are sterile fragments, both smooth and spiny/hairy, showing K- and H-branching types as well as clustered configurations. This, compared with later *Zosterophyllum myretonianum*, suggests early acquisition of the capacity for rhizomatous growth and the "turfing" of larger areas by the same taxon (Tiffney and Niklas 1985). Such a growth habit was probably typical of the plants (e.g., *Gosslingia*) recorded in the major diversi-

fication in the Late Lochkovian of southern Britain. In contrast, very limited evidence for the *Cooksonia* group [e.g., sporangia preserved at one level in Long Mountain marine sediments, and the single specimen of *Cooksonia bohemica* (Schweitzer 1983)] suggests a basal clustering of axes.

The increases in sporing capacity and duration mentioned in (2) are particularly evident in the zosterophylls, in species of *Zosterophyllum*, whose strobili are first recorded in the Lochkovian, and axes even earlier [e.g., in Australia (J. Tims, pers. comm. January 1998)], and in *Gosslingia* and *Deheubarthia*.

4. Some representatives of the *Cooksonia* growth habit remained small, into the Lower Devonian, being constrained by the "adaptive impasse" of determinate growth, but they underwent major diversification in sporangial and spore characters, as described here in mesofossils. It seems likely that such plants produced the equivalent of extensive "moss forests," probably coexisting in this ecological niche with the dyad and tetrad producers.

Caveat: The size of the fragmentary fossils at the north Brown Clee Hill locality is clearly controlled by the facies. They occur in the very fine grained siltstones at the top of the cyclothem. We know that elsewhere in the area, *Zosterophyllum* grew, and its spikes are preserved in slightly coarser grained rock. Thus it would bc logical to deduce that the small dimensions and fragmentary nature of the fossils result from, first, the breakup of much larger specimens during transport, and then the selective burial of their tips or isolated sporangia. However, we have never found isolated sporangia of *Zosterophyllum* (which could be identified by their spores) among the mesofossils, and while the terminal sporangia on short lengths of branching axes obviously derive from more profusely branched plants, it seems unlikely to us that they were the tips of plants of large stature. This is based on the absence of wider axes in coarser sediments elsewhere in the region, but more importantly on comparative size (axis diameter relative to branching and sporangial diameter) and frequency of branching in coalified fossils in coeval localities (e.g., Targrove) and older ones. We conclude that the north Brown Clee Hill plants are "scaled down" versions of the isotomous branching/terminal sporangia growth habit.

5. There is no direct evidence for a "rooting system" in any of the axial, isotomously branching plants. Indeed, in Silurian examples, preservation in marine settings is indicative of considerable transport. Evidence from the slightly younger Rhynie Chert suggests that surface creeping or buried axes, usually with rhizoids confined to the presumed lower surface, were not extensive and if subterranean would not have penetrated the thin soils to any depth (see chapter 13).

6. The appearance of *Baragwanathia* in Ludlow rocks of Australia marks the advent of much higher productivity per plant and potential for greater impact on substrate. Although the Silurian representatives have never been described in detail, Tims and Chambers (1984: 277) mention fragments at least 400 mm long, and although they place them in the Lower Devonian taxon *B. longifolia*, they record that they were "not identical in form." Garratt (1978) illustrated two fine Silurian examples, one of maximum width 40 mm inclusive of leaves and the other with a prominent stem about 5.0 mm wide. Such fragments probably derived from plants of considerable size and it seems not unlikely that they would have borne the adventitious roots recorded in younger examples (J. Tims, pers. comm. January 1998). In the present day northern hemisphere, the *Drepanophycus spinaeformis* first recorded in the Upper Gedinnian of Germany (Schweitzer 1980b) had a similar growth habit, with roots recorded *in situ* in fluvial sediments in later Emsian examples (e.g., Rayner 1984), where they extend through several centimeters of sediment and exhibit dichotomous branching.

GLOBAL PHYTOGEOGRAPHY AND PHYTOPROVINCIALISM

Ordovician–Early Silurian

Although reports of the interval from Ordovician to Early Silurian are relatively few, they are stratigraphically and geographically wide-spread, and they indicate that cryptospore assemblages are remarkably constant in composition (both temporally and spatially) throughout the Ordovician and Early Silurian (Gray 1985, 1991; Richardson 1996a; Strother et al. 1996; Wellman 1996) (figure 2.5). These data indicate that the vegetation was cosmopolitan but of limited diversity, with little evolutionary change during an interval of at least 40 million years. Gray et al. (1992) suggested that variation in the paleogeographic distribution of envelope-enclosed permanent tetrads (exemplified by type of envelope ornament) provides the first evidence for phytogeographic differentiation among early land plants during the Ordovician—Early Silurian. They suggest that a cool Malvinokaffric Realm and a warmer extra-Malvinokaffric Realm existed. However, certain reported data from Bohemia (e.g., Vavrdova 1989), Libya (Richardson 1988), and Turkey (Steemans et al. 1996) conflict with their findings, and more paleogeographically extensive recording is required to clarify this matter (Wellman 1996).

Wenlock to Prídolí

In the interval from Wenlock to Prídolí (figures 2.6, 2.7), there is considerable variation in the database. There are substantial differences in the number and quality of reports and the extent of their paleogeographic coverage (e.g., dearth of latest Prídolí occurrences in Laurentia, absence of spores in critical Australian deposits) for different time intervals. This creates problems when attempting to identify phytogeographic variation—problems that are further enhanced by difficulties associated with recognizing the effects of small-scale geographical variation dependent on local environmental control, variation due to subtle facies effects, and possibly noncontemporaneity. Nonetheless, certain patterns are beginning to emerge and are summarized here.

Wenlock to Prídolí spore assemblages are generally well documented from Laurentia, Avalonia, and Baltica paleocontinents, which were at this time closely associated and

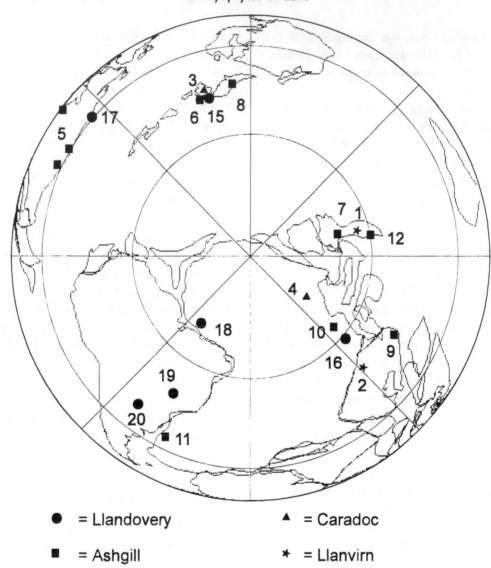

● = Llandovery   ▲ = Caradoc

■ = Ashgill   ✶ = Llanvirn

Figure 2.5.
Distribution of spore assemblages from the Ordovician and Llandovery on base map for
Late Ordovician (based on Eldridge et al. 1996; 450 myr). This gives a general idea of the
geographical distribution of localities in the Ordovician southern hemisphere but does not
take into account the postulated, quite extensive, and highly debated "polar wandering"
during the time intervals. Numbers refer to table 2.3. The Chinese Xinjiang locality has been
omitted, as we are uncertain of its position (possibly on Kazakhstan plate). Australia is
missing from this projection.

straddling the equator, and from northern
Gondwana, which lay further to the paleosouth
and separated by a relatively large ocean.
Records from elsewhere in Gondwana and high
northern paleolatitudes are sparse. Sequences
of spore assemblages from Laurentia, Avalonia,
and Baltica are notably similar in composition,
suggesting that these land masses contained a
flora representing a single paleophytogeographical realm, and subtle differences probably reflect small-scale paleogeographical varia-

tion. Sequences of spore assemblages from
northern Gondwana (e.g., Spain and north
Africa), although sharing many taxa, also
exhibit marked differences, suggesting differentiation of a further paleophytogeographical
realm. The limited megafossil record for Gondwana indicates the widespread occurrence of
plants in Late Wenlock to Ludlow times
[rhyniophytoids in southern Bolivia (Toro et al.
1997), *Cooksonia* (Morel et al. 1995) in the upper
Ludlow/Prídolí on its southwestern margin,

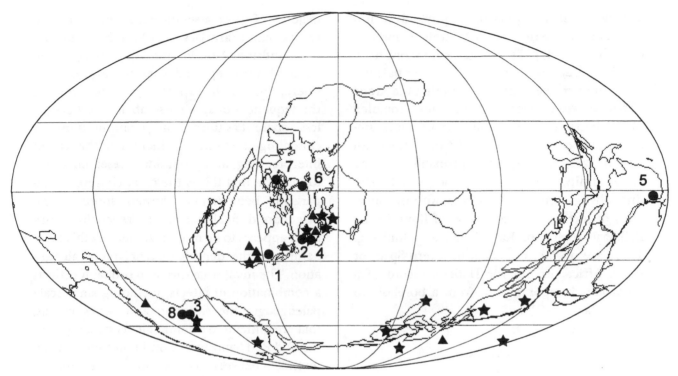

Figure 2.6.
Distribution of Silurian spore and megafossil assemblages (excluding Prídolí), on base map redrawn from Eldridge et al. (1996; 425 myr). *Numbers* refer to table 2.1. *Solid circles plus numbers* refer to megafossil localities in table 2.1. *Stars* indicate Wenlock/Ludlow microfossils; *triangles* indicate Prídolí microfossils.

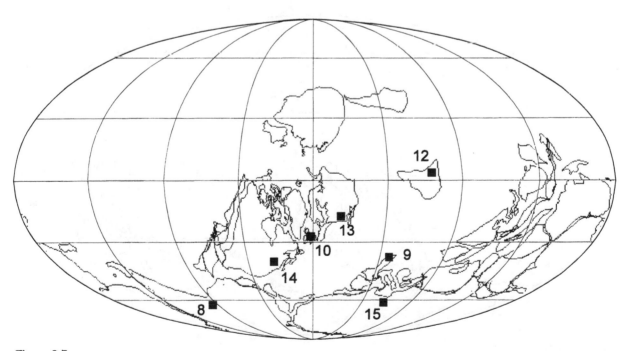

Figure 2.7.
Distribution of Prídolí megafossil assemblages on a base map redrawn from Eldridge et al. (1996; 408 myr). *Numbers* refer to table 2.1.

and *Baragwanathia* plus rhyniophytoids at low paleolatitudes to the east in Australia (Tims and Chambers 1984)]. The megafossil record from the Late Ludlow—Přídolí of Laurentia indicates uniform vegetation, except that the Arctic Canadian assemblage contains more complex plants than in the coeval southern Britain, mainland Europe, Greenland, and North American assemblages, although they probably have the same affinity. Edwards (1990) suggested a distinct phytoprovince in Australia, with the possibility of further differentiation of floras in the Balkhash region of Kazakhstan and Xinjiang. There are no megafossils in northern Spain or north Africa, except for Daber's record of a *Cooksonia*-like sporangium from a borehole in Libya (Daber 1971).

Lochkovian

By Lochkovian times (figure 2.8), Laurentia, Avalonia, and Baltica were all essentially part of the same large continent (the "Old Red Sandstone Continent"), now not far removed from Gondwana, the distance between them having diminished through the Late Silurian. The vast

majority of spore assemblages known from the Lochkovian (Early Devonian) are from the Old Red Sandstone Continent and northern Gondwana, with few reported assemblages from elsewhere. One might expect differences between the sequences of spore assemblages to become less apparent as the two megacontinents moved closer together offering increased chances of interbreeding, and preliminary research seems to indicate that this is the case. However, comparisons between Lochkovian spore assemblages are difficult because more localized variation appears to be more prevalent by this time, hampering the identification of larger-scale variation. The small-scale variation probably reflects a combination of effects, including small-scale paleogeographic variation and facies effects, that are difficult to disentangle. For example, in the Old Red Sandstone Continent there are differences between coeval spore assemblages from upland intermontane and lowland floodplain deposits (Richardson et al. 1984; Wellman 1993b; Wellman and Richardson 1996). It is unclear to what extent these reflect subtle facies effects or variation (small- or large-scale) in the distribution of flora due to variation in environ-

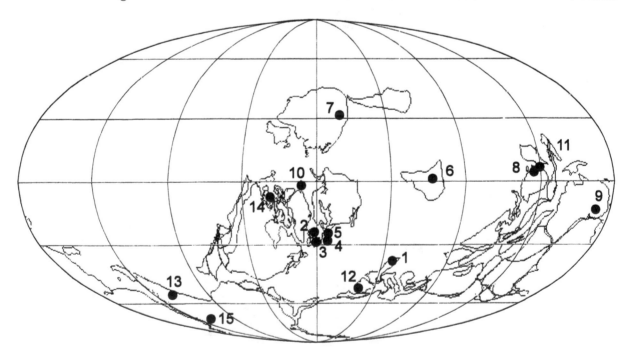

Figure 2.8.
Distribution of Lochkovian mega- and microfossil assemblages based on an equatorial projection produced by Eldridge et al. (1996; 408 myr). The PALEOMAP Project. *Numbers* refer to those on table 2.2.

ment [e.g., altitude, climate, substrate for *Aneurospora/Cooksonia* (Edwards 1990)]. However, it is clear that cryptospores are far more abundant in the lowland floodplain deposits than the upland intermontane deposits, possibly because the former offered a greater number of wetter habitats more suited to the cryptospore-producing plants. Megafossil assemblages on the Old Red Sandstone Continent are too few, and they clearly show major taphonomic/facies bias (e.g., the mesofossils of north Brown Clee Hill) to be of use in solving problems posed by the microfossil record.

On a global scale, megafossils show a similar distribution to the Silurian, but a highly diverse assemblage recorded from the Kusbass basin, Siberia (Stepanov 1975), holds the key to the possible differentiation of a northern high-latitude phytoprovince, in addition to those of Laurussia, Kazakhstan (where further bizarre diversity is recorded), and also Australia. A limited and relatively poorly dated assemblage from Argentina contains no present northern hemisphere fertile taxa although they are at a similar level of organization.

SUMMING UP

Our brief in this paper was to document the evidence for the nature of vegetation and its global distribution during the earliest stages of terrestrialization by embryophytes, as a background to the many interdisciplinary studies covered in this book. We have therefore omitted any deliberations on phylogenetic relationships (see, e.g., Kenrick and Crane 1997a) or changes in diversity through the time interval (Edwards and Davies 1990). It is, however, relevant to comment on the extent to which the fossil record is representative of global vegetation.

The inadequacies of the plant megafossil record are well documented (e.g., Edwards 1990 for this time interval; Bateman 1991 for the Paleozoic) and we have attempted to remediate this by including some information on palynomorph assemblages. These are indispensable to an analysis of the earliest stages of terrestrial-

ization (being the only record!) and more rarely to extending paleogeographical ranges. In addition to the limitations shared with megafossils (e.g., facies and author bias, overrepresentation of the present circum-Atlantic area), the microfossil record has problems of its own. Thus in pre-Wenlock records, where assemblages comprise for the most part morphologically simple spores, ultrastructural studies are beginning to reveal that the same dispersed taxon may have had a number of producers. It has also been noted, although conversely, that different morphotypes (monads, dyads, and tetrads) possess identical envelopes and may have been produced by the same parent plant taxon. Finally, envelope-enclosed cryptospores may have lost their envelopes (during transportation or diagenesis), thus confusing naked and envelope-enclosed forms, although Johnson (1985) provides evidence indicating that this is unlikely.

For the post-Wenlock, quantitative studies of lists of dispersed spore species with numbers of megafossil taxa are few (e.g., Fanning, Richardson, and Edwards 1991), although the former clearly outnumber the latter. This has been attributed to production of spores by plants growing outside the "catchment area" resulting in preservation, or to the plants themselves being of low fossilization potential (see, e.g., Edwards and Richardson 2000). The occurrence of megafossils containing spores from the latest Silurian—earliest Devonian allows a more critical examination of the relationship between the dispersed spore and plant megafossil record. Findings suggest that spore formation was usually simultaneous, producing spores attributable to only one taxon in any sporangium (see summaries in Fanning, Richardson, and Edwards 1991; Edwards and Richardson 1996). However, Fanning et al. (1988) found that morphologically identical plant megafossils may contain different spore types; they suggested that for *Cooksonia* this reflected reticulate evolution, in which changes in spore form were rapid compared with those in gross morphology, although morphological convergence in sporangium shape (e.g., in *Salopella*) of plants probably not closely related also occurred.

To a small extent, this is counterbalanced by examples where morphologically similar spores were produced by different plant types [e.g., hilate cryptospores (Wellman, Edwards, and Axe 1998b) and triradiate spores with micro-granulate ornament]. Despite this progress in identifying the spore producers, there are major outstanding deficiencies, particularly relating to the sources of patinate and trilete spores and those with proximal radial muri (*Emphanisporites* spp.), and, especially important, the types of spores produced by the earliest lycophytes.

Progress in these areas, together with quantitative assessments of geographic variation on local and global scales, will play an important role in elucidating details of early terrestrial vegetation, as will the discovery of new assemblages of both mega- and microfossils.

3

Rustling in the Undergrowth:
Animals in Early Terrestrial Ecosystems

William A. Shear and Paul A. Selden

The study of animals in early terrestrial ecosystems as represented in the fossil record, long a neglected field, has undergone a striking renaissance in the past decade and a half. Spurred on at first by the recognition of unconventional kinds of fossil remains (Shear et al. 1984), we have now seen the exploration of informative new sites (Jeram et al. 1990; Shear et al. 1996), the development of new trace fossil evidence (e.g., Retallack and Feakes 1987; Banks and Colthart 1993; Trewin and McNamara 1995; Wright et al. 1995), attempts to integrate the animal evidence with that of plants (Scott 1991, 1992; Chaloner et al. 1991; Stephenson and Scott 1992; Edwards, Selden, et al. 1995), and frequent reviews of progress (Selden and Jeram 1989; Shear and Kukalová-Peck 1990; Shear 1990, 1992; Scott et al. 1992; Edwards and Selden 1993; Gray and Boucot 1994). Book-length treatments of the subject have also appeared [Little 1990; Behrensmeyer et al. 1992 (in part); Gordon and Olson 1995].

In this chapter, we propose to examine the presently available fossil evidence from two viewpoints: that of faunistics and that of trophic relationships, limiting our discussion to the Devonian and earlier. It is our hope that yet another review of this rapidly expanding field may usefully synthesize information old and new.

THE ECOLOGICAL SETTING

Earth's terrestrial environment in the mid Paleozoic must have presented a strikingly different picture from that with which we are familiar today. We now recognize that the greening of the landscape by early autotrophs was a protracted affair, which can be separated broadly into four or five overlapping phases (Gray 1985, 1993; Edwards and Selden 1993; see Kenrick and Crane 1997b, for a slightly different arrangement).

The first of these phases, which began in the Precambrian, consisted of microbial mats and occurs on bare mineral surfaces today. The second was a bryophytic phase, which began in the Ordovician and ended in the Early Devonian. This was overlapped by the third phase, which started in the Late Silurian. It consisted of small plants (rhyniophytoids) with axial organization and terminal sporangia, such as *Cooksonia*. The earliest good faunal evidence comes from this phase: the Ludford Lane biota. It is possible to imagine the Ludford Lane land animals (predatory centipeds and trigonotarbids) creeping through a salt-marsh rhyniophytoid sward and occasionally becoming commingled with the aquatic eurypterids, fish, and annelids during storms.

The fourth phase possibly began as early as in the Late Ludlow in Australia, or as early as the Gedinnian on the Laurentian continent, and represents the first major diversification of land plant life, such as zosterophylls, drepanophycaleans, and trimerophytes, and *Baragwanathia* in the southern hemisphere. All other early terrestrial faunas belong here (e.g., Gilboa, Alken-an-der-Mosel, and Rhynie), and the beginning of complex terrestrial ecosystems can be discerned. The decline of the fourth phase is difficult to determine. If the phase is recognized on plants alone, whether it includes the lycopsid/sphenopsid/pteridosperm forests of the Late Paleozoic depends on whether such plant groups are included with their extinct progenitors or their modern descendants. It is possible that the advent of trees (i.e., *Archaeopteris*) in the Middle to Late Devonian contributed to the end of this phase and to the modernization that was to take place in the Late Paleozoic.

In ecological terms, we do not know how widespread herbivory was in the Late Paleozoic. Early terrestrial ecosystems were decomposer-based (see later), but just when modern herbivory became commonplace and signalled the start of phase five is unknown.

For 50 years after the discovery of the Rhynie Chert animals in the 1920s (Hirst 1923), until the description of the Alken-an-der-Mosel and Gilboa faunas in the 1970s (Størmer 1970, 1972, 1973, 1974, 1976; Shear et al. 1984), no new early terrestrial faunas were found. During the last 25 years, our knowledge of early terrestrial faunas increased enormously, as the Alken and Gilboa faunas were described in detail, and the Ludford Lane, Red Hill, South Mountain, and eastern Canada faunas came to light. Nevertheless, until quite recently, our understanding of early terrestrial animal life was based almost entirely on information from these few Lagerstätten.

It is satisfying that the faunal lists from these localities broadly match, so that a consensus view of early terrestrial animals is possible. In the last few years, some new localities in the Catskill delta area, the Welsh Borderland, and Canada have turned up, which may prove to be more informative in the future.

FAUNISTICS

This section will attempt to summarize the fossil evidence of early land animals from both a chronological and a systematic perspective. We will begin with an overview of the kinds of evidence available for the emergence and evolutionary development of land animals, then turn to a chronological, site-by-site survey of the fossil record.

Trace Fossils

Trackways and Traces

Prior to the appearance of more or less complete body fossils of animals (arthropods) in the Late Silurian (Jeram et al. 1990), trace fossil evidence and tiny fragments attributed to animals carry us back well into the Ordovician.

The Borrowdale Volcanics

Johnson et al. (1994) described a series of trackways attributed to the ichnogenera *Diplichnites* and *Diplopodichnus* from strata within the Caradocian Borrowdale Volcanic Group of northwestern England. They argued persuasively for the subaerial nature of the deposit bearing the trackways, but conservatively stopped short of attributing them to terrestrial animals. However, the trackways are consistent with ones made experimentally by modern millipeds (Diplopoda) and later-occurring trackways that seem definitively terrestrial (Trewin and McNamara 1995). The animals that made the Borrowdale trackways may have been capable of limited excursions on land, or indeed could have been terrestrial.

Potters Mills

Retallack and Feakes (1987) described meandering, subvertical burrows in an Upper Ordovician paleosol in Pennsylvania. These burrows have bilaterally symmetrical backfilling resembling that found in the burrows made through decaying wood by some living millipeds. However, it is presently impossible to say with any certainty whether these burrows were made by early millipeds (which do not appear as body fossils until the Late Silurian), or even by arthropods. They remain a tantalizing bit of evidence of possible, very early colonization of the land by large animals.

Millerstown

Utilizing the technique of hydrofluoric acid maceration, Gray and Boucot (1994) recovered intriguing fragments of animals from the Lower Silurian (Llandovery) Tuscarora Formation, near Millerstown, Pennsylvania. Not formally trace fossils, they are treated here because at the present time it is impossible to attribute these fragments to specific taxa, even at the phylum level. They include pieces of bristles or setae which may be arthropodan (though they do not appear to be hollow, as are nearly all arthropod setae; annelid setae are not hollow), trachea-like tubes, and patches of cuticle with at least two kinds of spines. The evidence is strong that the Tuscarora Formation in Pennsylvania, at least the basal part examined by Gray and Boucot, was deposited as an alluvial fan. It is far from any contemporary paleoshoreline and lacks any marine fossils, including marine microfossils. Plant spores are abundant. The Tuscarora fragments suggest that some forms of freshwater or terrestrial animals were present by the Early Silurian; the trachea-like tubes are especially important, as they cannot be attributed to plants and have at least some of the attributes of the tracheae of air-breathing animals.

Newfoundland

Late Silurian trackways attributable to terrestrial or semiterrestrial arthropods are relatively abundant. Recently described examples include *Diplichnites* from Newfoundland, Canada (Wright et al. 1995). Wright and colleagues suspected the arthropleurid *Eoarthropleura* of being the maker of the Newfoundland trackway, a contention strengthened by the discovery of *Eoarthropleura* body fossils in the Lower Devonian of New Brunswick (Shear et al. 1996) and the Upper Silurian of Shropshire, England (Shear and Selden 1995). However, their argument was based on the anatomy of the leg tip of the much later (Carboniferous) *Arthropleura*, and recently

discovered material of the legs of *Eoarthropleura* show a strikingly different structure.

Western Australia

As many as eleven different species of arthropods (not all of them terrestrial) may have been responsible for a rich diversity of trackways and burrows in the Upper Silurian Tumblagooda Sandstone of Western Australia (Trewin and McNamara 1995). *Diplichnites* trackways range in width from 10 to 300 mm, some traceable for several meters across the exposures, crossing and recrossing similar trails. Other trails include drag marks between the footprints, as well as marks made by paddle-like limbs. Trewin and McNamara attributed the trackways to myriapods, eurypterids, scorpions, and euthycarcinoids; while some were made under water, a large number clearly represent animals walking on sediments exposed to air, and possibly on dry, wind-blown sand. These Australian trackways have a great deal in common with Lower Devonian ichnofossils from the Taylor Group of Antarctica (Bradshaw 1981; Gevers and Twomey 1982), and similar Devonian trackways are known from the Old Red Sandstone of Britain (e.g., Pollard and Walker 1984).

Tetrapod Trackways

Clack (1997) has reviewed the trackway evidence for terrestrial locomotion by Devonian tetrapods. She found the evidence, most of it from Famennian localities, to be ambiguous, and not clearly indicative in any one case of land-going behavior by any of the tetrapods now known from Devonian skeletal material. While some of the tracks, particularly those from the Genoa River Beds in New South Wales, Australia, undoubtedly represent those of tetrapods, there is no published independent support for the subaerial nature of the deposit; the animal that made them could have been swimming.

At the present time, most of the trackways that can be clearly attributed to tetrapods occur in regions that were, in the Devonian, far from the sites that have yielded skeletal material, and it is difficult to connect any of the known Devonian tetrapods to these trackways.

Coprolites

Yet another line of trace fossil evidence comes from coprolites. Upper Silurian and Lower Devonian coprolites attributable to micro-arthropods have been studied by Sherwood-Pike and Gray (1984) and Edwards, Selden, et al. (1995). The pellets described by Sherwood-Pike and Gray from the Silurian of Gotland, Sweden, contained fungal hyphae, attributed to the predominantly terrestrial Ascomycetes, while those studied by Edwards and her colleagues (1995) consisted of masses of spores of land plants. The coprolite evidence will be discussed in detail in a later section.

Injuries to Plants

Damage to plants, probably caused by arthropods, has been reviewed by Chaloner et al. (1991). Controversy still remains over the damage to Rhynie plants (Lower Devonian) first noted by Kidston and Lang (1921a) and reported in more detail by Kevan et al. (1975). Kidston and Lang initially attributed the lesions in Rhynie plants to damage from volcanic cinders or other purely physical agents, perhaps unaware of the presence of animal fossils in the deposit. However, in the light of data on the details of the kinds of damage caused by the attacks of insects and other animals on modern plants (Labandeira and Phillips 1996), the arguments of Kevan and his coauthors are quite convincing, although none of the known animals from Rhynie possess the right kind of feeding apparatus, nor are any of their modern analogues feeders on living plant material (Edwards and Selden 1993).

Banks and Colthart (1993) examined in detail damaged specimens of *Psilophyton* from the Lower Devonian (Emsian) of Gaspé, Québec, Canada. They found several types of wounding and wound reaction, suggesting attack by animals and subsequent response by the plants. Possible coprolites were also found, both within and outside of plant stems. Animal fossils have been reported from the same formation (Battery Point Formation), but none of

them are potential candidates for having caused the wounding (Labandeira et al. 1988; Shear et al. 1996). No further evidence of damage to plants by animals during the Devonian period has been found, but data on plant–animal interactions of this type from the Carboniferous onward are substantial (Chaloner et al. 1991; Scott 1992; Scott et al. 1992; Stephenson and Scott 1992) and growing (Labandeira and Phillips 1996).

Body Fossils

This evidence is organized chronologically and by fossil site. In subsequent sections, the major groups of animals will be treated in more detail as to their mode of life and ecological role.

Body fossils of Silurian and Devonian land animals occur in two preservational contexts: as organically preserved cuticle that can be extracted from the matrix using hydrofluoric acid (HF) maceration (Shear et al. 1984; Jeram 1994a; Braun 1997), and as impression fossils. Only lately has it been recognized that both types can occur together (Shear et al. 1996).

Silurian Period

The oldest reported fossil of a putatively terrestrial animal is that of the supposed milliped *Archidesmus loganensis* Peach, 1899, from the Llandovery of Scotland. However, both Rolfe (1980) and Almond (1985) have examined the specimen and are inclined to support the contention of Ritchie (1963) that it is probably a misinterpreted algal fossil. Almond (1985) noted myriapod-like arthropod fossils from the Wenlock Fish Beds of Lesmahagow, Scotland, but thought them best considered *incertae sedis*. From the Stonehaven Group of Scotland, other myriapod-like fossils have been recorded, including *Kampecaris obanensis* Peach, and a convincing, as yet unnamed, diplopod (Almond 1985; Shear 1998). Marshall (1991) has argued on the basis of spore evidence that the true age of the Stonehaven Group is Late Wenlock to Early Ludlow. From the setting, the terrestrial nature of these animals cannot be fully established. Most of these Silurian myriapod-like fossils are typical impressions, but some also preserve cuticle (Rolfe 1980).

Ludford Lane. A basal Prídolí (c. 414 Ma) locality, Ludford Lane was well known for many years as an outcrop of the Ludlow Bone Bed, a concentration deposit containing vertebrate teeth and scales in a thin, rusty-weathering sandstone. The best-known outcrop of the Ludlow Bone Bed is at the junction of Whitcliffe Road (formerly known as Ludford Lane) and the main road to Leominster, just south of the bridge over the river Teme at Ludlow. Here, the Ludlow Bone Bed Member is a 0.21 m siltstone containing, in addition to the main basal bonebed, a number of other thin bone-beds (see Bassett et al. 1982 for details).

Many years of eager collecting of the basal bone-bed, latterly using long chisels, had resulted in an overhang that threatened to collapse into the lane. The Nature Conservancy Council stepped in to remedy the situation and removed the overhanging rock before tidying up the site and preserving it. It was this removed material that was subjected to hydrofluoric acid treatment and that produced the Ludford Lane biota (Jeram et al. 1990).

The half-meter of material overlying the Bone Bed Member (the Platyschisma Shale Member) consists of storm-dominated silts with some minor concentration layers. Some of these lags consist of very fine, unoxidized mud, which is easily washed away. Hydrofluoric acid digestion of these sediments produced a diverse biota of marine (e.g., scolecodonts), brackish (e.g., eurypterids), and terrestrial components. It has been suggested that the storms washed components of a terrestrial ecosystem into the estuarine environment, where they were mixed with shallow marine fauna. The animals may have lived in a rhyniophytoid salt marsh. Sedimentology of the Platyschisma Shale Member reveals that the biota was deposited in lags within dominantly intertidal and storm-generated deposition. In some parts of the Platyschisma Shale, plant- and animal-rich blocks of finely (tidal?) laminated siltstone form breccias within channels, giving the impression of storm-induced erosion of a tidal salt marsh. Salt marshes are more stable than bare muddy, sandy, or rocky littoral zones because of the ameliorating effect of the plants. Salinity, temperature, and water supply may fluctuate as much as on the bare shore, but at root, litter, and substrate level these fluctuations are lessened. Interestingly, salt marshes are providing a route for talitrid amphipods, crabs, and prosobranch and pulmonate gastropods to colonize the land at the present day, as well as for recolonization of the littoral by terrestrial insects and spiders (Little 1990).

The animal fossils consist of organically preserved cuticle and were recovered by HF maceration; only a single reasonably intact individual, a trigonotarbid, has been found (Dunlop 1996a). Fragments of eoarthropleurids (Shear and Selden 1995), scutigeromorph centipeds (Shear et al. 1998), and scorpions have been identified. Numbers of other specimens cannot reliably be assigned to taxa but may include spiders and archaeognath insects.

Early Devonian Period

Rhynie. Perhaps the most famous (but comparatively little-studied) Devonian locale for early terrestrial animal fossils is the hot spring chert deposit at Rhynie, Scotland (Trewin 1994). Here, silica-saturated waters from a thermal spring flooded and silicified nearby vegetation, with dense growth of vascular plants, algae, and fungi.

Significant numbers of animals, all arthropods, were also preserved. The quality of preservation is unparalleled anywhere in the fossil record, with many of the animals three dimensionally embedded in the translucent chert and showing minute setae, cuticular microsculpture, and the impressions on the cuticle of individual epidermal cells. Although the technique has not been much used, some material may also be isolated from the matrix using HF.

The age of the chert has been set at Siegenian (Late Lochkovian–Early Pragian) by means of spores, while argon dating has produced a figure of 396±8 mya, which spans from the Late Gedinnian through the entire Siegenian and Emsian to Early Eifelian (Trewin 1994); thus the oft-cited date of 404 mya is at the lower limit of possibilities. In any case, the site can be considered to date from the middle to late Early Devonian.

The fauna includes hexapods [Collembola (Greenslade and Whalley 1986)], mites (Hirst 1923; Dubinin 1962), trigonotarbid arachnids (Hirst 1923; Shear et al. 1987) and a freshwater crustacean, as well as a possible eurypterid and an enigmatic pair of mandibles that may be insectan (Hirst 1923; unfortunately, this specimen seems to have been lost). In 1993, Andrew Jeram found part of the leg of a scutigeromorph centiped in a sawn slab of Rhynie Chert, the first new animal to be discovered in the chert since 1923 (Shear et al. 1998). Current faunal studies by N. Trewin and L. Anderson (Aberdeen University) are also yielding new animals.

Alken-an-der-Mosel. The arthropod fauna of the Nellenköpfchen-Schichten of Alken-an-der-Mosel, Germany, was described by Størmer in a series of papers that culminated in a discussion of the paleoecology of the site (Størmer 1970–1976). Recently, Braun (1997) described some new material from the Alken site.

The Alken section comprises two horizons of plant-rich shales separated by about 20 m of sandstones and siltstones. During the Lower Emsian (upper Lower Devonian), there was an island bordered by tidal sand-flats with temporary lagoons. Marine- and brackish-water animals inhabited the lagoons, which were bordered by mangrove-type vegetation. Therefore, as at Ludford Lane, there is a commingling of terrestrial and aquatic organisms in the Lagerstätte.

Psilophytes were common, both terrestrial (and possibly partly submerged) and true aquatics. Størmer (1976) suggested that euryhaline bivalves were the permanent inhabitants of the lagoon. At flood times, marine animals were washed in. Some swimming animals probably wandered in and out of the lagoon: *Parahughmilleria, Jaekelopterus,* and other eurypterids, and the ostracoderm *Rhinopteraspis.* Xiphosurans, stylonuroid eurypterids, and scorpions were probably amphibious. Terrestrial animals recorded at Alken are trigonotarbids and the arthropleurid *Eoarthropleura;* the latter could have been amphibious (Størmer 1976).

Braun (1997) isolated trigonotarbid and spider cuticular remains from Emsian sediments of the Eifel region, and he also demonstrated that cuticle could be obtained from the Alken matrix.

Sites in Eastern Canada. Labandeira et al. (1988) macerated a single fragmentary specimen of an archaeognath insect, *Gaspea paleoentognatha* Labandeira, Beall, and Hueber 1987, from a beach cobble probably derived from the Battery Point Formation. Jeram et al. (1990) argued that this specimen may not be a fossil, but a contemporary archaeognath that crawled into a crack in the rock. They pointed out that the three-dimensional preservation is unique, that there is no associated cuticle fauna, as generally found at other sites, that the specimen lacks the usual taphonomic signs (e.g., impressions of matrix granules), and that the morphological differences between *Gaspea* and living archaeognaths noted by the authors are not clear. Until more material is found, the true nature of *Gaspea* remains in doubt; as of 1999, nothing additional concerning this specimen has appeared in print.

Unquestionable terrestrial arthropod fossils were reported from Gaspé by Shear et al. (1996). Several specimens of a large, cylindrical milliped were found in the Battery Point Formation. They also briefly described a scorpion and an eoarthropleurid from the Campbellton Formation (Emsian) of New Brunswick, Canada. The scorpion specimen was notable because book lung tissue was preserved, the earliest evidence of air-breathing in scorpions. Extensive assemblages of vascular plants have been recorded from both the Campbellton and Battery Point Formations (Gensel 1982; Gensel and Andrews 1984; and see chapter 11).

Other Sites. Nearly contemporaneous with Rhynie are the milliped fossils briefly reported from the Early Gedinnian (Lochkovian) Lake Forfar by Trewin and Davidson (1996), and millipeds and kampecarids from elsewhere in Scotland (Almond 1985). Kampecarids and a variety of diplopod-like fossils appear sporadically in the Lower Devonian rocks of Britain; some of these represent new high-level taxa (Almond 1986; Shear 1998).

Middle and Late Devonian Period

Blenheim-Gilboa. The extensive exposures of Middle and Upper Devonian rocks found in New York State and Pennsylvania have yielded a great deal of evidence of the development of terrestrial animal life. The best-studied site at this time is mid-Givetian Blenheim-Gilboa of the Schoharie Valley of New York State (Shear et al. 1984), where a thick, plant-bearing lens of black shale was uncovered during excavations for a pumped storage power project. Before the reservoir was flooded, a large amount of matrix was collected by paleobotanists James Douglas Grierson and Patricia Bonamo.

The Gilboa (Brown Mountain) host rock is a dark gray shale that occurs as lenses within the thick Panther Mountain Formation. This formation forms part of the Catskill Clastic Wedge, a delta complex developed on the western side of a mountain chain, close to the equator, during the Devonian period.

Latest palynomorph data (Richardson et al. 1993) give an Eifelian/Givetian (382 Ma) age for the Lagerstätte, slightly older than suggested in previous literature. A tropical savannah climate with alternate wet and dry seasons has been postulated on paleogeographical and paleobotanical evidence (Shear et al. 1987). The stratigraphy associated with this deposit is complex (Woodrow 1985; Miller and Woodrow 1991), and it is difficult to place in relation to a constantly fluctuating paleoshoreline. However, *in situ* evidence suggests that the particular location of the Blenheim-Gilboa lens was probably an interdistributary pond on a prograding delta.

Subjecting this material to HF maceration to release plant fossils, Grierson and Bonamo discovered a complex fauna of terrestrial arthropods, including trigonotarbids (Shear et al. 1987), centipeds (Shear and Bonamo 1988), mites (Norton et al. 1988), spiders and amblypygids (Selden et al. 1991), scorpions, pseudoscorpions (Schawaller et al. 1991), arthropleurids, archaeognaths, and other as yet undetermined taxa.

The fossils consist of organically preserved cuticle, and while the level of preservation is not as good as at Rhynie, it surpasses that at Ludlow Lane, and a larger number of more complete specimens have been recovered than have been found at other Middle and Lower Devonian cuticle-producing sites.

The main floral component of the Lagerstätte is the lycopsid *Leclercqia*, which apparently grew on stream banks and around ponds. This occurs in the rock in dense mats of stems interlocked by means of their hooklike, divided leaves. Other lycopods and shrubby progymnosperms are also found. Both *Leclercqia* and the animals in this Lagerstätte were perhaps transported to the site of deposition (allochthonous), possibly with the mats of drifted vegetation providing filters that trapped the animal debris during flood episodes (Shear et al. 1984). However, it is possible that the animals also lived among *Leclercqia* stems in life.

Other Catskill Sites. Other sites in the region have been explored and are yielding similar faunal elements, but much more remains to be done. Following up on the report by Kjellesvig-Waering (1986) of a scorpion from a shale quarry near Blenheim-Gilboa at South Mountain, Shear and Selden (1995) determined that instead of a scorpion, the cuticle fragments isolated using HF maceration by paleobotanist F. M. Hueber were from an eoarthropleurid. Subsequent exploration of this quarry has revealed one or more small lenses similar to the fossil-bearing rock at Blenheim-Gilboa, and maceration has produced more eoarthropleurid remains, trigonotarbids, spiders, mites, archaeognaths, and scorpions.

The age of the deposit was placed by Hueber and Banks (1979) at Early Frasnian, earliest Late Devonian, but recent examination of the spores by J. Beck (pers. comm., 1998) suggests a somewhat earlier late–latest Givetian to Early Frasnian. Thus, this site is clearly younger than Blenheim-Gilboa. While there are some similarities between this site and the *Leclercqia*-dominated Blenheim-Gilboa, the major floral elements yet described are psilophytes, liverworts, and archaeopterid trees (Hueber and Banks 1979). Lycopods, the dominant element at Blenheim-Gilboa, are apparently absent. The faunal differences noted may represent real differences as opposed to sampling errors, since the flora is so strikingly different. The differences in flora may be due either to the ages of the two deposits, or to ecological differences.

From the upper Delaware Valley of New York State and Pennsylvania have come a few tantalizing specimens, including a flat-backed milliped resembling *Archidesmus,* and a trigonotarbid closely related to those found at the earlier Alken-an-der-Mosel site in Germany (Shear 2000). The authors were not successful in relocating the outcrops or any corresponding lithology during the field season of 1996, however, so the true provenance of these specimens remains in doubt. The exposures in the region range over much of Frasnian and Famennian time.

Red Hill, a latest Famennian site in central Pennsylvania, has produced material of the tetrapod *Hynerpeton bassetti* Daeschler, Shubin, Thompson, and Amaral 1994. Plant material, mostly derived from archaeopterid trees and *Protolepidodendron,* is unusually abundant and well preserved at this site; HF macerations yield mats of leaves and stems strongly reminiscent of the lower layers of modern forest litter. Brown in color and translucent, the fossilized litter is possibly secondarily oxidized, or perhaps was never reduced and carbonized to the extent of coeval deposits elsewhere. Terrestrial arthropod remains have also been discovered, including both large impression fossils and organically preserved cuticle. Elements found so far include trigonotarbids, scorpions, and fragments that cannot be assigned to known taxa.

Into the Carboniferous

A 23-million-year gap separates the modest Devonian terrestrial faunas from the extensive and complex ones of the Upper Carboniferous. At this time, there are no known Tournaisian records for terrestrial animals; but a single trigonotarbid specimen (*Pocononia whitei* Ewing 1930) has been collected from the Pocono Mountain Formation of Virginia, probably located near the Tournaisian/Visean boundary (Dunlop 1996b). A significant Visean site for terrestrial animals, including arachnids, myriapods, and a rich fauna of tetrapods is found at East Kirkton, West Lothian, Scotland (Rolfe et al. 1994).

FAUNAL ELEMENTS

Chelicerata

Scorpionida

The earliest evidence of air-breathing in scorpions comes from the Emsian of New Brunswick, Canada (Shear et al. 1996). Among preserved scraps of scorpion cuticle is an abdominal plate with attached lung tissue. Mesoscorpions are a prominent element in the preserved cuticle faunas of the Silurian and Devonian, appearing at Ludford Lane, Gilboa, South Mountain, and Red Hill. It is difficult to establish terrestriality for scorpions without having certain body parts (lungs, leg-tips) present, but Jeram (in Shear et al. 1996) saw evidence for the terrestriality of all mesoscorpions. Contemporaneous aquatic Devonian scorpions are known from Alken (Brauckmann 1977), but the unique respiratory arrangements of *Waeringoscorpio* require reexamination; the supposed feathery gills of this scorpion are unlike the book gills of earlier, aquatic Silurian forms.

All extant scorpions are predatory (Polis 1990). The known morphology of fossil scorpions is consistent with these feeding habits. The size range of Devonian scorpions is not easy to establish. The New Brunswick Emsian scorpion may have been about 9.5 cm long (Shear et al. 1996), while the size of fragments found at Gilboa and South Mountain suggest animals less than 1 cm in length, although the maturity of any of these specimens cannot be assessed. Carboniferous terrestrial scorpions reached 30 cm (Jeram 1994b).

Trigonotarbida

Trigonotarbids are superficially spider-like arachnids lacking silk-spinning organs and with the abdomen enclosed in segmented armor. They range from Late Silurian to Early Permian (Dunlop 1996a). The single Silurian record is of *Eotarbus jerami* Dunlop 1996, from Ludford

Lane. Devonian records are scarce. Brauck-mann (1994) reported two partial specimens from Emsian strata near Waxweiler, Germany, and found them similar to the ones previously recorded at the nearly contemporaneous Alken site (Størmer 1970), while Schultka (1991) assigned an Emsian trigonotarbid from the German Bensberger Shale to a Carboniferous genus, *Trigonotarbus*. The exquisitely preserved Rhynie trigonotarbids have been restudied by Dunlop, but the results have not yet been published (Dunlop, pers. comm. 1995). Shear et al. (1987) described three genera, containing seven species, from the cuticle fauna of Blenheim-Gilboa, although one of these is now known to be a spider (Selden et al. 1991). Trigonotarbid cuticle has been found at South Mountain, evidently not from any of the Blenheim-Gilboa species. As yet undescribed material, both from isolated cuticle and from a single complete impression fossil, is now known from the Late Famennian of Red Hill, Pennsylvania, and another new impression fossil is probably from the Upper Devonian of the Delaware Valley, New York. These are important because with *Poconoinia* of the Early Carbonifer-ous, they bridge the enormous time gap between the Middle Devonian and Late Carboniferous trigonotarbids.

Dunlop (1996c) attempted to untangle the classification and phylogeny of the trigonotar-bids, an enterprise that is beyond the limited scope of this chapter.

Two ecomorphotypes of trigonotarbids can be recognized among the Devonian material. "Paleocharinids" are generally small (1–6 mm long), lightly sclerotized animals with remnant lateral eyes (a pleisomorphy) that probably were inhabitants of the litter. They are limited to the ?Silurian and Devonian. (The question mark is needed because the assignment of *Eotarbus jerami* to one or the other of these groups is uncertain.) "Aphantomartids" are larger, more heavily sclerotized animals without lateral eyes (apomorphic), postulated to be surface-dwellers or possibly arboreal; they range from ?Silurian to Permian. As far as can be deter-mined, the functional morphology of the feed-ing apparatus of both types is that of a carnivore;

prey was probably chewed with the chelicerae and liquid remains sucked up (Dunlop 1994).

Araneae

A single species of Devonian spider, *Attercopus fimbriunguis* Shear, Selden, and Rolfe 1987 has been described from Blenheim-Gilboa (Shear, Palmer, et al. 1989; Selden et al. 1991; see the lat-ter paper for discussion of other putative records of Devonian spiders). This is the oldest known fossil of this ecologically dominant arachnid order, and it appears to be a represen-tative of a taxon forming the sister-group to all other spiders. However, undiagnostic scraps of cuticle from Ludford Lane (Upper Silurian) may be from spiders. Abundant new material of *Attercopus* has appeared in macerations from South Mountain, in somewhat younger (Frasn-ian?) rocks.

Attercopus possesses a well-developed spinning apparatus and was evidently capable of making silk constructs, although analogies with other primitive spiders make us suspect that it was a burrower and used silk only to line its burrow or possibly to construct simple trip-lines. All spi-ders so far studied are predatory and the anatomy of *Attercopus* does not differ signifi-cantly from living spiders in this respect.

Amblypygi

Ecchosis pulchibothrium Selden and Shear, 1991, known only from leg podomeres from Blenheim-Gilboa, may be an amblypygid (Selden et al. 1991) on the basis of the strongly ornamented trichobothrial base on the patella. However, the patella is differently shaped, lacking the adapta-tions found in living Amblypygi for reflexion of the leg, and the supposed tibiae have rows of large sockets bearing robust, striated, bifid spines unlike any known from modern amblypygids (but which do occur in liphistiomorph spiders!). It is possible that the material represents an undiagnosed order of arachnids, or of a very primitive amblypygid ancestor lacking the extreme flattening of the body found in Carboniferous–modern forms. This flattened habitus may have developed as an adaptation for hiding under loose tree bark, a habitat only just available in the Middle Devonian. Che-licerae are unknown for *Ecchosis,* but all living amblypygids are predatory.

Pseudoscorpionida

The only known Devonian pseudoscorpion is *Dracochela deprehendor* Schawaller and Shear 1991 from Blenheim-Gilboa (Shear, Schawaller, and Bonamo 1989; Schawaller et al. 1991).

Pseudoscorpions are tiny predators living in soil, litter, or crevices in bark, or under stones (Weygoldt 1969). *Dracochela* is not very different from modern neobisioid pseudoscorpions and undoubtedly had a similar lifestyle.

Acari

The earliest mite fossils are specimens from the Rhynie Chert (Hirst 1923). Originally placed in but a single species, they may actually represent as many as four distinct families (Dubinin 1962) but require detailed reexamination (Norton et al. 1988). Two oribatid mites (Norton et al. 1988) and one alicorhagiid (Kethley et al. 1989) have been discovered at Blenheim-Gilboa. A few poorly preserved specimens of oribatids similar to one of the Gilboa oribatid families (Devon-acaridae) have been recovered from the younger South Mountain site.

Oribatids are detritivores or fungivores in modern ecosystems; alicorhagiids are known to prey on nematodes. Mites are undoubtedly underrepresented in the record because of their low fossilization potential (Shear 1990). They play a major role in soil and litter communities today and doubtless did so in the past.

Atelocerata

Diplopoda

Milliped fossils may reach as far back as the Wen-lock (Almond 1986; Marshall 1991), depending on the dating of the Stonehaven Group of Scot-land, which contains myriapod-like fossils. If Marshall's proposed Wenlock–Ludlow bound-ary age for these strata is correct, or if Almond's Lesmahagow Fish Bed (Wenlock) specimens are in fact millipeds, these would be the oldest body

fossils of land-dwelling animals. With the excep-tion of these specimens, *Archidesmus macnicoli* of the Scottish Gedinnian is the earliest milliped. In an unpublished thesis, Almond (1986) has comprehensively studied the few other Devon-ian milliped fossils and diagnosed a number of new taxa, some of which do not seem very close to modern millipeds and hint at an early diver-sity for this group.

Until very recently all records for Devonian millipeds were from Britain; Shear et al. (1996) have now described cylindrical millipeds from the Emsian of Canada. These specimens are very distinct from the British representatives and are similar to Carboniferous and modern ana-logues. Two specimens of flat-backed millipeds have recently turned up from the Upper Devon-ian of the Delaware Valley. Tesakov and Alekseev (1992) described possible milliped diplotergites from the Devonian of Kazakhstan, but this mate-rial is somewhat ambiguous. The absence of mil-lipeds from any of the discovered cuticle faunas is curious, but it probably results from a preser-vational bias. The cuticle of millipeds is impreg-nated with calcium carbonate, which would readily disappear in the acid preservational envi-ronments with which cuticle faunas are associ-ated. The organic fraction of the cuticle, thus weakened, would probably disintegrate.

With very few exceptions, all living millipeds are detritivores—a few rarely attack live plant material, and still fewer may scavenge carrion (Crawford 1992; Hopkin and Read 1992).

Chilopoda

Leg segments attributable to scutigeromorph centipeds have been found at Ludford Lane (Silurian), Rhynie (Lower Devonian), and Blenheim-Gilboa (Middle Devonian). These distinctive podomeres are pentagonal in cross section and bear marginal setae and serrations (Shear et al. 1998). Shear and Bonamo (1988) diagnosed a new centiped order, Devonobio-morpha, on the basis of *Devonobius delta* Shear and Bonamo, 1988, from Blenheim-Gilboa. Both of these groups are heterotergous, an adaptation for fast running.

The scutigeromorphs, with their extraordinarily long legs, were probably surface-running, rather than litter inhabitants, but the much smaller *Devonobius* (about 10 mm long) seems to have been cryptic. *Devonobius* bears well-developed poison claws with which to seize prey; today's scutigeromorphs are also predators.

Arthropleuridea

Arthropleurids are an extinct class of atelocerates, ranging in the fossil record from Upper Silurian to Upper Carboniferous. Respiratory structures are not known for any arthropleurids, but it is widely assumed because of the preservational context and other anatomical features that they were at least partly terrestrial. Of the three orders (one as yet undescribed), two are known from the Silurian and Devonian (Shear 1998). Eoarthropleurids are a part of the Ludford Lane (Upper Silurian; Shear and Selden 1995) fauna, and they also occur at Alken and in New Brunswick (Lower Devonian; Størmer 1970; Shear et al. 1996) and South Mountain (Upper Devonian; Shear and Selden 1995). Ranging in size up to 15 cm long, they probably resembled large, flat-backed millipeds, but fully articulated specimens have never been found, so nothing is known of their heads or how many segments may have been in the trunk. Later Carboniferous forms were detritivores (Rolfe and Ingham 1967).

An unnamed order of arthropleurids is prominent in the Blenheim-Gilboa and South Mountain faunas. These animals were probably less than 5 mm long, with but 9 or 10 trunk segments. The structure of the head is very similar to that of millipeds, except that eyes and antennae seem to be lacking, and suggests detritivory as a feeding habit (Shear 1998; Wilson and Shear, in press). At Blenheim-Gilboa, this animal is one of the most abundant, and thus it was probably a significant portion of the prey available to the trigonotarbids, spiders, and centipeds.

Hexapoda

Parainsecta are represented in Devonian faunas by Collembola (springtails) from Rhynie. Greenslade and Whalley (1986) and Greenslade (1988) claimed that *Rhyniella praecursor* Hirst, 1923,

belongs to the living family Isotomidae, and that other species of springtails remain to be described from the chert, including neanurids.

Collembola are entognathous detritivores/fungivores that occur in modern soils in enormous numbers. A few species are also found in aquatic to semiaquatic situations, feeding on algal mats; this may have been the original habitat and may have carried them onto land very early (Shear and Kukalová-Peck 1990).

The evidence for Devonian insects is equivocal. We have already alluded to the problems with the supposed earliest insect fossil, *Gaspea paleoentognatha*; more material must be discovered and the original specimen described in detail before it can be accepted as genuine. Meanwhile, pieces of cuticle closely resembling that of modern archaeognaths have been found at a number of cuticle-bearing sites. The cuticle is thick, often poorly preserved, and bears short, arched rows of oblong sockets for the insertion of scales. There is a single scrap of this distinctive cuticle from Ludford Lane, large sheets have appeared in macerations from Blenheim-Gilboa, and a few pieces from South Mountain still have the scales in place. In addition, a series of antenna-like fragments and compound eye scaffolding from Blenheim-Gilboa may signal the presence of insects; particularly intriguing are ?antennomeres with unevenly bifurcate setae, known from living thysanurans.

Several unusual fossils from South Mountain are very similar to the cibarial or frontal pumping apparatus that occurs on the heads of modern insects with sucking mouthparts. This is particularly intriguing, given recent evidence of suctorial damage to Lower Devonian plants (Banks and Colthart 1993).

Thus, while the evidence is suggestive, unquestionable fossils of insects have yet to be found in Devonian rocks; the earliest known insect at this time is *Delitzschala bitterfeldensis* Brauckmann and Schneider, 1996, from the lower Namurian A of Germany. This becomes the first known Lower Carboniferous insect fossil courtesy of a redefinition of the Lower Carboniferous to include the lower part of the Namurian A.

Living archaeognaths are microherbivores (subsisting on algal and fungal crusts and films) or detritivores.

Chordata

Vertebrata, Tetrapoda

All evidence of Devonian tetrapods is limited to the Late Frasnian and Famennian. No less than eight genera of tetrapods have been named, coming from localities in Greenland, Scotland, Pennsylvania, Australia, Russia, and Latvia (Ahlberg and Milner 1994; Daeschler and Shubin 1995). The ecological setting of most sites seems clearly to be deltaic or lacustrine, though there are indications of a shallow marine environment for *Tulerpeton* (Clack 1997).

Vertebrates were comparative late-comers to land, and the evidence suggests that they retained a strong connection to the aquatic habitat throughout the Late Devonian. An approximately 23-million-year gap in the record during the Early Carboniferous makes it very difficult to relate these early tetrapods to later forms (Carroll 1992).

The evidence tells us that by the time of their first appearance in the fossil record in the Frasnian and Famennian, tetrapods had already achieved considerable diversity and worldwide distribution, so their actual origin must be placed at some time earlier in the Devonian. The earliest record of limb bones (Ahlberg 1991) is from the Frasnian of Scat Craig, Scotland, of *Elginerpeton pancheni* Ahlberg 1995. *Obruchevichthyes*, from Latvia and Russia, is of a similar age (Ahlberg and Milner 1994). Famennian tetrapods, in additional genera, come from Australia, Greenland, and North America (Daeschler and Shubin 1995). The phylogenetic relationships of these genera and the abundant and diverse tetrapods of the upper Early Carboniferous are not well understood (Thomson 1991; Carroll 1992; Ahlberg and Milner 1994).

The ecological setting and habits of the early tetrapods have been the subject of some controversy. Bendix-Almgreen et al. (1990) pointed to evidence suggesting that *Acanthostega* of Greenland lived in active channels in fluvial-dominated environments. Work by Clack and her colleagues (summarized by Zimmer 1995) has shown that the forelimbs of *Acanthostega* may have been paddle-like, and that the animal retained internal gills. They suggest that *Acanthostega* inhabited weed-choked river channels, where it lay in wait for prey. It is not clear whether this form was primarily or secondarily aquatic, but it probably spent little, if any, time on land.

The exact sedimentary context from which the large number of *Ichthyostega* specimens comes has not been established (Clack 1997) and the animal has been reconstructed by Jarvik (1996) as much more of a terrestrial creature, albeit with a fishlike tail supported by fin rays. Clack (1997), on the other hand, pointed to a number of anatomical details that suggest a primarily aquatic lifestyle. In particular, she argued for massive, powerful forequarters, much smaller, paddle-like hind limbs, and a relatively rigid, barrel-shaped body, leading her to postulate a seal-like lifestyle. The large, sharp, recurved teeth of *Ichthyostega* would have served well for catching and holding fish.

Conversely, the teeth of other early tetrapods, such as *Elginerpeton* and *Acanthostega*, are relatively small and closely set. The flattened skulls of all three genera, with eye sockets located dorsally, suggest a shallow-water habitus and the taking of prey from on or near the surface. It is possible that *Acanthostega*, at least, fed on insects or other land dwellers that fell into streams.

Thus, the early tetrapods of the Devonian seem to have been carnivores based in aquatic environments, playing only a limited role, if any, in the terrestrial food chain.

TROPHIC RELATIONSHIPS

Co-occurrences of Animals with Plants

Early terrestrial plant and animal fossils often occur commingled at the same localities. However, there are many sites that have yielded only plants, not animals, since plants are generally more abundant as fossils. The co-occurrence of plant and animal fossils is mainly the result of paleobotanists finding the animal fossils in passing, while looking for plants (this is certainly true of Rhynie and Gilboa), and later searches for animal fossils rely on finding good plant fossils, which indicate that animals may also be present.

Nevertheless, there seems little doubt that plants and animals were associated ecologically in early terrestrial ecosystems. What is the nature of the relationship? Presumably, plants occupied the producer trophic level on which the remainder of the ecosystem is based. In this section, we examine the evidence for direct interaction between plants and animals in early terrestrial ecosystems.

Animal–Plant Interactions

Animal–plant interactions in modern ecosystems can be classified into feeding, shelter, transport, reproduction, and coevolution (Scott et al. 1992); examples of all of these have been found in the fossil record, but evidence only of feeding and coincidental spore dispersal in the Siluro-Devonian.

Feeding by animals on plants includes not only direct herbivory but also digestion of dead plant material and its decomposing microflora by detritivores. Fossil evidence for animals feeding on plants comes from four sources: (1) plant morphology (anatomy and pathology), (2) animal morphology, (3) direct associations between animals and plants, and (4) coprolites.

Evidence from Plant Morphology and Pathology

Indirect paleobotanical evidence for plant–animal interaction in early terrestrial ecosystems comes from plant morphology—in the form of adaptations for defense from animal attack, and pathological reactions to attack—in the form of healed wounds. Both are presumably related to herbivory.

Defensive Adaptations.

Spines and Enations. Many early terrestrial plants bore spines, which suggests they may have been defensive adaptations against herbivory. However, other reasons for the spines have been proposed. Kevan et al. (1975) suggested that the

short, upward-pointing, scalelike enations of early vascular plants facilitated upward (but not downward) climbing of plant axes by arthropods; this would favor spore-eaters, which might then jump or glide off the sporangium after feeding, thus aiding dispersal.

Outgrowths from the axis in early vascular plants would undoubtedly increase surface area for photosynthesis, and possibly some of the outgrowths contained chlorophyll, although this is difficult to prove given the state of preservation (Lyon and Edwards 1991). Dense, silvery hairs, on the other hand, are used by plants in exposed areas today to prevent excess transpiration (e.g., hairy alpines and cacti). Another reason suggested for axial lateral branches is to enable support and separation of axes in dense stands, or for a scrambling habit.

Nevertheless, some of the spines on early terrestrial plants look like sharp, defensive spines, and some occur on sporangia (Fanning, Edwards, and Richardson 1991), although a nutritional role has been postulated by these authors in addition to the possible defensive one. Finally, spines on spores, which commonly have bent or bifid tips, may not be defensive but could have attached to the bodies of animals for dispersal (Kevan et al. 1975).

The main problem with a defensive role for spines on the axes of early terrestrial plants is that they are too large to have been an effective deterrent against the animals that are known to have occurred with the plant fossils. While smaller animals could conceivably have been overlooked in the rock matrix, one would expect larger animals to be preserved. (On the other hand, processing rock in acid without looking at the surfaces could result in larger animals being disintegrated into unrecognisable pieces. For this reason, it is always advisable to search rock surfaces carefully before maceration.)

Evidence for Chemical Defenses. Darkly pigmented residues within spinous cellular hairs on aerial axes of the zosterophyll *Trichopherophyton* from the Rhynie Chert were considered by Lyon and Edwards (1991) as possible evidence for the formation and/or storage of substances that were toxic or distasteful to herbivores. These authors argued, however, that this suggestion was not in accord with the lack of evidence for herbivore fossils in this Lagerstätte.

Spore Coats. Silurian spores are simpler, with no, fewer, or less complex spines and other ornament, than those of the Devonian and later. An increase in complexity continued throughout the Devonian so that a great many palynospecies are recognizable by the end of the period (see Kevan et al. 1975: pl. 55). These authors postulated that such a radiation must be adaptive, since energy is involved in the formation of ornament. Reasons for spore ornament include increased dispersal resulting from increased buoyancy in air or water or attachment to a dispersal vector. Since dispersal is a primary function of spores, such function is clearly useful. However, complex ornament would also help deter spore-eating organisms, for which the spore coat is nutritively valueless in comparison with the internal protoplasm. The thick spore coat of sporopollenin would also deter palynophagy (spore feeding) in addition to preventing water loss. Kevan et al. (1975: Appendix) discussed these and other adaptations of spore coats in great detail.

It is difficult to be sure how much coat thickness and ornament are directed against adverse physical, animal, or pathogen attack and how much toward dispersal mechanisms, including animal dispersal externally or in guts. Nor do we know how much coat thickness and ornament are related to the intrinsic needs of the spores (e.g., resistance to desiccation and increased buoyancy). There may also be constraints on spore structure that are essentially nonadaptive.

Wounds in Fossil Plants. Wounds in *Psilophyton* from the Emsian (394–387 Ma) of Canada were described by Banks (1981), Banks and Colthart (1993), and Trant and Gensel (1985). Banks and Colthart (1993) described wounds apparently produced by chewing and piercing in *Psilophyton* and an unnamed trimerophyte from Gaspé. The surface (presumed chewing) wounds on *Psilophyton* axes are generally small (<3 mm in diameter) and elliptical. They apparently occurred on erect axes 1 to 3 feet above ground level, and there is evidence, in the form

of cell proliferation, for wound repair. Hence, Banks and Colthart (1993) concluded that the wounds were formed while the plants were alive. Piercing wounds were identified by their cone-like form, the apex extending into the axis.

Lesions with a distinct tissue reaction were recognized in Rhynie plant axes by Kidston and Lang (1921a). These lesions (figured also by Kevan et al. 1975) commonly consist of dark areas extending from the epidermis into the cells in a wedgelike form, in most cases not reaching the vascular tissue, and outward to form a blister-like shape. Clear cells surrounding the wound apparently show an elongate growth reaction. The dark material ("opaque organic matter" of Kevan et al. 1975) is presumably some sort of healing exudate. Kevan et al. (1975) distinguished three types of lesions, which differ in the regularity of the wound and the amount of opaque organic matter.

Kidston and Lang (1921a,b) speculated that prolonged exposure to siliceous water or water vapor from the hot springs may have been the cause of these lesions. Kevan et al. (1975), citing Tasch's (1957) study, which concluded that the Rhynie plants did not grow in "soil saturated with hot silicic waters" (Tasch 1957:17), suggested the alternative possibility: that the injuries could be attributed to sap-feeding animals. They pointed out that the juicy phloem-equivalent tissue of *Aglaophyton* was close to the surface of the plant. Labandeira and Phillips (1996) noted that wound reactions, such as cell proliferation, a dark exudate around the damaged area, and stylet insertion paths (some to vascular tissue) point to deliberate piercing by animals with sucking mouthparts rather than accidental injury. These have not been demonstrated in Rhynie plants.

It seems to us that there are a number of ways in which lesions could have been produced in early land plants. Chemical irritation, as originally suggested by Kidston and Lang (1921a,b), may well have been present in the hot-spring environment, but its effect probably would not have been to produce deep, thin lesions more suggestive of stylet wounds. Physical wounding can be caused accidentally, not just for the purpose of extracting nutriment from the plants. Again, hot water is unlikely to have produced the type of wounding seen; wind-blown sand grains are a possibility, but perhaps improbable too. As previously mentioned, some of the early vascular plants were spinose, and the spines may well have been able to puncture adjacent axes. It is conceivable that spinose stems gave their bearer an advantage over naked stems, which were thus vulnerable to wounding.

Deliberate wounding by animals in the form of chewing was suggested by Banks and Colthart (1993); this seems a less sophisticated method of feeding than sap sucking, since presumably only the cell contents could be digested, not the cellulose walls. However, it may well have been more efficient than sap sucking, since many cells would be attacked simultaneously; unless the sap sucker penetrated vascular tissue and tapped a continuous supply of fluid, the stylet would be able to puncture only one cell at a time. But the question arises whether just taking the occasional nip from a plant stem would be a very rewarding way of feeding. Perhaps the plants were deliberately wounded to allow sap to escape, which could then be lapped up.

Though not conclusive, the evidence for sap sucking in the Rhynie Chert and Gaspé plants is more than suggestive. Cell fluids are more easily digestible than cell walls, so sap feeding could have evolved easily, and there is the possibility of pre-adaptation to fluid sucking as a feeding method in aquatic precursors.

Evidence from Animal Morphology. A general survey of the animals found in early terrestrial ecosystems was given earlier in this chapter; here, we concentrate on the evidence from animal morphology for plant–animal interactions.

Animals, particularly arthropods, exhibit a wide variety of methods of feeding on plants, including biting, sawing, cutting, chewing, rasping, mining, boring, mopping exudate, swallowing whole, and piercing and sucking juices. Some of these are particular to certain plant organs (e.g., boring wood and seeds, leaf mining, and swallowing whole seeds, pollen grains, and spores). Animal mouthparts can be a good clue to their mode of feeding—the lepidopteran "tongue," for example. However, piercing and sucking mouthparts occur in parasites of both plants and animals, and chewing mouthparts may be considered generalized. So, morphology alone is often not good enough, and comparison with modern relatives of the systematic group to which the animal belongs is normally necessary to determine the probable method of feeding.

Arachnids Largely Predatory. Earlier in this chapter, the feeding methods of the various arachnid groups known from early terrestrial Lagerstätten were briefly discussed. All modern arachnids are predators with the exception of some mite groups (see later). Reports exist of spiderlings ingesting pollen grains (possibly by accident) and of nectar feeding by some adult spiders (Taylor and Foster 1994).

The oldest known mites (Acari) occur in the Rhynie Chert. Hirst (1923) thought the specimens were conspecific; he named them *Protacarus crani*, and he placed them, with some doubt, in the modern family Eupodidae. Dubinin (1962) considered they represented five species belonging to four families: *Protacarus crani* (Pachygnathidae), *Protospeleorchestes pseudoprotacarus* (Nanorchestidae), *Pseudoprotacarus scoticus* (Alicorhagiidae), and *Paraprotacarus hirsti* and *Palaeotydeus devonicus* (Tydeidae). John Kethley (Field Museum of Natural History, Chicago) restudied the specimens and questioned the alicorhagiid affinity of *Pseudoprotacarus scoticus* because of its pretarsal morphology (Kethley et al. 1989). He considered all to belong to the family Pachygnathidae (pers. comm. in Norton et al. 1988) with the exception of the nanorchestid (a family that is nevertheless included in the superfamily Pachygnathoidea).

Little is known about the food preferences of living pachygnathoid mites, but Krantz and Lindquist (1979) reasoned that pachygnathoids probably feed by sucking fluid from algal cells; they pointed to the sharply pointed mouthparts of these mites, thought by Trägårdh (1909) to be a piercing organ, and work by Schuster and Schuster (1977), who observed nanorchestids feeding on algal mats and refusing animal food.

Feeding Methods of Collembola and Archaeognatha. Devonian Collembola are known only from the Rhynie Chert (Greenslade and Whalley 1986). In his comprehensive review of the biology of Collembola, Christiansen (1964) noted that very little is known about the food preferences of collembolans with piercing and sucking mouthparts, but suggested that fungal juices might form their diet. What information is recorded about other Collembola is patchy and not always reliable. For example, some will eat almost anything when in captivity but are highly selective in the wild. Among the foods listed by Christiansen (1964) are fungal hyphae, bacteria, decaying plant and animal material, frass, algae, and spores. Some forms will eat living plant roots and seedlings and undecayed leaves, while others are carnivores preying on smaller animals (Protura, rotifers, tardigrades, and other Collembola).

Archaeognathans, known from Gilboa, South Mountain and perhaps Ludford Lane, have milling mouthparts that are used to scrape algae and lichens from rocks (Ferguson 1990).

Kraus and Kraus (1994) gave a brief review of food preferences of Tracheata, in which they concluded that a food niche common to all the main tracheate taxa was piercing/sucking. For example, Symphyla feed on plant rootlets, other soft tissues of plants, and small arthropods. Pauropoda and Protura apparently suck the juices of fungal hyphae. Some members of Zygentoma, Pterygota, and Collembola also feed in this way. These authors pointed out that relatively few, unrelated, tracheate groups have developed new (i.e., apomorphic) feeding mechanisms, and that these correlated with a body size increase, as in many Pterygota.

Chilopoda Predatory. Modern centipeds are generally unspecialized predators. While the Devonian Devonobiomorpha have no modern representatives, their morphology points to them as undoubted predators. Scutigeromorpha are among the earliest known land animals, found in the Silurian at Ludford Lane, and the Devonian of Rhynie and Blenheim-Gilboa (Shear et al. 1998). Modern scutigeromorphs are fast-running predators; *Scutigera coleoptrata* is a common house centiped in the Mediter-

ranean region and has been introduced into houses widely throughout warm parts of the world, where it can be seen chasing insects attracted to lights at night.

Diplopoda and Arthropleurida Detritus Feeders. Millipeds are principally detritivores, although some use gut flora to digest living plant tissue. In modern millipeds, the density of teeth on the gnathal lobes determines the size of food particles ingested; smaller species have smaller teeth, and the smaller the particles, the greater the proportion of food assimilated (Köhler et al. 1991).

Millipeds gain nutriment not only directly from the detrital material but also by digesting saprotrophic microorganisms present on the detritus, which may proliferate during the passage of food through the gut. They eat a variety of foods but when provided with choice show a clear preference for partly decayed leaves from particular species (Kheirallah 1979; Piearce 1989).

Coprophagy is common among litter feeders, and autocoprophagy is essential for the milliped *Apheloria montana* (McBrayer 1973). It may be that this species has to continually reinfect its gut with appropriate microbes. Enzymes are excreted by salivary and midgut epithelial glands and pass through the peritrophic membrane to digest the food. One function of the peritrophic membrane, which occurs in a variety of arthropods and annelids in addition to millipeds, is to protect the gut epithelium from abrasion by coarse detritus. Martin and Kirkham (1989) have shown how the membrane wraps around individual gut items and may also be involved in the digestive process. Broken-down peritrophic membrane passes out with the feces, and the membrane is renewed at ecdysis in arthropods. In Diplopoda, anal glands secrete a gluelike substance that binds fecal particles together (Schlüter 1982, 1983).

Arthropleurids appear to have had mouthparts similar to millipeds, and a Carboniferous *Arthropleura* specimen found with gut contents (Rolfe and Ingham 1967) showed it ate plant detritus. We presume that the Devonian eoarthropleurids were detritivores, too.

Direct Associations of Plants and Animals.
Trigonotarbids in Sporangia. A good example of

a direct association between plants and animals in an early terrestrial ecosystem is the trigonotarbids found within axes and dehisced sporangia in the Rhynie Chert (Kevan et al. 1975: pl. 56; Rolfe 1985: pl. 1, fig. 1). Kevan et al. (1975) reported the suggestion of W. D. I. Rolfe, who brought these specimens to their attention, that the trigonotarbids could have gotten there actively, for feeding or to avoid adverse physical conditions (e.g., temperature, humidity, windspeed), or passively by postmortem transport. Kevan et al. (1975) rejected the latter on the basis of the autochthonous/parautochthonous nature of the deposit, and preferred the idea of trigonotarbids entering the plant cavities actively, possibly to feed on spores.

Since trigonotarbids are almost certain to have been carnivores, it is more likely that they sought shelter there, perhaps for molting (Rolfe 1980). Indeed, Rolfe (1980) reported that none of about 15 dehisced sporangia still in growth position examined by Dianne Edwards showed any trigonotarbid remains, and the geopetal infill of the trigonotarbid carcass and the plant axis shown in Kevan et al. (1975: pl. 56, fig. 1) indicated that the sporangium containing the trigonotarbid was not in life position.

Many occurrences of trigonotarbids in the Rhynie Chert can be shown to be of molted animals (Rolfe 1980); one interesting example shows a trigonotarbid leg inside a trigonotarbid abdomen within a plant axis (Kevan et al. 1975: pl. 56, fig. 1). This lends support to the idea that trigonotarbids used dead plant axes and empty sporangia as refugia for molting rather than for palynophagy. We note here that Bonamo and Richardson examined the contents of more than 150 *Leclercqia* sporangia but found no animal remains (reported in Shear et al. 1987).

Coprolites. Coprolites bearing plant remains provide direct evidence of plant–animal interaction in early terrestrial ecosystems. However, this evidence needs careful evaluation. Coprolites are proving to be not uncommon in macerates of early terrestrial sediments, although they are not always recognised as such, especially if their constituents are poorly aggregated, but also if they are composed of spores that are all of the same type, in which case they show a superficial resemblance to sporangia (Edwards, Selden, et al. 1995). The main occurrences of coprolites in Siluro-Devonian sediments are reviewed here. By Carboniferous times, coprolites had become relatively common (Scott and Taylor 1983).

Silurian. Possible coprolites containing fungal hyphae were described from the Late Silurian (Ludlow) Burgsvik Sandstone of Gotland, Sweden, by Sherwood-Pike and Gray (1985). The possible coprolites are subcylindrical, 62 to 260 mm long and 18 to 40 mm in diameter, and consist of hyphae in an amorphous groundmass. These authors discussed the possible producers of the fecal-like masses, and fungivorous arthropods seemed to be the most likely candidates. Nevertheless, they conceded that the hyphae may have been later invaders of the coprolites, a view echoed by Scott et al. (1992).

Judging from the photographs and drawings in their paper (Sherwood-Pike and Gray 1985: fig. 2G, fig. 5F,G), the hyphae appear to be continuous, and not broken into segments as might be expected if the producer was feeding on the contents. However, the walls of the hyphae in the supposed coprolites appear to be more irregular than those of the hyphae found loose in the macerates.

Sherwood-Pike and Gray (1985:9) discussed and dismissed the possibility of modern contamination as the source of the fungi in their samples; in contrast, Selden (1981: fig. 23m) illustrated fungi etched from Silurian limestones of the Baltic region, which he concluded were modern fungi growing on the surface of weathered rock before etching. Devonian fungi are well known from the Rhynie Chert (Taylor, Remy, et al. 1995) and Gaspé (Banks and Colthart 1993).

If the hyphal aggregates described by Sherwood-Pike and Gray (1985) are indeed coprolites, then they represent an early example of presumably terrestrial coprolites, but not necessarily, in our opinion, of fungivory.

Devonian of Gaspé. Banks and Colthart (1993) described possible coprolites or regurgitates among material of *Psilophyton* and other trimerophytes from the Lower Devonian (Emsian) of Gaspé, Canada. A larger mass (3.0 × 1.5 mm) consisted of an unstructured mix of collenchyma and xylem tissues, while smaller pellets of amorphous material occurred within stems. The authors concluded that these two types of coprolite must have had different producers: The larger type seems to have resulted from indiscriminate chewing of a variety of tissues, while the smaller pellets seem to have been produced by animals munching their way along the inner cortex of the plant stems. It seems to us that the larger coprolite could have been produced by an animal eating living or dead plant tissue—that is, a herbivore or a detritivore.

Spore-bearing Coprolites from the Welsh Borderland Silurian and Devonian. Edwards, Selden, et al. (1995) described coprolites stuffed with numerous types of spores from a number of Silurian and Devonian sites in the Welsh Borderland. Because they contain spores, the coprolites were initially thought to be sporangia, but their regular shape, lack of a sporangial wall, and the presence of more than one spore type (up to nine in some specimens) and other debris ruled out this possibility.

Eighty-five coprolites range from 0.95×0.74 mm to 3.3×1.27 mm in size. Some are regularly cigar-shaped while others are truncated at one end. Consistency in shape, and the production of fecal pellets of a similar size and shape by some modern millipeds, suggested that the truncated examples are complete. Less regular shapes were interpreted as coprolites on their composition. Spore-dominated examples have smooth contours except where impregnated by pyrite. In contrast, those with cuticle, unidentifiable plant debris, and sporadic spores are less regular, with voids separated by draped debris.

Ninety-five percent of the coprolites contain spores, which occur in varying proportions with cuticle, tubes, and unidentified plant material. At least 25 different spore types were distinguished. Nonspore constituents include a featureless film, quite distinct from macerated cuticle, possibly solidified mucilage; featureless sheets interpreted as cuticle; occasional examples of sterome and tracheids; and possible nematophyte tubes.

Edwards, Selden, et al. (1995) discussed whether the coprolites were derived from terrestrial or aquatic animals, and whether they were feeding on living or dead plant tissue. The coprolites had clearly drifted into a fluvial overbank flood deposit. Some evidence favored aquatic producers—for example, the presence of cuticles of possible aquatic animals in the same sedimentary sequence: eurypterids, scorpions, and kampecarid myriapods occur there, although evidence suggests that some or all of these animals may have been terrestrial or amphibious in the Devonian. There is evidence for a land origin for the coprolites in the presence of predominantly land-derived debris, with the same preservational characteristics as in the productive samples, and the lack of identifiable remains of unequivocal aquatic organisms in the coprolites.

Extant herbivores exploit gut fungi and bacteria to degrade cellulose, but even with their assistance the nutrient value of living vegetative tissues would have been low; thus palynophagy is an attractive possibility. Because sporopollenin-impregnated spore walls would be largely impermeable to enzymes, a spore-eating animal would have to crack the spores to digest them; the evidence suggested that the few damaged spores seen in the coprolites were broken as a result of post-depositional phenomena, not deliberate cracking. Another possibility is that herbivores fed on peripheral exospore layers or locular fluids, or on sporangial contents before sporopollenin deposition occurred. Immature sporangia would have provided an energy-rich nutrient source of presumably relatively high nitrogen content.

Edwards, Selden, et al. (1995) concluded that the coprolites indicated an animal that was ingesting spores and spore masses, but there is no evidence of the spore contents having been digested. Since the coprolites consisted entirely of organic matter, and no sedimentary particles, the producer was unlikely to have been a soil or sediment feeder. Litter feeders eat a variety of foods, and studies of fecal contents of detritivorous invertebrates showed that a variety of matter passes through the gut, including spores, which remain undigested. The paucity of recognizable plant matter other than spores in the coprolites is consistent with their production by a detritivore. Digestible matter in litter is quickly broken down, in contrast to indigestible spores; many litter arthropods ingest their own feces or those of other detritivores, in which case a great deal of breakdown would already have occurred; indeed such a habit would tend to concentrate resistant items like spores in feces. A significant proportion of nutrition in detritivores is derived from the digestion of saprotrophic fungi and bacteria, so it is possible that the animal was selectively ingesting spores on which microorganisms were consuming the exospore—for example, the remains of extrasporal material such as demonstrated for *Cooksonia* sporangia.

Overrepresentation of spore-containing coprolites might reflect the lack of recognition of non-spore-bearing coprolites. Feeding experiments with millipeds show that fecal pellets dominated by plants are less regular and so, in fossils, could be overlooked or discarded as very poorly preserved fragments of plant axes. The available evidence suggests that the coprolites were produced by detritivores that fed on litter rather than a mixture of litter and sediment, and that ingested spores and spore masses, which were abundant in the debris. Digestive processes, coprophagy, and selective feeding were given as possible explanations for the high concentration of spores seen in the coprolites.

An impression of the size of the producer was deduced from the size of the coprolites and knowledge of the sizes of modern animals, so most collembolans, mites, nematodes (too small), and earthworms (too large) were excluded from consideration. The coprolites predate the earliest fossils, and the presumed origin, of terrestrial isopods and gastropods. Enchytraeid oligochaetes are common components of the soil fauna at the present day, but their casts would not resemble the coprolites. This leaves millipeds (or other myriapods) and possibly large collembolans, both of which are known as fossils in Devonian terrestrial biotas. Kampecarid myriapods occur in the same beds as the coprolites, but little is known of their biology. It is most likely that the animals producing the coprolites were terrestrial detritivores feeding on litter and were possibly arthropods similar to modern millipeds.

Undescribed Coprolites from the Devonian of Rhynie. Undescribed coprolites from the Rhynie Chert were brought to the authors' attention by Hass and coworkers. These are recognizable as coprolites from their shape and because identifiable contents are mainly fragmentary. They contain very few spores, and of the other identifiable contents, some dense material may be sporangium wall. Possible chitinous (i.e., arthropod) cuticles may also be present. In some preparations fungal hyphae appear to pass through the coprolites. The presence of possible arthropod cuticles would suggest a detritivore, since predatory arachnids would suck fluid out of prey and leave no cuticle-bearing coprolites. However, the rejectamenta of arachnids that chew their prey could possibly resemble coprolites.

TROPHIC RELATIONSHIPS IN EARLY LAND ECOSYSTEMS

Little Evidence for Herbivory

We have shown that most of the terrestrial animals found as fossils in early terrestrial biotas belong to primarily carnivorous groups. Some mites and collembolans could be microherbivores (fungivores or sap suckers) or decomposers; some of the myriapod taxa were probably detritivores. There is no evidence of animals that eat living, growing plant material.

The preponderance of carnivores in the Rhynie Chert was discussed by Kevan et al. (1975), who suggested three possible explanations: (1) small, soft-bodied prey animals were not preserved; (2) some of the arthropods were facultative herbivores; or (3) some of the predators were amphibious and returned to the water to feed.

Rolfe (1980) saw no problem with the lack of herbivores; he advanced the hypothesis that herbivores followed plants onto land and were succeeded by carnivores—a trophic-level succession with time. However, Rolfe (1985) rightly emphasized that the greatest proportion of primary production in the terrestrial ecosystem passes through the decomposer chain at the present day.

Shear et al. (1987:9) mentioned the "striking predominance of predatory arthropods" in both the Rhynie and Gilboa faunas, and they gave as potential reasons the small sample, the differential preservation of arthropod cuticles, and the possibility of soft-bodied herbivores. Selden and Edwards (1989) mentioned two possible reasons for the lack of herbivores in early terrestrial ecosystems: poor preservation/recovery of herbivores because of their probable small size, and the greater importance of the decomposer food chain.

In a review of early terrestrial ecosystems, Shear (1991) pointed out that the fossil record should be taken at face value (particularly since new localities preserving early faunas showed similar suites of arthropods). Siluro-Devonian ecosystems may have been based on detritivores and microherbivores, herbivory evolving only much later (Carboniferous?) when animals had developed gut microfloras to aid in the breakdown of lignin, other recalcitrant materials, and toxins, a view echoed by Edwards and Selden (1993).

A New Definition of Herbivory

Biologists are as guilty as laymen in relating more to large, conspicuous bushes, trees, birds, and mammals than to the unseen microbes and microarthropods with which terrestrial ecosystems teem. Price (1988) termed this "Noah's Ark Ecology." In his review of ecosystem development over time, Price (1988) pointed out the importance of the substrate, be it marine mud or terrestrial soil, and of the availability of nutrients in decaying organic matter. Other authors, such as Seastedt and Crossley (1984), have emphasized that, even at the present day, cycling of nutrients through the decomposer niche has a greater impact, in most cases, than herbivory in terrestrial ecosystems.

For reasons that will become clear in the discussion to come, we wish to propose a much narrower definition of herbivory than has been current in the literature. While herbivores have traditionally been considered to be animals feeding on any plant parts, we believe that at least three subgroups can be recognized, and we will limit our concept of herbivory to only one of them.

Herbivore processing of plant material is inefficient. An examination of the feces of any animal that feeds primarily on live plant material (aside from spores, seeds, or fruits) often reveals identifiable fragments of the plants eaten. Large numbers of other organisms have adapted to exploit these feces, which speaks volumes about the nutrient value still remaining.

The absence of endogenous cellulases and lignases, for example, from nearly all herbivorous animals strongly suggests that herbivory could not have been a primary adaptation of animals. Instead, plant-eating animals adopt various strategies that suggest that herbivory is a much later adaptation, developed through the acquisition by animals of mutualistic microbes that do most of the digestion for them. We will examine the reasons for thinking this is so.

Plants as Food

This section relies heavily on the information from the book edited by Abrahamson (1989).

Allelochemicals

Allelochemicals are substances produced by plants that offer some degree of protection against herbivores (Fraenkel 1959; Rosenthal and Janzen 1979). These are known as secondary compounds, because they have no role in the primary metabolic pathways of plants. The presence of such chemicals is nearly universal in vascular plants, with some being so poisonous that they are virtually immune to herbivory. Since plants have no excretory systems, these often-toxic chemicals must be sequestered. The longer a plant part is in existence, the more toxic it becomes as molecules of these compounds accumulate.

Nor is toxicity the only effect. Allelochemicals can inhibit feeding, act as repellents, reduce digestibility, and even mimic animal hormones, speeding up or slowing down developmental changes (Weis and Berenbaum 1989; Lindroth 1989).

Ephemeral parts of plants, such as flowers, are therefore not likely to accumulate large amounts of these substances, nor are seeds or spores. Fruits, which are designed to be eaten, very rarely contain appreciable amounts of toxins.

Strategies available to herbivores to circumvent allelochemical defenses include avoidance, detoxification by endogenous enzymes (Brattsten 1979), detoxification by gut microorganisms (Lindroth 1988), and sequestering. At least a few insects have co-opted plant toxins for their own defense (Duffey 1980).

Low Nutrient Content

Animals require not only calories, but specific nutrients, such as certain amino acids and minerals. Thus, the value of their food depends not just on the available calorific content, but on the amounts of micronutrients and protein available. Seeds, fruits, pollen, and spores may not only have a very dense calorific content but also be high in proteins and contain little indigestable material.

Leaves and shoots of plants, on the other hand, are not particularly nutritious. Much of the carbohydrate biomass is fundamentally indigestible by animals, consisting of such recalcitrant molecules as cellulose, hemicellulose, and lignin. While high in calorific value, these substances cannot be broken down by the endogenous enzyme systems of animals.

The protein content of plant foliage is very low, as measured by percent nitrogen. A number of studies have demonstrated that nitrogen content is an important regulator of insect feeding; nitrogen-poor food must be consumed in much greater quantities (Weis and Berenbaum 1989). Amino acid balance is also highly significant, and particular plants may be deficient in amino acids required by herbivores.

All parts of plants may be low in micronutrients; the best example is sodium, which is found in very low concentrations in plant tissues. Vertebrate herbivores may have to seek out other sources to overcome sodium deficiencies, hence the well-known phenomenon of the salt lick. The analogous process in some insects is "puddling," in which Lepidoptera, in particular, gather at animal feces, urine, or carrion to lap up sodium-rich fluids (Arms et al. 1974).

Furthermore, the quality of plant food changes drastically over time, perhaps by as much as an order of magnitude, so animal digestive systems must be prepared to adapt to such changes (Scriber and Slansky 1981).

Problems for Sap Suckers

Feeders on plant juices not only face most of the problems just outlined but in addition may incur problems of water balance. In particular, xylem sap is so low in nitrogen content that insects feeding on it must process up to 1,000 times their body weight in sap each day (Horsfield 1978). Both xylem and phloem sap are available only if the requisite tissues can be reached, and the channel must be kept open.

Our Concept of Herbivory

First, we consider as herbivores only those animals that feed on the living tissues of plants; animals that process dead or decayed plant material are detritivores. Their food is nutritionally very different from living plant material because it has already been attacked by fungi and bacteria, which render it more digestible and also enhance the nutrient content by converting some of the carbohydrate to protein.

We also consider as special cases animals that feed primarily on seeds, spores, and pollen. These plant parts, while living, are nutritionally very different from the vegetative parts of the plant; in particular, they may be high in calorific value and contain abundant lipids and protein. While seeds may be heavily defended by recalcitrant shells and coats, they usually contain little indigestible material and the rewards for cracking the defenses are high. Spore and pollen coats are largely indigestible and must be cracked physically or chemically to be digested.

The discussion will be more illuminating if we focus instead on the use of the vegetative parts of plants as food, because this draws into consideration all the problems of plants as food briefly surveyed here. *Thus, for us, true herbivores are animals that routinely feed on leaves, shoots, and roots.*

Where feeders on plant fluids fit in the scheme (or how feeders on plant fluids are defined) depends on the case; while avoiding the problem of indigestibility, most sap or cell content feeders are confronted by the same difficulties of nutritional deficit and allelochemicals.

Fundamental Adaptations of Herbivores

Because of the low nutrient content of their food, herbivores have only a limited number of options. One is to consume enormous quantities of low-quality foods and to extract from it what they can; this is a common strategy among lepidopteran larvae and orthopterans. The gut becomes extremely large to handle the mass of food, and conversion values are low. This strategy is probably not available to larger herbivores, such as mammals.

A second strategy, adopted by nearly all vertebrate herbivores and many arthropods, is to form mutualistic relationships with microorganisms that live in the gut and process the food. The host herbivore then either digests the mutualists or lives on their by-products. The coadaptations of the participants can become extremely detailed, such that the herbivore host regulates conditions in its gut to enhance the growth of the mutualists, which in turn may tune their life histories to changes in the host, signalled by hormonal changes.

Microorganisms play a vital role in digestion in litter-feeding invertebrates; for example, it is unlikely that arthropods can digest cellulose without the aid of gut microflora (Swift et al. 1979). In millipeds, there is no evidence that the gut microorganisms exist in a symbiotic relationship, as they do in termites (Hopkin and Read 1992), but the flora does include soil and intestinal bacteria, actinomycetes, and fungi (Szabo et al. 1992, and references therein). Indeed, the digestive efficiency of millipeds is reduced if microorganisms are excluded from the diet, and the microorganisms proliferate within the warm, moist conditions of the gut (Anderson and Ineson 1983; Tajovsky 1992). Desert spirostrepsid millipeds bask in the sun to increase body temperature and thus aid digestion (Crawford et al. 1987), an activity that renders the animals susceptible to water loss and predation. The microorganisms themselves form an important part of the diet of millipeds, as shown by the experiments of Bignell (1989, reported in Hopkin and Read 1992). The passage time of gut contents for three milliped species was measured as 2 to 5 hours during constant feeding, and 11 to 12 hours without food (Brüggl 1992). Assimilation efficiency has been measured in many ways by a number of authors (Brüggl 1992: table 3), and Hopkin and Read (1992) advocated caution when assessing figures based on subtraction of weight of feces from weight of food ingested, because of changes in moisture content and gut passage times. It is likely that a maximum efficiency of 30 percent is typical for millipeds.

The milliped example may illustrate how the mutualistic strategy evolved. For soil detritivores, the soil itself has been described as a vast, external rumen. Not only is new plant matter that hits the ground eaten, but fecal pellets are ingested and reingested as long as they have any nutritional content; they are worked on by the gut flora of the detritivores as well as by free-living soil bacteria and fungi. Indeed, there may be no clear line of demarcation between the two.

When detritivores ingest soil, plant remains, and microorganisms, the biochemical activity of the microorganisms continues in the gut, and some forms will colonize the gut cavity and adapt to life there. This, in turn, would lead to adaptations on the part of the host to retain and "culture" them. With a highly adapted gut flora, animals could attack fresher and fresher plant material, perhaps eventually shifting to parts of live plants.

This is, of course, but one hypothetical path to herbivory. Others must lead through the route of sap sucker and seed or spore eating.

CONCLUSIONS

The fossil animal evidence shows a preponderance of carnivores, the presence of some detritivores, and the absence of herbivores. Hence it points strongly toward a food chain based in detritivory. This is common in soils today, and note, too, that modern herbivory involves gut floras that evolved presumably from detritivorous ancestors.

This suggests that either we are sampling a soil ecosystem, or that the food chain, even for surface dwellers, was detritivore based. Although Devonian evidence is scarce, no definitively herbivorous terrestrial animals have been found, even aside from the major cuticle-bearing Lagerstätten (Shear et al. 1996). If the latter is true, then trophic relationships in early terrestrial ecosystems were radically different from those in today's herbivore-dominated world.

Plant evidence presents a somewhat different picture. There is evidence (previously summarized) for damage to plants, and it is reasonable to hypothesize that at least some of this damage was caused by animals in attempts to feed on living plants. The perpetrators of the damage are entirely unknown, but the pattern suggests feeding on plant fluids, either by piercing vascular tissue or by damaging the plants and lapping up any fluids that leaked out. Conclusive evidence for animals (insects) eating the vegetative parts of plants does not appear until the Late Carboniferous (Labandeira and Phillips 1996), and we can infer from anatomy that herbivorous vertebrates appeared even later, near the Carboniferous–Permian boundary.

4

New Data on *Nothia aphylla* Lyon 1964 ex El-Saadawy et Lacey 1979, a Poorly Known Plant from the Lower Devonian Rhynie Chert

Hans Kerp, Hagen Hass, and Volker Mosbrugger

The Lower Devonian chert from the Rhynie locality, Aberdeenshire, Scotland, has become famous as one of the finest sources of information on early life on land, for the flora as well as for the fauna (Trewin 1994; Cleal and Thomas 1995). Some of the plants originally described by Kidston and Lang in their classical monograph series (1917, 1920a,b, 1921a,b) are the best known and most detailed fossil land plants [e.g., *Aglaophyton major* (D. S. Edwards 1986; Remy and Hass 1996)]. Other taxa remain very incompletely known. These include *Nothia aphylla*, a taxon for which only aerial axes, sporangia, and small pieces of epidermis have been described.

When Kidston and Lang (1920b) found small leafless axes and detached "pear-shaped" sporangia in close association with *Asteroxylon mackiei* in two blocks of chert, they regarded them as belonging to this species. Lyon (1964) published a short note on the leafy fertile axes of *A. mackiei*. The leafless axes and sporangia originally attributed to *A. mackiei* appeared to belong to a different plant, which he called *Nothia aphylla*. This name was a nomen nudum, since he gave no diagnosis. Høeg (1967) gave a brief discussion and refigured some of Kidston and Lang's specimens and two of Lyon's photographs, but he neither gave a proper diagnosis nor did he indicate a type. The most detailed account of the monotypic genus *Nothia* (El-Saadawy and Lacey 1979)

included aerial axes, the xylem, smaller pieces of epidermis and sporangia, plus the first diagnosis of the genus and the species and a designated holotype. Most of El-Saadawy and Lacey's observations were based on cellulose acetate peels. Although they documented a number of previously unknown features, notably of the fertile parts of the plant, their photographs generally do not show much anatomical detail. Although the species is relatively abundant in some chert blocks, *Nothia aphylla* is commonly regarded as one of the least known higher land plants of the Rhynie Chert (Gensel and Andrews 1984; Taylor and Taylor 1993), in that only the aerial parts have been described to date. Here we give a description of the whole plant, including the previously unknown rhizomes.

Only aerial parts of most fossil plants are known. Although roots and rootlike structures are not uncommon, these are often difficult to correlate with aerial parts of the plants. The evolutionary significance of basal parts of early land plants and their role in the early colonization of terrestrial habitats are still poorly understood. The rhizomes described here and the basal parts of the aerial axes of *Nothia* are preserved in life position and in all anatomical details. Detailed anatomical studies enable the functional interpretation of the various tissue types and ecological adaptations. The material therefore offers a unique opportunity to reconstruct the growth form and the life history of an early land plant in an exemplary completeness. Moreover, the ecological and adaptive relations to the whole environment including the substratum can now be documented.

According to Niklas (1997), the greening of the terrestrial landscape by plant life was really more an invasion of air than of land. The example of *Nothia* shows that greening of the land was also an invasion of the substratum. Root-bearing land plants played an important role in the formation of soils (Algeo and Scheckler 1998). However, *Nothia* documents that plant—substratum interactions were already well developed in the Early Devonian, even when these plants still did not have real roots.

LOCALITY AND MATERIAL

The material was collected at Rhynie, 14 km south of Huntly, in the Grampian region of Scotland. For further details of the locality, see Kidston and Lang (1917, 1920a,b, 1921a,b), Trewin and Rice (1992), Trewin (1994, 1996), Cleal and Thomas (1995), and Rice et al. (1995). The Rhynie Chert has been dated as Pragian/ Siegenian (Richardson 1967; Richardson, in House et al. 1977). Rice et al. (1995) provided a radiometric age of 396±12 Ma years for the cherts, confirming the earlier palynological data.

Preservation and Taphonomy

The rhizomes of *Nothia* were found in two blocks of cherty sandstone. The figured aerial organs (figure 4.1, and see figures 4.11D,F,G, 4.12B–E, 4.15A,B, and 4.18) are from other blocks of chert—that is, a so-called massive to vuggy chert (Trewin 1996). The rhizome-bearing rocks consist of an irregular alternation of sandy, coaly, and partly silicified layers with strongly decomposed, compressed plant litter. Up to 5-cm-long and 1.5-cm-thick chert nodules occur randomly within the sandstone. These cherts contain uncompressed silicified plant remains, some showing excellent preservation.

Anatomically preserved *Nothia* rhizomes were found isolated within the decomposed and compressed plant debris and as the most abundant plant remains in the chert nodules. Preservation ranges from excellent in the debris layers and the nodular cherts to variable, as several stages of decomposition may occur in the litter layers. Rhizomes were not found in organic connection with aerial stomatiferous *Nothia* axes. The erect portions of the rhizomatous axes, particularly their distal parts, are rather strongly decomposed and compressed, and the anatomy is preserved in the lowermost few millimeters only. In both sandstone blocks, aerial axes of *Nothia* constitute a considerable proportion of the decomposed plant debris, as is evidenced by cuticle remnants. Compressed aerial axes of *Nothia* virtually always directly overlie the anatomically preserved rhizomatous portions.

In both sandstone blocks, some hundred rhizomatous axes of *Nothia* occur, all similarly oriented. The fact that all rhizoidal ridges and rhizoids occur in the same position indicates that these rhizomatous axes are preserved *in situ.* Within the nodular cherts, *Nothia* rhizomes are frequently associated with roots and rootlets of *Asteroxylon mackiei,* and less frequently with rhizomatous tubers of *Horneophyton lignieri* and axes of *Aglaophyton major. Asteroxylon* roots and *Horneophyton* tubers show the same orientation as the *Nothia* rhizomes. This further supports an *in situ* occurrence of all the rhizomatous organs in both sandstone blocks. *Asteroxylon* roots may show the same excellent preservation as the *Nothia* rhizomes. However, *Horneophyton* tubers are usually more strongly decayed and *Aglaophyton* axes are mostly preserved as "straws" only (cf. Trewin 1996: figure 2D,E). *Asteroxylon* rootlets and *Nothia* rhizomes both have been found penetrating tubers of *Horneophyton* and "straws" of *Aglaophyton,* but *Asteroxylon* rootlets and *Nothia* rhizomes are never penetrated.

METHODS

Specimens were studied in thin sections prepared by cementing slices of Rhynie Chert to standard microscopic slides with Lakeside cement, and then grinding the rock to a thickness of 50 to 150

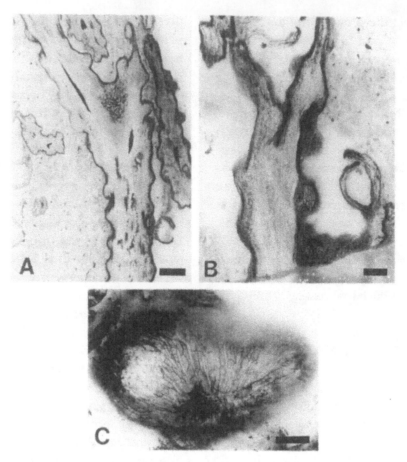

Figure 4.1.
Nothia aphylla, aerial axes and sporangia. **A:** Longitudinal section through a subdistal portion of a furcating aerial axis with numerous emergences. Slide P 2432. Scale bar = 1 mm. **B:** Longitudinal section through the distal portion of an aerial axis with two stalked sporangia. Slide P 2430a. Scale bar = 500 µm. **C:** Adaxial surface of a sporangium seen from below; note the epidermal pattern. Slide P 1646. Scale bar = 300 µm.

µm using silicon carbide abrasive powder according to the methods outlined in Hass and Rowe (1999). Slides are not covered with a cover slip. All specimens and thin sections are kept in the Remy Collection at the Abteilung für Paläobotanik, Westfälische Wilhelms-Universität, Münster. Collection numbers are given in the figure captions.

Specimens were photographed on 25 ISO panchromatic film with a Wild M400 Makrophot and an M20 microscope. In many cases, the use of incident light instead of transmitted light had the advantage that cell outlines, especially of thick-walled cells, are more visible, under lower as well as under higher magnifications. Slides were covered with immersion oil; for higher magnifications, immersion objectives were used directly on the surface of the rock.

SYSTEMATICS

Nothia aphylla Lyon 1964 ex El-Saadawy et Lacey 1979

Emended Diagnosis (Partly Adopted from El-Saadawy and Lacey 1979: 138)

Clonal plant with several subsequent generations of plantlets borne on rhizomes. Each plantlet consists of a rhizomatous, more or less horizontal axis bending upward several millimeters from its base and continuing as an erect aerial axis with several three-dimensional bifurcations. Last two orders of branches bear sporangia.

Aerial axes thin, rarely exceeding 2.5 mm in diameter, and covered with low, elongated emergences, each bearing a single central stoma. Emergences composed of spongy tissue. Stomata anomocytic with six to eight neighboring cells. Periclinal walls of guard cells thickened. Sporangia stalked, inserted laterally and terminally, spirally to semiverticillately arranged. Sporangial stalks adaxially recurved. Sporangia more or less reniform, stomatiferous, eusporangiate, and homosporous, opening apically by a transverse dehiscence slit formed in an apical strip of tiny cells. Spores approximately 65 μm in diameter, smooth to very finely reticulate distally, with a conspicuous proximal trilete mark, approximately 30 to 40 μm wide.

Rhizomes thinner than aerial axes, less than 2 mm in diameter, with only lateral branches, which represent the basal parts of the next generation of plantlets. Lateral branches arise perpendicularly from the lowermost erect parts of rhizomes. Rhizomes lack stomata but bear a median rhizoidal ridge on the ventral surface. The rhizoidal ridge is supported by bands or meshes of "thick-walled cells," extending from central xylem core of the rhizomes to the ridges through a special connective. Mature unicellular rhizoids unbranched with swollen and thick-walled tips.

Vascular core near the bases of the plantlets consists of irregularly mixed, vascular tissues. Distal parts, including the aerial axes, with a haplostele. Central xylem core of aerial axes consists of fusiform cells with uniform wall thickenings. The epidermis of basal parts of plantlets often meristematic. Epidermis of the aerial axes is possibly multilayered above the uppermost lateral branching, comprising elongate giant cells alternating with longitudinal rows of short cells. Giant cells less than 1.6 mm long and usually very narrow at the outer surface; the maximum width of the giant cells below the epidermal surface is approximately 200 μm. Short epidermal cells in longitudinal files; all cell walls thickened in aerial axes but external cell walls in sporangia are not thickened.

Possible male gametophyte: *Kidstonophyton discoides* Remy et Hass 1991 (see Remy et al. 1995).

Rhizomes

Gross Morphology

Nothia rhizomes consist of laterally branched systems comprising several orders of branching (figures 4.2A,B, 4.3A, 4.4). Individual axes depart more or less horizontally before they curve to become vertical. They are here termed primary or rhizomatous. The horizontal portions of the primary axes, here named horizontal axes, are characterized by the presence of rhizoidal ridges. The upright portions of the axes are termed erect or vertical axes. These vertical axes are always incomplete; only the lowermost, up to 8-mm-long, parts are preserved; they most probably represent the basal parts of the already known aerial axes (El-Saadawy and Lacey 1979: plate II).

The horizontal axes are dorsiventrally symmetrical in cross section with a ventral rhizoidal ridge (figure 4.5B–E). Vertical diameters range from less than 1 mm up to 2 mm, and horizontal diameters vary from less than 1 mm up to 2.3 mm. Horizontal axes branch only rarely; lateral branches depart from the upper flanks of the axes. Some of the horizontal axes are not bent upward but end blindly.

Most primary axes are bent upward at 2 to 25 mm from their insertion, usually at (nearly) right angles (figure 4.3A). The erect axes often slightly widen distally and they are less than 1 mm to 3.3 mm in diameter. Rhizoidal ridges are generally absent above the insertion of the uppermost lateral branch; these axes are then circular in cross section. Most branching is found within the lowermost 3 mm of the erect axes. Branches depart at right angles (figure 4.6A,B). The primary axes may develop one to four lateral branches or buds. Lateral branches are formed at different levels, on all sides of the primary axes. However, they mostly arise from the side that forms the continuation of the lower surface of the horizontal axis. These lateral branches generally grow in the same horizontal direction as the parent axes (figure 4.6A,B). Further lateral branches grow in all directions.

The simplest rhizomatous systems show a single lateral branch in each successive order,

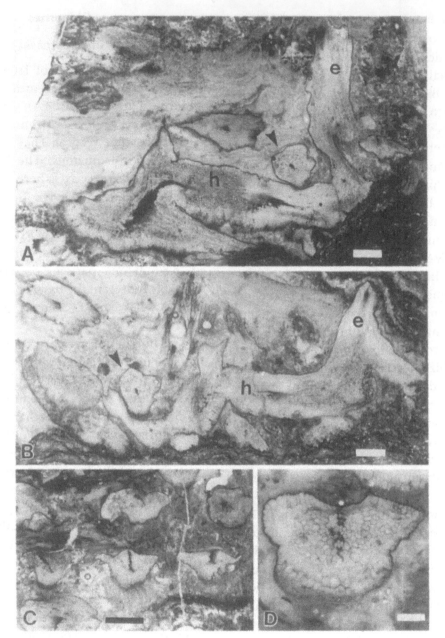

Figure 4.2.
Nothia aphylla, primary (rhizomatous) axes. **A:** Longitudinal section through two
subsequent generations of primary axes. Slide P 2986. Scale bar = 1 mm. (h = horizontal axis; e = erect axis). **B:** Same specimen as in (A), distal continuation of (A).
The axis indicated with an *arrow* is the same as the one with an arrow in (A). Note
the basal horizontal and erect portions of the primary axes. In both (A) and (B),
four subsequent generations of primary axes are visible. Slide P 2987. Scale bar =
1 mm. **C:** Cross section through two layers of primary axes in regular arrangement.
Slide P 2860. Scale bar = 1 mm. **D:** Cross section through a young primary axis.
Slide P 2841. Scale bar = 250 µm.

thereby forming steplike unidirectional axial
systems (figure 4.2B). In the most complex systems, each erect axis can bear up to four
branches at different sides; most of these
branches do not follow the main growth direction but they form knot- or ball-like aggregates
(figure 4.3C,D). Larger knots may be over 2 cm
in diameter and consist of more than 20 primary
axes. Within these knots, the axes can be regularly distributed in two or three vertical layers

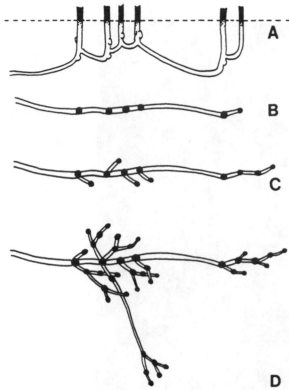

Figure 4.3.
Nothia aphylla. Reconstruction of the clonal growth pattern. **A:** A side view of a young specimen with soil surface (*dashed line*). **B:** The same seen from above (aerial axes, *black*). **C,D:** Further growth stages of the clonal plant seen from above.

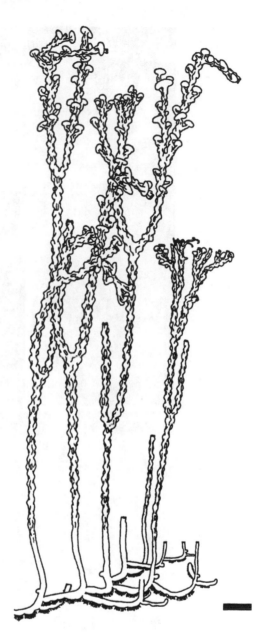

Figure 4.4.
Nothia aphylla. Reconstruction of a younger part of the clonal plant, partly adopted from El-Saadawy and Lacey (1979). Scale bar = 1 cm.

(figure 4.2C). Individual axes or simple systems radiate from the large knots, some of them forming smaller knots at distances of 1 to 3 cm (figure 4.3D). These radiating axes include the longest horizontal portions that have been observed, reaching a length of up to 2.5 cm before they bend upward. Numerous examples of transitions between both branching systems have been observed; the simple ones apparently represent young parts of the plant, whereas the complex knotlike aggregates are obviously older portions.

General Anatomy of Primary Axes

Primary axes are dorsiventral in cross sections, showing a ventral rhizoidal ridge (figures 4.5C–E, 4.7F). From the outside to the center, the tissues consist of an epidermis, a hypodermis, a cortex, and a vascular core (figure 4.7F). Ventrally, the rhizoidal ridge is connected

with the vascular core by a special tissue bridge interrupting the cortex; this tissue bridge is here termed the connective. Therefore, the ventrally interrupted cortex is horseshoe shaped in cross section (figures 4.7F, 4.8). The connective differs from the cortical tissue by its lack of intercellular spaces and is therefore reminiscent of "phloem" or hypodermal tissues. Small strands of "thick-walled cells" extend ventrally from the

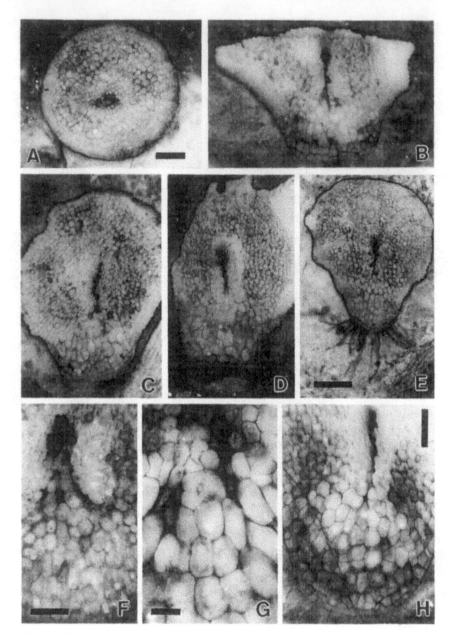

Figure 4.5.

Nothia aphylla, cross sections through primary axes. **A:** Basal portion near departure; note the circular outline and the lack of a rhizoidal ridge. Slide P 2811. Scale bar = 500 µm. **B:** Compressed, flattened axis, near the middle of a horizontal portion. Slide P 2936. Same magnification as (A). **C:** Axis approximately 2 mm after point of insertion showing bandlike xylem core. Slide P 2809. Same magnification as (A). **D:** Distal part of a horizontal portion showing the dark-colored cortex, ventrally interrupted by the connective. Slide P 2923. Same magnification as (A). **E:** Distal part of a horizontal portion showing a well-developed rhizoidal ridge with rhizoids. Slide P 2868. Scale bar = 500 µm. **F:** Detail of thick-walled vascular parenchyma ventrally extending into connective. Slide P 3016. Scale bar = 200 µm. **G:** Detail of connective showing parenchyma and thick-walled cells. Slide P 2972. Scale bar = 50 µm. **H:** Rhizoidal ridge, connective, and vascular core. Slide P 2843. Scale bar = 200 µm.

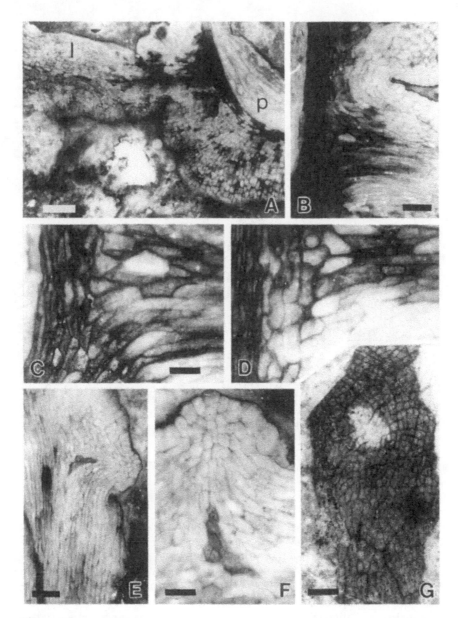

Figure 4.6.

Nothia aphylla, primary axes. **A:** Longitudinal section showing lateral branching to the left (p = parental axis; l = lateral axis). Note the perpendicular insertion of the branch and the vascular tissue. Slide P 2926. Scale bar = 500 μm. **B:** Longitudinal section showing two superimposed lateral branches at the right. Slide P 2896. Scale bar = 250 μm. **C:** Detail of (B) showing the lower lateral branch inserted perpendicularly with the connection of the vascular core to that of parental axis. Scale bar = 100 μm. **D:** Detail of upper lateral branch in (B) showing the arrangement of xylem cells of parental axis (*left*) and vascular parenchyma of branch (*right*). Same magnification as (C). **E:** Longitudinal section through an erect portion of a primary axis showing a developing bud. Slide P 2944. Scale bar = 250 μm. **F:** Detail of (E) showing the apex of the developing bud and the first xylem-like cells. Scale bar = 100 μm. **G:** Surface view of a horizontal primary axis showing the meristematic epidermis and a bud (*white spot above*). Slide P 2922. Scale bar = 200 μm.

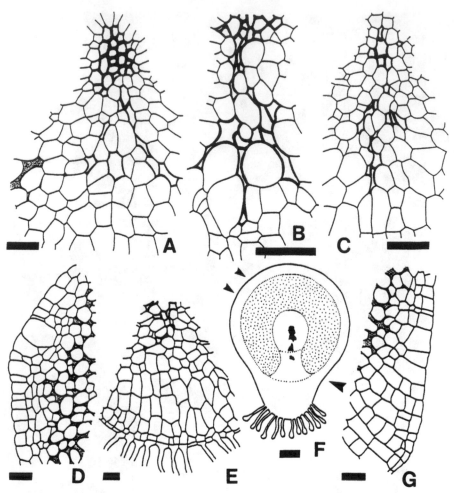

Figure 4.7.

Nothia aphylla cross sections through primary axes. **A:** Xylem core (*above*), ventral band of thick-walled cells and connective (*below*). Intercellular spaces of adjacent cortex *stippled.* Slide P 3016. Scale bar = 100 μm. **B:** Ventral meshes of thick-walled cells enclosing thin-walled cells. Slide P 2893. Scale bar = 100 μm. **C:** Young xylem core and ventral thick-walled cells. Slide P 2854. Scale bar = 100 μm. **D:** Dorsal side of the rhizome showing the epidermis, hypodermis, and cortex. Intercellular spaces *stippled.* The approximate location of this section is indicated in (F) with two small *arrows* (*left above*). Slide P 3040. Scale bar = 100 μm. **E:** Rhizoidal ridge with rhizoids, hypodermal tissues, and the connective with thick-walled cells (*above*). Slide P 2869. Scale bar = 100 μm. **F:** Schematic cross section though a rhizome. Outer zone consists of hypodermis and epidermis. Horseshoe-shaped cortex (*stippled*) interrupted by the connective. Vascular core with xylem (*black*) and thick-walled cells (*black*). Slide P 2868. Scale bar = 0.5 mm. **G:** Lateral side of the rhizome with the transition to the rhizoidal ridge. Intercellular spaces in adjacent cortex *stippled.* The approximate location of this section is indicated on (F) with an *arrow* (*right below*). Slide P 3040. Scale bar = 100 μm.

central xylem core through the "phloem" and the connective to the rhizoidal ridge (figure 4.7A–C).

Parts of axes lacking a rhizoidal ridge are circular in cross section with a clear radial symmetry (figure 4.5A). The vascular core of a primary axis shows marked, ontogenetic variations from the base to the erect part (figure 4.9). Serial sections show that it changes from a mixed organization to a haplostele.

Dorsal Surface of Primary Axes: Epidermis

Primary axes show a large variation in epidermal structure, ranging from meristematic areas to apparently mature regions. Primary axes that bend at a short distance from inception contain

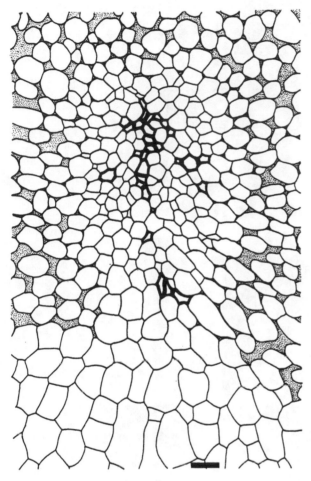

Figure 4.8.
Nothia aphylla cross section through a primary axis. Detail from the center of the rhizome, showing the cortex with prominent intercellular spaces (*stippled*), basally interrupted by the connective. Oval vascular core consists of xylem and thick-walled vascular parenchyma (*black*), sieve cells, and thin-walled vascular parenchyma. Slide P 2805. Scale bar = 50 μm.

a relatively high proportion of meristematic epidermal tissue. In longer horizontal axes, meristems usually occur basally as isolated patches surrounded by more or less mature epidermal tissue.

At least three different types of meristematic tissues can be distinguished on the basis of cell shape and division mode. The first consists of patches of very large prismatic cells, interpreted as initials (figures 4.2D, 4.10B). These cells are extremely deep, 140 to 190 μm, and they are square in surface view, 70 to 80 μm long and 70 to 80 μm wide. Segmentation by anticlinal walls is rare; occasional unequal periclinal divisions result in the formation of very small internal cells (figure 4.10B). This type of tissue strongly

resembles the dormant meristems with large initial areas that are known from extant plants [e.g., *Selaginella* (Cusick 1954)].

The second type consists of much smaller, uniformly shaped cells arranged in regular blocks of two to four cells (figure 4.6G). In surface view, cells are square to rectangular; in cross section they are prismatic (figure 4.7D). The cells are 35 to 70 μm long, approximately 35 μm wide and 60 to 70 μm deep. Their shape, size, and arrangement suggest that they were formed by very regular anticlinal and periclinal segmentations of larger initial cells. This tissue occurs as large, more or less rectangular areas in the basal portions of primary axes, and as dome-shaped outgrowths on vertical portions in which the cell blocks are arranged more or less concentrically. Cross and longitudinal sections through such outgrowths show that internal tissues consisting of thin-walled elongate cells, or even a few thick-walled (?xylem) cells, are present (figures 4.6E,F). The thick-walled cells are not connected to the vascular tissues of the parent axis. The internal organization of the outgrowths suggests that they are young buds, and therefore they are considered to represent the first developmental stages of new primary axes. However, in all the buds examined the covering meristem consists exclusively of rectangular cells, and distinct apical cells seem to be absent.

Adjacent to this tissue, a third type of meristematic tissue may occur (figure 4.6G). In surface view, the meristematic cells occur as irregular polygonal blocks of cells. Within the blocks, cells are bounded by long, slightly sinuous, perpendicular anticlinal walls. The cells are triangular to polygonal in shape in surface view and small, 30 to 50 μm long, 30 to 50 μm wide, and 60 to 100 μm deep.

In many primary axes, the last two types of meristems gradually pass into the apparently mature epidermal tissues. In surface view, the cells of the mature epidermal tissues are elongate, and 100 to 350 μm long. Zones of wider and longer epidermal cells (up to 350 μm long, about 60 μm wide, and about 60 μm deep) alternate with zones of shorter and narrower epidermal cells (up to 200 μm long, about 35 μm wide, and about 35 μm deep) (figure 4.10A).

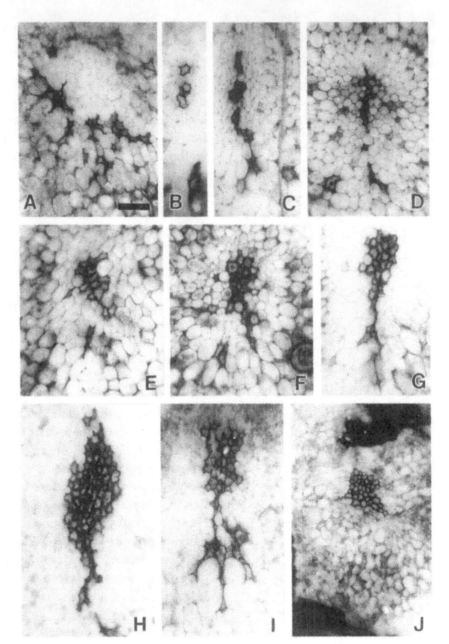

Figure 4.9.
Nothia aphylla, vascular cores in various parts of the primary axes, from the departure to the erect portion. All cross sections are at the same magnification. **A:** Vascular core at the departure of the primary axis from its parental axis; vascular core of parenchyma only. Slide P 2822. Scale bar = 100 μm. **B:** Detail of the vascular core at approximately 1 mm from the departure showing only four thick-walled cells of vascular parenchyma. Slide P 2906. **C:** Vascular core at approximately 2 mm from the departure showing first stage of xylem core, forming a ventrally directed band. Slide P 2809. **D:** Same as (C), "phloem" and connective well preserved. Slide P 2805. **E:** Vascular core at approximately 3 mm from the departure showing an increase of the xylem core. Note the irregular outlines of the xylem cells. Slide P 2940. **F:** Vascular core at approximately halfway from the horizontal part, showing the cortex (*right*) and the connective (*below*). Slide P 2972. **G:** Same as (F); xylem and the ventral thick-walled cells. Slide P 2924. **H:** Vascular core in the lowermost part of the erect portion showing the massive xylem core just below a lateral branching. Slide P 2871. **I:** Vascular core at the point just before the axis attains a vertical position, showing a distinct xylem core and meshy ventral bands of thick-walled vascular parenchyma. Slide P 2893. **J:** The trapezoidal xylem core in an erect axis above the uppermost lateral branching. Slide P 2921.

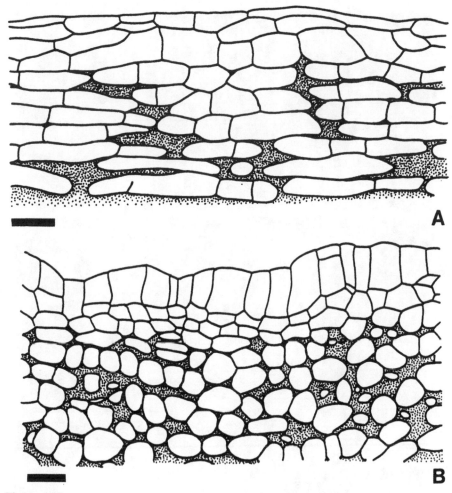

Figure 4.10.
Nothia aphylla longitudinal sections through dorsal side of primary axes. **A:** Long rhizome with mature epidermis (*above*), a one- to two-layered hypodermis and a typical cortex (intercellular spaces *stippled*). Slide P 2924. Scale bar = 100 µm. **B:** Short rhizome with meristematic epidermis consisting of large prismatic cells. Cells of hypodermis and cortex unusually short (intercellular spaces *stippled*). Slide P 2986. Scale bar= 100 µm.

Above the uppermost lateral branch of primary axes, this epidermal pattern can change abruptly into an alternation of isolated giant cells with two to four longitudinal files of short cells (figures 4.11A–C, 4.12B). A similar epidermal pattern is also found in the aerial axes and sporangia of *Nothia*. The terms *short cells* and *giant cells* are used in an informal way to describe the different types of epidermal cells of *Nothia* and do not denote similarities or analogies to similarly named cells in other groups of plants (e.g., Poaceae).

The giant cells usually occur isolated, but sometimes short rows of giant cells are found. The giant cells are 250 to 500 µm long, 70 to 80 µm wide, and 95 to 120 µm deep. The short cells are 45 to 70 µm long, 15 to 80 µm wide, and only 25 to 60 µm deep. In surface view, the giant cells seem to be much narrower than short cells, being only 15 to 25 µm wide. However, transverse sections reveal that short epidermal cells partly overlie the giant cells (figures 4.11H, 4.12A), so that they appear flasklike.

Some developmental stages of this epidermal pattern were observed in cross sections of the basal parts of the aerial axes (e.g., figure 4.11E). In some sections, the epidermis consists of very large, 100- to 120-µm-wide and 95- to 150-µm-deep cells regularly alternating with narrow cells of the same depth. In other specimens, these narrow cells are segmented by one or two periclinal walls, thus forming two- to three-layered radial cell files (figure 4.11E). Epidermal cells of the young files may widen and start to overgrow

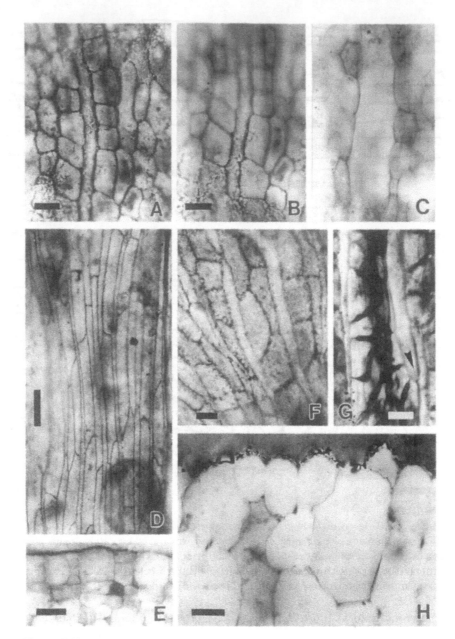

Figure 4.11.

Nothia aphylla, epidermis of primary axes, aerial axes, and sporangia. **A:** Erect primary axis showing a giant cell and the adjacent files of short cells. Slide P 2923. Scale bar = 50 μm. **B:** Detail of (A) showing the limited width of the giant cell at the axis surface. Scale bar = 50 μm. **C:** Same as (B) but at a different (deeper) focusing level, showing the internal widening of the giant cell. Same magnification as (B). **D:** Epidermis of an aerial axis, area between the stomata. Note the narrow giant cells and longitudinal files of short cells. Slide P 1644. Scale bar = 100 μm. **E:** Cross section though an erect portion of a primary axis showing an early developmental stage of the epidermal pattern, giant cells alternating with radial rows of short cells. Slide P 2824. Scale bar = 100 μm. **F:** Main zone at the adaxial surface of a sporangium showing giant cells alternating with files of short cells. Slide P 1646. Scale bar = 50 μm. **G:** Transition zone at the abaxial surface of a sporangium showing two longitudinally adjacent giant cells (*arrow*) at lower right. Note the wall thickenings of short cells. Slide P 2430b. Scale bar = 50 μm. **H:** Cross section through the erect portion of a primary axis showing the mature epidermis with a giant cell (*right*) and swollen short epidermal cells. Slide P 3042. Scale bar = 25 μm.

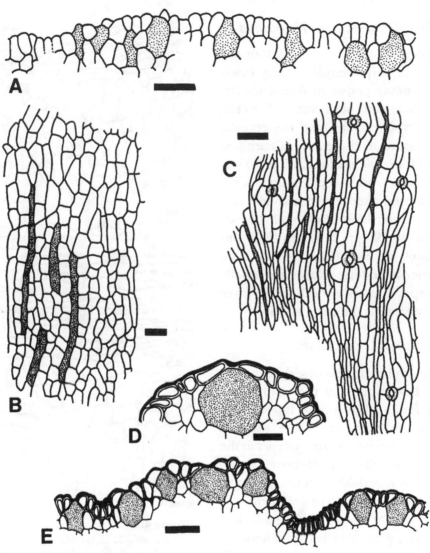

Figure 4.12.
Nothia aphylla, epidermis of primary and aerial axes. Giant cells *stippled*. **A:** Cross section through an erect portion of a primary axis showing giant cells alternating with short epidermal cells. Same specimen as in figure 4.11H. (Slide P 3042). Scale bar = 100 µm. **B:** Primary axis, erect portion; epidermal surface; alternation of giant cells and short cells in lower part only. Same specimen as in figure 4.11A. Scale bar = 100 µm. **C:** Aerial axis; epidermal surface; stomata at top of emergences; short cells on emergences and adjacent to giant cells are widened. Slide P 1638. Scale bar = 200 µm. **D:** Cross section through an emergence of an aerial axis with an extremely large giant cell, largely covered by widened short cells. Slide P 2312. Scale bar = 100 µm. **E:** Cross section through an aerial axis with giant cells more or less regularly alternating with short epidermal cells; short cells are very narrow in the depression; cell walls in the subepidermal layer are partly thickened. Slide P 2430. Scale bar = 100 µm.

the adjacent large cells on both sides. This suggests that the giant cells are homologous to the epidermal and subepidermal layers of intervening cells and that all three cell types may constitute a heterogeneous and multilayered epidermis.

The cell walls of the giant cells and the short cells are not thickened in the rhizomes. No stomata have been observed. Except for the giant cells, all the mature epidermal tissues of the rhizomatous dorsal sides or flanks resemble the epidermal tissues in the aerial axes and sporangia in having convex outer cell walls in cross section.

Dorsal Surface of Primary Axes: Hypodermis

A continuous one- or two-layered, occasionally even three-layered, hypodermis lacking intercellular spaces is developed in the dorsal and lateral regions of basal parts of rhizomes. The cells of the outer hypodermal layer underneath meristematic regions of the epidermis are low and bricklike, 60 to 100 μm long, 35 to 70 μm wide, and 25 to 40 μm deep (figures 4.7D, 4.10B). Mature elongated epidermal cells are underlain by an outer layer of elongated hypodermal cells, up to 190 μm long, up to 70 μm wide, and 50 to 60 μm deep (figure 4.10A). The cells of the inner hypodermal layer are usually longer and deeper than those of the outer layer and are the same size as the cortical cells.

Cortex of Primary Axes

In primary axes, the cortex is horseshoe shaped and ventrally interrupted by the connective. The cortex is four to fifteen cell layers thick, depending on the diameter of the axes. In horizontal axes, the cortex is often deeper at the dorsal side. Cortical cells normally are elongate, 100 to 300 μm long, and 50 to 85 μm in diameter. In cross sections, all cortical cells are rounded; intercellular spaces are large. In longitudinal sections, cortical cells form longitudinal files with rounded or tapering end cells (figure 4.13B). The cortical cells may be unusually short to even globular in horizontal axes, with large proportions of meristematic tissue (figure 4.10B).

Stout hyphae of a glomacean fungus occur abundantly in the intercellular spaces of the cortex; these give the tissue a brownish appearance (figure 4.5C,E). Fungal hyphae usually do not intrude into cortical cells. However, in erect axes, above the uppermost lateral branching, cortical tissues are frequently completely disintegrated. There, fungal hyphae are even more abundant and chlamydospore-like bodies are occasionally observed.

Rhizoidal Ridges: General Features

Each rhizome has a median rhizoidal ridge on the ventral surface. This ridge is 750 to 1,100 μm

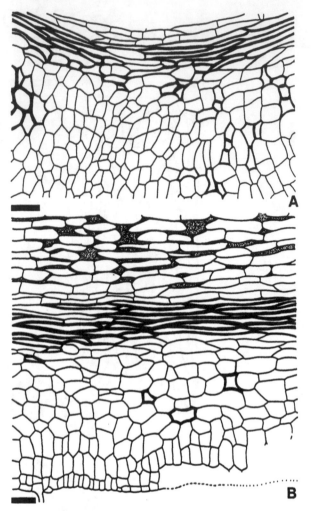

Figure 4.13.
Nothia aphylla longitudinal sections through primary axes. **A:** Section from the vascular core (*above*) to the rhizoidal ridge (*below*). Note longitudinal files of thin-walled cells (*top*) and dispersed thick-walled cells ventral to the xylem core. Slide P 2926. Scale bar = 100 μm. **B:** Section from the cortex (*above*) to the rhizoid-bearing epidermis (*below*). Cortex with intercellular spaces (*stippled*). "Phloem" consisting of nonthickened fusiform cells and longitudinal files of parenchyma cells. Xylem of thick-walled fusiform cells. Below the xylem, two layers of phloem, four layers of the connective, and a four- to six-layered rhizoidal ridge. Note dispersed thick-walled cells in the connective. Slide P 2921. Scale bar = 100 μm.

wide and 300 to 650 μm deep; in cross section the ridge is flat to dome shaped. The ridges are sometimes more than 10 mm long, although they are often interrupted by narrow transverse incisions about 100 μm wide and up to a few hundred micrometers deep. Many of the rhizomatous axes are broken along these incisions. The rhizoidal ridges consist of an epidermis and radial cell files of a strongly developed hypodermis that is internally bounded by the connective (figure 4.5H).

The hypodermis of the rhizoidal ridges is composed of cells that are extremely short but radially deep. Most of the epidermal cells of the ridges bear rhizoids that depart radially (figures 4.5E, 4.14A). In some chert horizons, nearly all epidermal cells bear young or mature rhizoids. In other horizons, the epidermis and rhizoids of most ridges have already deteriorated but their hypodermal tissues are still intact.

Epidermis of the Rhizoidal Ridges

The epidermal cells on the lateral sides of the rhizomes are elongate and up to 200 μm long; on the ridges, they are short (30 to 50 μm long) (figure 4.13B). Intermediate cells are found in a narrow, four- to seven-layer-wide strip on the lateral margins of the rhizoidal ridges (figure 4.5C, left). In this strip, epidermal cells ventrally decrease in length but increase in depth. By sub-

sequent series of anti- and periclinal divisions, the epidermal cells then produce the radial cell files of the ridges and the rhizoid-bearing epidermis (figure 4.7G).

The rhizoid-bearing epidermal cells are almost cuboidal cells (30 to 50 μm long, 35 to 50 μm wide, and 35 to 60 μm deep; figure 4.15F,G), or wide and shallow (35 to 50 μm long, 60 to 95 μm wide, and only 15 to 35 μm deep; figure 4.14D). The cuboidal cells occur predominantly on young ridges and near the margins of mature ridges, whereas shallower cells are found in the central parts of mature ridges (figures 4.14D, 4.15F).

The outer periclinal walls of the rhizoid-bearing epidermal cells are well cutinized; the rhizoids penetrate this cuticle (figure 4.14D). Rhizoids are unbranched, up to 1.5 mm long, and about 30 μm in diameter. Very young rhizoids, less than 100 μm long, may have tapering tips (figure 4.15F); older ones have enlarged or swollen tips with cap-like wall thickenings (figure

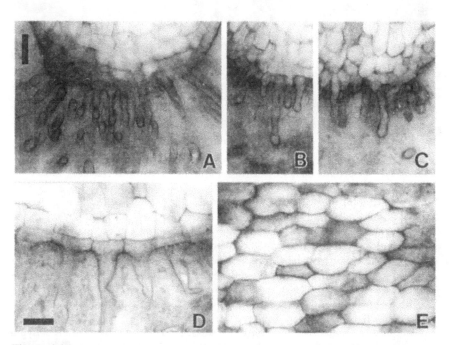

Figure 4.14.
Nothia aphylla, rhizoidal ridges. **A:** Cross section showing radiating mature rhizoids. Slide P 2869. Scale bar = 100 μm. **B:** Cross section showing rhizoids with swollen tips. Slide P 2838. Same magnification as (A). **C:** Cross section showing young rhizoids. Slide P 2837. Same magnification as (A). **D:** Cross section showing mature rhizoids with shallow epidermal cells. Slide P 2957. Scale bar = 50 μm. **E:** Tangential longitudinal section (lateral sides of the ridge to the *left* and the *right*). Hypodermal cells are extremely short but widened. Slide P 2825. Same magnification as (D).

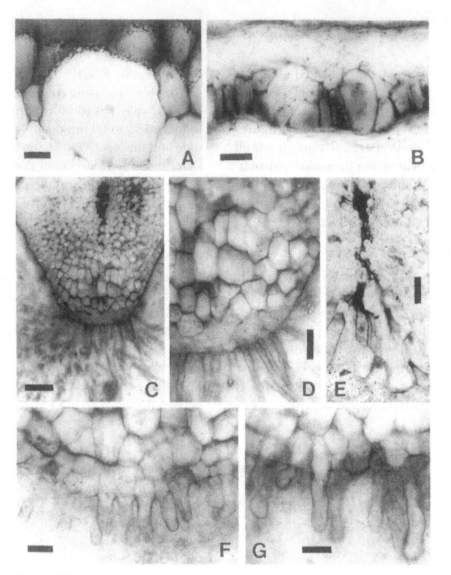

Figure 4.15.

Nothia aphylla, cross sections through the epidermis of the aerial axes, sporangia and rhizoidal ridges of the primary axes. **A:** Aerial axis showing a giant cell and short epidermal cells. Note the outlines of the short cells overlying the giant cell and the wall thickenings in the outer two layers of short cells. Slide P 2430. Scale bar = 25 µm. **B:** Epidermis of a sporangium (surface below), two giant cells alternating with short cells. Slide P 2428. Scale bar = 50 µm. **C:** Primary axis with mature rhizoidal ridge. Slide P 2808. Scale bar = 250 µm. **D:** Detail of (C) showing the decay of epidermal cell walls and thickened periclinal walls in hypodermis of the rhizoidal ridge. Scale bar = 100 µm. **E:** Primary axis detail of rhizoidal ridge with deteriorated epidermis and external protrusions of the outer hypodermal layer (*below*). Slide P 2886. Scale bar = 100 µm. **F:** Detail of the rhizoidal ridge showing early developmental stages of the rhizoids with partly thickened periclinal walls of hypodermal cells. Slide P 2957. Scale bar = 50 µm. **G:** A slightly later stage than in (F), with cuboidal epidermal cells. Slide P 2837. Scale bar = 50 µm.

4.14B). Tips may be up to 70 µm long and up to 50 µm in diameter.

In many rhizomes, the rhizoids and epidermal cells of the rhizoidal ridges show various stages of disintegration possibly caused by fungal infections. First the anticlinal epidermal cell walls disintegrate and only the cuticle and the rhizoids remain (figure 4.15C,D). Then rhizoids and finally the cuticle disappear. In this final stage, the rhizoidal ridges are enclosed only by hypodermal tissues. The cells of the outermost hypodermal layer then may form large, outwardly directed protrusions (figure 4.15E), indicating that the tissue was still alive after the epidermis decayed and that the outer hypodermal layer may, at least to some extent, have served for water absorption.

Some of the periclinal and anticlinal walls of the hypodermal cells of the rhizoidal ridges are thickened. These thickened walls enclose blocks of cells with thinner "inner" walls (figure 4.15D,F). These blocks of hypodermal cells may be infected by fungi and then disintegrate, as described for the epidermis, starting with the decomposition of the nonthickened internal anticlinal walls. The fact that thickened walls of the hypodermal cells resist this disintegration may indicate they had a protective function.

The decay of epidermal and hypodermal cells is frequently accompanied by, or may even be caused by, a massive occurrence of a small fungus. This fungus, as well as the glomacean fungus in the inner cortex, will be described in detail at a later time.

Hypodermis of the Rhizoidal Ridges

The rhizoidal ridges mainly consist of a special, three- to seven-layered hypodermis of parenchymatous cells lacking intercellular spaces. The hypodermal cells of the rhizoidal ridges are arranged in radial files (figures 4.5C, 4.7G), and clearly all the cells within a file were formed from a single epidermal cell. Young and mature rhizoidal ridges may have the same number of layers of hypodermal cells, but cell shapes differ, being shallow in young rhizoidal ridges and deeper (>200 μm deep) in mature ones.

In most rhizoidal ridges, outer, middle, and inner hypodermal layers differ in cell size (figure 4.7E). Cells of the outer layers of the hypodermis are short and shallow (20 to 50 μm long, 50 to 105 μm wide, and 50 to 80 μm deep; figure 4.14C). The decreased cell length in these layers is most obvious in longitudinal sections (figures 4.13B, 4.14E) and is compensated for by a marked increase in the number of cells. The middle hypodermal layers of rhizoidal ridges show the largest cells. They are 35 to 85 μm long, 50 to 105 μm wide, and 120 μm to over 200 μm

deep (figure 4.15D). The innermost hypodermal layers adjacent to the connective consist of more or less cuboidal cells (70 to 120 μm long, 70 to 85 μm wide, and 60 to 120 μm deep). These inner layers form the transition to the cells of the connective, which are longer and arranged in longitudinal cell files (figure 4.13A,B).

Connective

The connective forms a bridge between the hypodermis of the rhizoidal ridge and the tissues of the vascular core (figure 4.5B,D,H). The connective is laterally bordered by cortical tissue and it is slightly wedge shaped in cross section, three to nine cells wide and three to five cell layers deep. Intercellular spaces are absent as in the hypodermis of the rhizoidal ridges and in the vascular tissues. No special boundary layers between the connective and the adjacent tissues have been observed, but the cells of the connective differ in shape, size, arrangement, and/or the lack of intercellular spaces from those of all adjacent tissues. The connective consists of thin-walled cells, intermingled with some thick-walled cells (figures 4.5C, 4.9F,I). The thin-walled cells are intermediate in size and in shape between the longitudinally elongated cells of the outer "phloem" region and the almost cuboidal cells of innermost hypodermal layers (figure 4.13A,B). Within the connective, the thin-walled cells shorten and become wider and deeper ventrally. These cells are 80 to 250 μm long, 30 to 70 μm wide, and 50 to 100 μm deep. Radially elongate cells predominate in cross sections (figure 4.9F). The cells of the connective are arranged in longitudinal files (figure 4.13B). In the outer layers, these files mainly consist of 80- to 140-μm-long cells, whereas the cells of the inner layer are usually up to 250 μm long. The thick-walled cells in the connective occur as isolated cells, in small groups, or in bands, the latter forming the continuation of similar cells departing ventrally from the xylem core.

Vascular Tissues of Primary Axes

The vascular core is more or less circular to slightly ventrally elongated in cross sections, and 300 to 500 μm or 10 to 20 cells in diameter. It consists of a uniform central xylem core surrounded by "phloem." Small strands of "thick-walled" cells extend from the xylem core down to the rhizoidal ridges. Intercellular spaces are generally lacking in all vascular tissues. The vascular tissues consist of at least four different cell types (figure 4.13A,B):

1. Fusiform, thin-walled cells with tapering or S-like ends
2. Short, "thin-walled" cells arranged in longitudinal files
3. Short cells with strongly thickened walls, arranged in longitudinal files
4. Fusiform cells with strongly thickened walls and tapering ends

According to their shape and distribution, we interpret the type 1 cells as possible sieve cells and the type 4 cells as xylem (water-conducting cells). The shape, arrangement in longitudinal rows (figure 4.13B), and location within the vascular core indicate that the cells of types 2 and 3 represent vascular parenchyma. Both types have similar cell shapes and may occur within a single individual cell file (figures 4.13A, 4.16C). However, the thick-walled cells are compressed by neighboring thin-walled cells. We therefore consider them different cell types and name them thin-walled cells and thick-walled cells.

The possible sieve cells are 300 μm to over 700 μm long and 15 to 45 μm in diameter (figure 4.13A,B). They are not arranged in longitudinal files. Clear sieve areas have not yet been observed in their walls; therefore, their true nature is still uncertain. The thin-walled cells are short, 100 to 300 μm long, and 15 to 45 μm in diameter. Thick-walled cells and sieve elements are not distinguishable in cross sections because they generally have similar diameters and thin walls.

The xylem cells are over 700 μm long, 10 to 35 μm in diameter and have long and tapering ends; the cell walls are uniformly approximately 2 to 3 μm thick (figure 4.16A). In shape and wall thickening, these cells rather resemble fibers or hydroids more than tracheids. In some specimens, the cell walls are split for most of their length, thus giving them a double-walled

appearance; however, they still adhere in the cell corners (figure 4.17F). In the primary axes, these cells first appear a few millimeters above the base of the axes. They become more frequent distally, forming eventually a central xylem core.

Thick-walled cells extend as small strands from the xylem core down to the connective (figure 4.7B,C). In longitudinal and some cross sections (figure 4.5F,G) the thick-walled cells appear as ventrally directed meshes that enclose portions of thin-walled tissue (figures 4.16B,E, 4.17G). All walls of the thick-walled cells are uniformly 2 to 3 μm thick; in this they resemble the fusiform xylem cells. All thick-walled cells are compressed by neighboring thin-walled cells. They are arranged in longitudinal files in which they often alternate with cells of thin-walled cells (figures 4.6D, 4.16C). Near the xylem core, the thick-walled cells are 70 to 150 μm long, 9 to 20 μm wide, and 35 to 45 μm deep, whereas those adjacent to the innermost hypodermal layer of the rhizoidal ridges are 50 to 60 μm long, 20 to 35 μm wide, and 100 to 110 μm deep. The changes in shape and dimensions of the thick-walled cells are therefore similar to those in the adjacent thin-walled cells.

Cross sections of the primary axes reveal great variability in anatomy of the vascular core (figure 4.9A–J). Serial sections show a whole series of developmental stages from the bases to the erect parts of the primary axes. The vascular core generally changes from a mixed type with irregularly distributed thick-walled cells at the bases, to a protostelic type in the erect portions of primary axes.

The vascular core of a primary axis is connected to its parental axis only by vascular parenchyma (figures 4.6A,B, 4.9A). The vascular core of the branch (i.e., the primary axis) is positioned perpendicularly to that of the parental axis; its center often is composed of a group of pyramidal cells (figure 4.6C,D). Shortly above the insertion, thick-walled cells occur irregularly in the entire vascular core of the branch (figure 4.17A–C). They either occur isolated, in short horizontal bands, or in groups (figures 4.16D, 4.17A,C). Farther away from the parental axes, the number of thick-walled cells

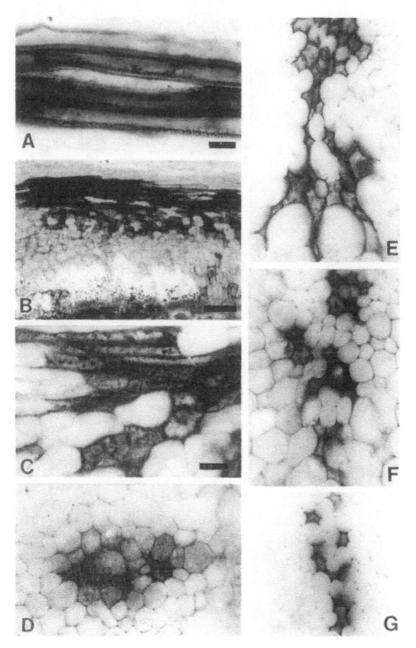

Figure 4.16.

Nothia aphylla, vascular tissues in horizontal portions of primary axes. **A:** Longitudinal section at the same level as figure 4.9I showing the central xylem core of fusiform cells with uniform wall thickenings. Slide P 3032. Scale bar = 25 μm. **B:** Central vascular core and meshy bands of ventral thick-walled cells. Slide P 3009. Scale bar = 250 μm. **C:** Detail of figure 4.6A. Longitudinal section through a newly erect portion showing thin- and thick-walled vascular parenchyma. Slide P 2926. Scale bar = 50 μm. **D:** Cross section near the departure from parental axis showing transitions between thin- and thick-walled vascular parenchyma. Slide P 2811. Same magnification as (C). **E:** Cross section showing details of the ventral meshy band of thick-walled cells in the "phloem" and the connective (*below*). Slide P 2893. Same magnification as (C). **F:** Cross section above the departure showing the young xylem core (*above*) and groups of thick-walled vascular parenchyma. Thin-walled cells of different sizes. Slide P 2841. Same magnification as (C). **G:** Cross section near the departure from the parental axis showing a mixed vascular tissue with only few thick-walled vascular parenchyma cells. Slide P 2846. Same magnification as (C).

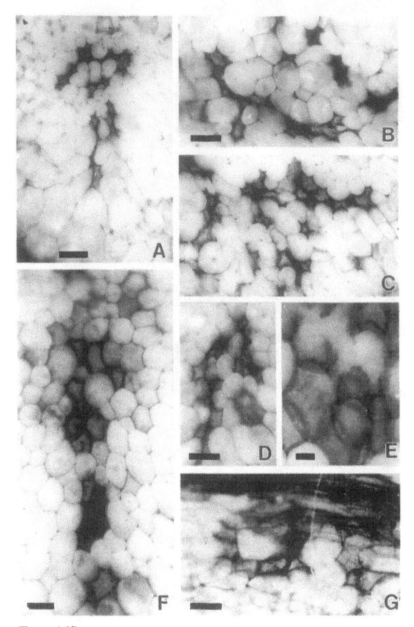

Figure 4.17.

Nothia aphylla, horizontal primary axes, details of vascular tissues. **A:** Cross section near the departure from the parental axis showing the mixed-type vascular core. Slide P 2970. Scale bar = 50 μm. **B:** Cross section just above the departure from parental axis showing the mixed vascular core. Slide P 2822. Scale bar = 50 μm. **C:** Cross section at the departure showing the horizontal arrangement of thick-walled vascular parenchyma. Same slide and same magnification as (B). **D:** Cross section approximately 2 mm above the departure. Note the transitions between thin- and thick-walled cells. Slide P 2808. Scale bar = 30 μm. **E:** Cross section showing a detail of the ventral band of thick-walled cells leading toward the rhizoidal ridge. Note the small cells of thin-walled vascular parenchyma alternating with thick-walled cells. Slide P 2854. Scale bar = 50 μm. **F:** Cross section showing a detail of the xylem core. Note the splitting of the cell walls. Slide P 2893. Scale bar = 10 μm. **G:** Longitudinal section showing a detail of the meshy ventral thick-walled cells. Slide P 3009. Scale bar = 70 μm.

decreases until eventually only two or three cells are left (figures 4.9B, 4.16G).

At 1 to 2 mm from the base of the axis, the first xylem cells appear in the center of the vascular core (figures 4.7C, 4.16F). In the first stages, up to 15 xylem cells form a central xylem core, one to two cells wide and irregular or bandlike in cross section (figures 4.5C, 4.8, 4.9C,D,). In these stages, cells with moderately thickened walls occur adjacent to the xylem cells (figure 4.17D).

Farther away from the insertion but still in the horizontal parts of the primary axes, the central xylem core increases to 20 to 30 cells and becomes more circular in cross section (figures 4.7A, 4.9E,F). The central xylem core further increases in diameter in the curved and erect portions of the primary axes, then being more or less circular to diamond shaped in cross sections (figure 4.9I). In this region, lateral branches are given off. Cross sections reveal that the xylem core is very variable in shape where the branches are inserted, and that the number of its cells can vary from approximately 40 to more than 100 (figure 4.9H), possibly including a number of thick-walled cells. The central core again becomes circular to diamond shaped in cross section above the departure of the uppermost lateral branch. All cells are xylem. Small central cells with diameters of approximately 10 μm can be enclosed by one or two layers of larger cells of up to about 35 μm in diameter (figure 4.9J). In these portions, the protostele consists of approximately 40 xylem cells with uniformly thickened walls, looking very similar to the xylem core of aerial axes of *Nothia* (El-Saadawy and Lacey 1979).

Aerial axes

For the general morphology and anatomy of aerial axes, and for general measurements, see El-Saadawy and Lacey (1979). However, some additional and more detailed information on epidermal pattern, the structure of the epidermis, and the cortical tissues is given here. These descriptions are based on several dozen fertile and vegetative axes (figure 4.1A,B), preserved in vuggy to massive cherts. Additional photographic documentation is given in Edwards, Kerp, and Hass (1998).

Epidermis of Aerial Axes

The epidermal pattern of the mature vegetative axes of *Nothia* is mainly determined by the following:

1. Basic epidermal pattern (i.e., the alternation of longitudinally orientated giant cells and files of short cells)
2. More or less distinct parastichy of stoma-bearing emergences, and the additional widening of epidermal cells on the emergences

The basic epidermal pattern of mature aerial axes is very similar to the overall epidermal pattern of the erect parts of the primary (rhizomatous) axes. Giant cells occur isolated and they appear extremely narrow in surface view because they are largely covered by overlying epidermal short cells (figure 4.11D). In cross sections, the giant cells are large and flask shaped to trapezoidal (figure 4.15A). The short cells are arranged in longitudinal files. Files of short cells alternate with giant cells in a more or less regular mode.

The epidermal cells of the aerial axes differ from those of the primary axes in their size and some cytological details. The giant cells of the aerial axes are up to 1.6 mm long, 90 to 110 μm and occasionally up to 200 μm deep, and only 10 to 20 μm wide or even less at the surface; their largest width is 70 to 120 μm and sometimes 200 μm (figure 4.12D,E). Their walls are not thickened. The short epidermal cells are 120 to 300 μm long, 30 to 80 μm deep, and 15 to 100 μm wide. Shorter and longer cells seem to be distributed irregularly. However, some axes show a predominance of cells of a particular length (e.g., 150 to 250 μm). Files of short cells can be more than 2.5 mm long and may comprise more than 15 cells (figure 4.12C). The end walls of the cells in these files are horizontal or oblique. Wider cells mainly occur adjacent to giant cells and partly overlie them. Narrow short cells usually occur at some distance from the giant cells. All periclinal and anticlinal walls of the short cells are strongly thickened, and the walls are bulged outwardly. The giant cells alternate with two to twelve files of short cells. Some axes show a more or less regular alternation of giant cells and two to four files of short cells, whereas in others the bands of short cells are formed by four to seven, occasionally even up to twelve, files. The number of giant cells is strongly reduced in some portions of the aerial axes.

The epidermis shows a further differentiation due to the occurrence of stomata-bearing emergences. The emergences are 250 to 350 μm deep, 700 to 1,200 μm long, and 300 to 500 μm wide, and elongated, elliptical to lens shaped in surface view. They comprise seven to ten longitudinal files of short cells plus two to four intervening giant cells (figure 4.12C; see also Remy and Hass 1991a: fig. 26K). The short and giant cells on the emergences are considerably widened, whereas those in depressions are narrow. In surface view, the emergences are characterized by 50- to 100-μm-wide short cells and the depressions by narrow, 10- to 30-μm-wide short cells (figure 4.12E). The stomata lie in the central files of short cells.

At some distance from the bifurcations, aerial axes often show a more or less regular parastichous pattern of the emergences. Emergences can be so densely positioned that they touch each other; occasionally they form irregular ridges. These regions of the aerial axes have approximately three stoma-bearing emergences per square millimeter. Near the bifurcations, emergences are often much shorter and only 400 to 600 μm long. Consequently, the stomata are more densely positioned, approximately five or even more per square millimeter.

The stomata are anomocytic and the guard cells are surrounded by six to eight epidermal cells that resemble normal epidermal cells in their size and arrangement (figure 4.12C). However, the lateral neighboring cells can often be extremely short and wide, and they thus appear as a radiating arrangement. The stomata on the aerial axes are sometimes longer than wide and sometimes wider than long, 75 to 105 μm long

and 70 to 95 μm wide. The areas delimited by the stomatal ledges are oval, 35 to 45 μm long and 25 to 40 μm wide. All the observed stomatal pores are extremely narrow, 18 to 24 μm long and only 2 to 3 μm wide. The guard cells are nearly rectangular in cross section and the periclinal walls are strongly thickened (see Remy and Hass 1991a: figure 27G–R). A few specimens show guard cells in contact with internally isolated, irregularly shaped parenchymatous cells; usually the guard cells are directly in contact with a wide substomatal chamber.

Hypodermis of Aerial Axes

The hypodermis consists of one to three layers of cells without intercellular spaces below the thick-walled short epidermal cells. The hypodermal cells are similar in shape and size to the latter. In most specimens, the hypodermis is discontinuous and interrupted by the giant cells; in the better-preserved specimens, the innermost layer of the hypodermis underlies the giant cells. The cells of the outer hypodermal layer directly below the epidermis occasionally have thickened periclinal and anticlinal walls (figure 4.15A), as do the short epidermal cells. In other specimens, the inner periclinal walls are not thickened, or there is no wall thickening. The cell walls of the innermost one or two layers are never thickened. The similarities in cell shape, size, and wall thickenings of the outer hypodermal layer to the short epidermal cells suggest that the epidermis of *Nothia* may be multilayered as was suggested by the developmental stages of the epidermis and hypodermis in the primary (rhizomatous) axes.

Cortical and Vascular Tissues of Aerial Axes

Cortical tissues are strongly decomposed in virtually all the aerial axes examined. Description is based on the relatively few aerial axes with areas of partly decomposed cortical tissues. On the basis of different cell shapes, an inner and an outer cortex can be differentiated (Edwards, Kerp, and Hass 1998). The cells of the outer cortex are short or deeper than long, whereas those of the inner cortex are clearly elongated. The outer cortex is three or four layers thick, whereas the inner cortex can be more than ten cell layers deep.

The anatomy of the outer cortex differs beneath the emergences and depressions. Intercellular spaces are extremely large in the tissue underlying the emergences, resulting in a very spongy appearance of the tissue. They are small but well developed below the depressions. The cells of the outer cortex below the depressions are arranged in short longitudinal files. The cells are 60 to 80 µm long and 60 to 80 µm deep. In some areas, the cells are radially elongated and more than two times deeper than long (100 to 120 µm deep and only 35 to 60 µm long). End walls are often oblique and at an angle of approximately 60µm to the hypodermis. Such tissues strongly resemble the palisade-like assimilation tissues of vertically positioned assimilating shoots of certain extant plants (Meyer 1962:15).

In the spongy outer cortex underlying the emergences, cell shape is variable and can be spherical and approximately 70 µm in diameter with up to 50-µm-long cylindrical, armlike protrusions that touch those of adjacent cells. Others are up to 170 µm deep and only 50 to 60 µm long with short, 10- to 20-µm-long protrusions. Cells rapidly increase in depth toward the centers of the emergences where the largest intercellular spaces occur. These cells strongly resemble a three- to four-layer-thick spongy mesophyll. The outer cortex in the emergences shows no increase of the number of cells or cell layers in comparison to the outer cortex underlying the depressions, suggesting that the growth of emergences is not achieved by cell divisions but by cell expansion.

The inner cortex is homogeneous over all the aerial axes. Cells are arranged in longitudinal files; most have horizontal end walls. Intercellular spaces are well developed between the individual cell files, but they become smaller toward the center of the axis. The cells of the inner cortex are 35 to 70 µm in diameter and 100 to 150 µm long in the outer and 200 to 250 µm long in the inner layers.

No new information is available on the vascular tissues. El-Saadawy and Lacey (1979) de-scribed the "xylem" as approximately 250 µm in diameter and consisting of small central cells approximately 10 µm in diameter and larger surrounding cells less than 30 µm in diameter, which is very similar to the xylem in the erect parts of the rhizomes previously described.

Sporangia

The sporangia of *Nothia* are reniform (about 3 mm wide, 1.2 mm thick, and 1.5 mm long), eusporangiate, and homosporous. They open apically by a transverse dehiscence slit. Sporangia are borne on adaxially recurved stalks and inserted laterally and terminally in a spiral to semiverticillate arrangement.

New information on the epidermis of the sporangia complements the earlier descriptions of El-Saadawy and Lacey (1979). The sporangial epidermis shows an alternation of giant cells and files of short cells, a pattern very similar to that of the aerial and primary axes, apart from some significant differences in the arrangement and size of the cells and wall thickenings.

In surface view, three distinct zones (figure 4.18A) can be distinguished from the base to the distal region: (1) a main zone of mature epidermal cells, that extends from the base of the sporangium to the region with the last stomata, (2) a transitional zone, characterized by the rapid decrease in size of the short epidermal cells, that ends with the last giant cells, and (3) a distal dehiscence strip with only very small cells.

The main zone is characterized by stomata and mature short cells (figures 4.1C, 4.11F, 4.18B). The epidermal pattern of alternating giant cells and files of short cells is basically similar to that of aerial axes, although the giant cells are arranged more regularly. At its circumference, a single sporangium may have up to 50 giant cells, which alternate with one to six files of short cells. In the main zone, the short cells are arranged in distinct longitudinal files, where oblique end walls are more common than in the aerial axes. The giant cells in the main zone of the sporangia are up to 1.2 mm long, 15 to 20 µm wide at the surface, 80 to 130 µm wide below the surface, and 80 to 130 µm deep (figure

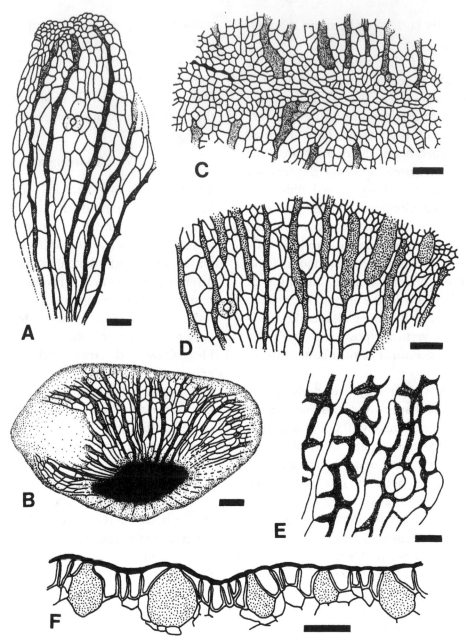

Figure 4.18.

Nothia aphylla, epidermis of sporangia. Giant cells *stippled*, except for (E). **A:** Lateral surface. Slide P 2324. Scale bar = 100 µm. **B:** Outer surface of the adaxial valve. Same specimen as figure 4.1C (slide P 1646). Scale bar = 200 µm. **C:** Top view showing the distal strip of small cells and the developing dehiscence slit (*black*). Slide P 1650. Scale bar = 100 µm. **D:** Transition zone of the adaxial valve showing short and wide giant cells. Note the two adjacent giant cells in the center. Slide P 2430b. Scale bar = 100 µm. **E:** Detail of the surface of an abaxial valve showing the main zone of the sporangium with distinct wall thickenings of the short cells. Slide P 1647. Scale bar = 50 µm. **F:** Cross section through the epidermis showing the more or less regular alternation of giant cells and short cells. Slide P 2428. Scale bar = 100 µm.

4.18F). Internally, they often seem to extend into the tapetal regions. The walls of giant cells are not thickened. The short cells in the main zone are 40 to 200 µm long, 20 to 90 µm wide, and 40 to 70 µm deep. The widest short cells lie next to the giant cells (figures 4.15B, 4.18F). The short cells normally show conspicuous thickenings of the inner periclinal and all anticlinal walls. In surface view, the 3- to 4-µm-wide thickenings of vertical anticlinal walls are often less prominent than those of horizontal anticlinal walls, which may be up to 8 µm thick (figure 4.18E). Cross sections show that the anticlinal walls are thickest at the surface and that they become thinner internally until they are 2 to 3 µm thick, as are the internal periclinal walls. Wall thickenings of short cells are most obvious in the upper parts of the main zone and in the transition zone. Toward the basis of the sporangium, the walls seem to become thinner. However, in all specimens the cell walls show various stages of decomposition.

The sporangia lack emergences and have only a few stomata. Sporangial surfaces are sometimes uneven, but a regular arrangement of protrusions or a relationship to the occurrence of stomata cannot be determined. The few sporangia that are almost complete show one to three stomata on the adaxial and three to six stomata on the abaxial surface. Stomata are anomocytic with six to eight neighboring cells that resemble, or are, normal epidermal cells (figure 4.18A,E). In one specimen, a giant cell lies directly next to a guard cell (figure 4.18D). The stomata of the sporangia are smaller than those of the aerial axes, being 60 to 75 µm long and 60 to 85 µm wide. They are usually circular in outline or wider than long; the stomatal ledges are oval, 30 to 40 µm long and 20 to 30 µm wide. The stomatal pores are poorly preserved and about 15 µm long. In cross sections, the guard cells show rectangular outlines and thickenings of the periclinal walls that are similar to those of the aerial axes. Substomatal chambers are wide and well developed (Edwards, Kerp, and Hass 1998).

The transition zone of a sporangium begins above the uppermost stomata and ends with the disappearance of the giant cells. It is about 300 to 400 µm long and mainly characterized by a decrease in size of the short cells (figure 4.18A). Some of the giant cells run from the base to the tip of the sporangium, whereas others in the transition zone are only 150 to 500 µm long. Short giant cells lying immediately next to each other with oblique end walls are occasionally present in the transition zone (figures 4.11G, 4.18D). A similar arrangement of giant cells has been found in the erect parts of the primary (rhizomatous) axes. In surface view, some of the giant cells of the transition zone are oval in outline and others show clublike distal widening (figure 4.18D). They may be up to 80 µm wide at the epidermal surface and in this respect they differ from those in the aerial axes and the main zone of the sporangia. In the transition zone, the giant cells alternate with longitudinal bands of short epidermal cells that are one to seven cells wide. The latter are often not arranged in distinct longitudinal files as is seen in the main zone. In the transition zone, the short epidermal cells decrease in length from approximately 100 µm in the basal part to 50 to 20 µm in the distal part. The cell width decreases from approximately 70 µm to 50 to 20 µm; the depth of the cells remains at 40 to 70 µm. The extremely wide giant cells in the transition zone are not, or hardly, covered by these small short cells. The short cells show the same type of wall thickenings as the cells in the main zone of the sporangium (figure 4.11G).

A distal strip covers the tip of the sporangium. It extends over the entire width of the reniform sporangia and is from about 2.5 to over 3 mm long (figure 4.18A,C). It comprises four to ten rows of very small, more or less uniformly shaped, isodiametric to elongate cells that are 15 to 30 µm in diameter. The distal strip thus varies from less than 100 µm to over 200 µm wide. Slightly elongated cells may be concentrated near the median line of the distal strip, where they are often arranged in short horizontal bands. The cells near the transition zone are about 40 µm deep and the depth decreases rapidly toward the distal margin of the sporangium; the median cells of the distal strip are only 15 to 20 µm deep. The anticlinal and internal periclinal walls of the cells are substantially thickened and up to 6 µm thick. The external periclinal wall is not thickened. The cell corners and often also the middle parts of the cell walls are the thickest. Cells may (?schizogenously) separate locally along the primary walls; this results in the formation of larger, more or less horizontal slits (figure 4.18C).

A one- to three-layered tissue, very similar to that in aerial axes, may be present underneath the epidermis in main and transition zones of sporangia. The cells are similar in shape to the epidermal short cells. As in the aerial axes, non-thickened cells of this tissue may underlie the giant cells.

DISCUSSION

Systematics

Organic connections between the rhizomes and the stomatiferous (aerial) axes of *Nothia aphylla* have not been found. However, the rhizomes and the aerial axes share unique features that unequivocally demonstrate their relationship. These include striking similarities in epidermal patterns and the wall thickenings of the xylem cells. The epidermis of the erect parts of the rhizomes and the aerial axes show the same basic pattern of longitudinal files of short epidermal cells alternating with giant cells. In both, the convex outer walls of short cells overlie the giant cells so that they appear very narrow in surface view. The giant cells are extremely deep, wide, and long, and they can extend deep into the cortical tissues. The organization and development of the subepidermal tissue suggest that the epidermis might be multilayered in the rhizomes and in the aerial axes. The epidermis of the two differs only in the thickening of the walls of the short cells, in the occurrence of stomata and emergences, and in the width of the short and giant cells in the emergences.

The erect parts of the rhizomes and aerial axes of *Nothia aphylla* both have a protostele, in which small, central xylem cells (about 10 μm in diameter) are surrounded by larger ones (up to 30 μm). The strongly thickened walls of the xylem cells are massive, uniform and lack differential thickenings; these fusiform cells therefore resemble fibers. These features are not known for any of the other Rhynie Chert plants.

Their co-occurrence in the same cherts is further evidence for the relationship of the rhizomes and aerial axes. The rhizomes in the cherty sandstones are found in life position and in close association with hundreds of strongly decayed fertile and vegetative axes of *Nothia aphylla* still showing characteristic features such

as the shape of sporangia, epidermis, emergences, and the xylem cells. In some horizons, rhizomes and aerial axes of *Nothia aphylla* form monotypic assemblages. Anatomically well-preserved aerial axes of *Nothia aphylla* are often found in life position in the so-called vuggy to massive cherts (figure 4.1A,B). These cherts are interpreted as silicifications of subaerial parts of the vegetation caused by surface flooding (Trewin 1996). *Nothia* rhizomes have never been found in these cherts but occur in cherty sandstones containing chert nodules that are interpreted as silicifications of substrates and plant debris by subsurface permeation. The apparent lack of rhizomes in the vuggy to massive chert containing aerial axes of *Nothia* can therefore be related to their subterranean growth environment.

Although new data on *Nothia aphylla* have become available and this taxon is now one of the best-described (in terms of detail) and probably also best-known early land plants, its taxonomic relationship remains unclear. The stalked reniform sporangia of *Nothia* consisting of two valves and occurring in lateral and in terminal position are most similar to those of the zosterophylls. However, the characteristic wall thickenings of the water-conducting cells typical of zosterophylls are absent in *Nothia aphylla*.

Unique Aspects of Anatomy Relative to Form and Function

Aerial axes and rhizomes of *Nothia aphylla* show some unique anatomical features worthy of discussion in relation to their functional morphology. These include the peculiar giant cells, the emergences and the distribution of the stomata, the outer cortex of aerial axes, and the connective and vascular parenchyma in the rhizomes.

Aerial Axes: Epidermal and Cortical Features

The most characteristic feature of the epidermis of *Nothia aphylla* is the presence of the enigmatic giant cells. El-Saadawy and Lacey (1979) reported these cells from sporangia and aerial axes, but they described them as cavities and suggested that they might be substomatal chambers. However, the end walls here documented between two successive giant cells clearly demonstrate that they are true cells. In extant plants, such giant epidermal cells often occur in connection with a multilayered epidermis, they are often partly covered by neighboring epidermal cells, and they often extend deeply into the mesophyll as in *Nothia*. The presence, morphology, and classification of such giant epidermal cells in extant plants has been discussed extensively in the literature. These large cells include excretion, crystal storage, and hinge cells. During contraction, tissues may fold along these hinge cells when the turgor pressure reduces. There is no unequivocal evidence for the function of the giant cells in *Nothia*. However, the large cell lumina may have served as large water reservoirs. In some cases, the outlines of the giant cells are concave, probably due to the loss of turgor. This suggests that an overall loss of turgor in the giant cells may have led to a considerable contraction of subhypodermal tissue, causing a folding of the axial surface. In the sporangia, such folding may have induced the opening of the sporangia.

The density of stomata-bearing emergences in *Nothia aphylla* is about three per square millimeter in aerial axes, but it may be higher below and above the bifurcations. This indicates that the plant had only a minor ability to regulate transpiration and a minor adaptation of the aerial axes to water stress, especially when compared with the often closely associated axes of *Asteroxylon mackiei* which have up to 30 stomata per square millimeter (Edwards, Kerp, and Hass 1998). The central position of the stomata on the emergences and the spongy nature of the underlying cortex obviously enhance transpiration rate and gas exchange. On the other hand, the possibly multilayered epidermis with two external layers of thick-walled cells may have had a protective function (e.g., against insolation but also against uncontrolled water loss). Together, these epidermal features may suggest that the aerial axes lived under relatively favorable conditions with high atmospheric CO_2 conditions and at least a seasonally humid climate. The here presumed short-lived nature of the aerial axes is consistent with such an interpretation.

Kenrick and Crane (1997a) described emergences ("protrusions") on the axes and on the sporangia. However, none of the about 50 sporangia we examined has regular projections with centrally positioned stomata, nor did El-Saadawy and Lacey (1979) figure such emergences. We want to emphasize here the unique structure of the emergences of *Nothia*. They are not formed by the production of additional cells by cell division but by an expansion of the already existing cells. In this respect, they differ from most trichomes and/or emergences of Early Devonian zosterophylls and trimerophytes. However, in most cladistic analyses of early land plants (e.g., Kenrick and Crane 1997a), such fundamental differences have not been taken into consideration.

In the aerial axes of *Nothia*, an inner and an outer cortex can be distinguished. The inner cortex is developed as a normal axial cortical tissue with vertically elongate cells, but the outer cortex, which seems to be the assimilation tissue, is of special interest. The position and the histology of the assimilation tissues of all Rhynie Chert plants are still conjectural, mainly because little attention has been paid to the outer regions of cortical tissues. Edwards, Kerp, and Hass (1998) first described them in some detail for *Rhynia gwynne-vaughanii*, *Aglaophyton major*, *Horneophyton lignieri*, and *Nothia aphylla*; all these plants have similarly shaped outer cortical cells that are arranged in the same way. The outer layer is bounded by the hypodermis but there is no sharp boundary with the inner cortex. In all these plants, the outer cortex is approximately three to five layers thick and consists of two types of cells occurring in a special arrangement. All four plants have an outer cortex consisting of spongy parenchyma around and beneath the stomata, and palisade parenchyma in regions between stomata. The individual cells of spongy parenchyma have armlike protrusions touching those of adjacent cells. In *Nothia*, these are deeper than in the other three species, which have a spongy tissue consisting mainly of elongated cells. The cells of the outer cortex between the patches of spongy tissue are isodiametric or in some areas deeper than long. The latter very short and deep cells in the outermost layer of the cortex, which make an angle of approximately 60° with the overlying hypodermis, resemble palisade cells. The cells of the assimilating palisade parenchyma of the erect aerial shoots of many extant plants have obliquely positioned upper and lower anticlinal walls that make a sharp angle with the overlying hypo- or epidermis (Meyer 1962). The spongy parenchyma occurring around and beneath the stomata therefore seems to have functioned for gas exchange. The palisade-like parenchyma of the Rhynie Chert plants was apparently the assimilating tissue.

Rhizomes: Connective and Vascular Parenchyma

The connective in *Nothia* rhizomes is a special parenchymatous tissue located right between, and forming a direct connection to, the rhizoidal ridge and the vascular core (figure 4.5D). Its special location and the lack of intercellular spaces suggest that this special tissue served as the optimal transport of water and nutrients from the rhizoids to the vascular core. Tissues similar to this connective are unknown for fossil and extant land plants. This may be related to the special evolutionary status of *Nothia* as an early land plant that still lacks real roots.

The vascular tissues of the *Nothia* rhizomes show large amounts of parenchyma (figures 4.7A–C, 4.8, 4.13A,B). Vascular parenchyma has not yet been described for aerial axes of Rhynie Chert plants, except for some rhizomatous parts of *Aglaophyton* (Remy and Hass 1996). So far, for these aerial axes, only two cell types have been described from vascular tissues: fusiform "phloem" and xylem cells. It is still uncertain whether vascular parenchyma was really restricted to rhizomatous axes only.

The vascular parenchyma of *Nothia* consists of two cell types—the thin- and thick-walled cells. The thick-walled cells are arranged in loose meshes extending from the xylem core down to the rhizoidal ridge (figure 4.16B), including very similar cells in the connective. The thick walls may have facilitated an apoplastic water movement (via the cell walls), as in extant roots. They are compressed by the thin-walled neigh-

boring cells. Therefore, it can be assumed that these cells were not living. Thick-walled vascular parenchyma of *Aglaophyton* rhizomes shows the same cell shapes and arrangement between the rhizoidal ridge and the xylem core as in *Nothia*. This further supports our interpretations regarding the function of this tissue in *Nothia*. However, in *Aglaophyton*, these cells show the same netlike wall thickening as its xylem cells (Remy and Hass 1996). This might even indicate that the thick-walled cells are not vascular parenchyma *sensu stricto* but xylem cells that have derived ontogenetically from the thin-walled cells of the vascular parenchyma.

In *Nothia*, both thin- and thick-walled vascular parenchyma forms the perpendicular vascular connection between the vascular core of the rhizome and its lateral axes (figure 4.6C,D). This resembles the vascular connection between main and lateral roots and between parental and adventitious axes in extant plants. In *Nothia* lateral axes, fusiform xylem cells first appear at some distance from their departure, suggesting that xylem is produced only in mature regions.

GROWTH HABIT AND ECOLOGY

The subterranean nature of *Nothia* rhizomes is evidenced by taphonomic data and by a number of anatomical/morphological features. The excellently preserved rhizomes are found in several subsequent horizons, all in life position. The rhizomes are occasionally associated with well-preserved rootlets of *Asteroxylon* also in life position. *Nothia* rhizomes and *Asteroxylon* rootlets penetrate axes of the associated plant debris. This indicates that *Nothia* rhizomes and *Asteroxylon* rootlets lived in a sandy soil containing plant litter. The presence of rhizoids and especially the lack of stomata in the *Nothia* rhizomes further substantiate their subterranean nature. The epidermis and the rhizoids of the ventral rhizoidal ridges of *Nothia* rhizomes can decay rapidly, but in the still-living rhizomes the hypodermal tissue then apparently takes over the function of an outer protective layer. This is very similar to the situation in true roots of numerous extant plants, where rapidly decaying root epidermis and root hairs are replaced by the outer layer of the exodermis. The steplike ascending growth of *Nothia* rhizomes, representing subsequent generations, may also indicate an adaptation to a habitat with accumulating substrate layers. In contrast to the radially symmetrical rootlets of *Asteroxylon*, the rhizomes of *Nothia* are dorsiventrally symmetrical. This may suggest that the latter evolved from prostrate aerial axes.

The modes of preservation of rhizomes and aerial axes of *Nothia* in the different chert types suggest that the rhizomes and the aerial axes had different life spans. In contrast to the excellent preservation of aerial axes of other taxa such as *Aglaophyton major* and *Rhynia gwynnevaughanii*, the cortical tissues of the aerial axes of *Nothia* are nearly always strongly decayed in the vuggy to massive cherts, even though the plants were preserved in erect life position. This suggests that the aerial axes of *Nothia*, which have a different anatomy, decayed more rapidly than those of the other taxa. However, the presence of some completely preserved aerial axes of *Nothia* shows that the usually poor condition is not directly related to its preservation potential. *Nothia* rhizomes are generally excellently preserved except for the uppermost erect portions, which probably represent basal regions of the typical aerial axes with emergences and stomata. In all *Nothia* specimens examined, the erect portions of the rhizomes are completely identical in preservation to those of the decorticated aerial axes in the vuggy cherts. In addition, the presence of meristematic tissues and young, just-expanding buds, indicates that the rhizomatous axes were still living when their erect parts had already decayed. The presence of meristems in older rhizomes, which show an overtopping by several subsequent generations of rhizomes, suggests a life span of more than only a single season for an individual rhizome. The aerial axes of *Nothia* terminate in sporangia (El-Saadawy and Lacey 1979), which might suggest determinate growth. The organization and preservation of the rhizomes and aerial axes thus seems to indicate that aerial axes lived only

for a growing season and subterranean rhizomes had a much longer life span.

These observations strongly suggest that *Nothia aphylla* was a geophyte. Extant geophytes are characterized by short-lived aerial parts that are only moderately adapted to seasonal climatic changes, and long living subterranean parts that can survive under unfavorable conditions like seasonal drought or coldness. Extant geophytes are successfully adapted to climates with distinct periodic, often seasonal, changes. A very different interpretation of *Nothia aphylla* was recently given by Kenrick and Crane (1997a:275), who suggested that it was a semiaquatic plant. This interpretation was based on the absence of tracheidal thickenings in the xylem cells and the relatively thin cuticles of the aerial axes lacking distinct cuticular flanges, in combination with general paleoecological interpretations of the Rhynie Chert. However, our material does not provide any indications for a semiaquatic growth habit, either taphonomically or anatomically. Moreover, it should be noted that the Rhynie Chert represents a rather wide variety of biotopes, ranging from aquatic via humid to better-drained sandy habitats. These can be reconstructed on the basis of sedimentological, taphonomic, and biological data. The latter include autecological and synecological data on various organisms ranging from fungi and algae to higher land plants (Remy et al. 1997). As argued previously, we interpret *Nothia aphylla* as a geophyte that colonized sandy substrates. However, it must be admitted that the aerial axes display features that may indicate humid conditions. Regarding the short life span of these axes and the longevity of the rhizomatous axes, we conclude that *Nothia aphylla* was well adapted to temporary, presumably seasonal humidity.

The organization of the rhizomes clearly shows that *Nothia* had a clonal growth. Each individual plantlet consists of a basal rhizoid-bearing portion (i.e., the primary axis) that passes distally into a bifurcating aerial axis. The aerial axes bifurcate several times and distally bear stalked, laterally and terminally positioned sporangia. The clonal growth in *Nothia* is restricted to the basal region of the erect parts of the rhizomes, each of which may bear one to four buds or lateral branches. Clonal growth seems to be very common in Early Devonian land plants. At least three other Rhynie Chert plants had a similar habit. Adventitious buds and lateral axes are formed by stomatiferous axes in *Aglaophyton* and *Rhynia*, whereas adventitious tubers are formed by older ones in *Horneophyton*. Johnson and Gensel (1992) discussed the abundance of arrested apices and similar structures in various Early Devonian plants. They suggested that these structures may have effected additional branching and irregular branching patterns. Although they did not explicitly regard this as clonal growth, it may be interpreted as such. In *Nothia*, clonal growth follows distinct patterns. In the simplest cases, axes of subsequent generations follow a unidirectional course in which longer rhizome portions may alternate with shorter ones (figure 4.3B). Local clusters of more densely positioned aerial axes are produced by additional branching (figure 4.3C,D).

Nothia is thus another example of a Rhynie Chert plant with clonal growth. Two Rhynie Chert genera spread with local, interconnected clusters of erect axes, one by stomatiferous axes

lying on the surface (*Aglaophyton*), the other by subterranean rhizomes (*Nothia*). The third example (*Horneophyton*) formed only dense clusters by subterranean tubers. The considerable variation in growth strategies observed in the few plants known from the Rhynie Chert indicates that such strategies, including different adaptations to the surface and/or substratum, played an important role in the establishment and development of a land flora.

ACKNOWLEDGMENTS

This study was supported by grant Mo 412/13-1 of the Deutsche Forschungsgemeinschaft to V. Mosbrugger, H. Kerp, and W. Remy. The constructive remarks of Dianne Edwards and Pat Gensel on an earlier draft of this paper are highly appreciated. We want to dedicate this paper to the memory of our late colleague and friend Prof. Dr. Winfried Remy (†1995) who initiated Rhynie Chert research in Münster and considerably contributed toward our understanding of early land plants and early terrestrial ecosystems.

5

Morphology of Above- and Below-Ground Structures in Early Devonian (Pragian–Emsian) Plants

Patricia G. Gensel, Michele E. Kotyk, and James F. Basinger

The Pragian–Emsian of the Lower Devonian is distinguished by plants that were much larger in axial diameter and height, were more complex in branching pattern, sporangial morphology, and internal histology, and exhibited greater variation in morphology and arrangement of reproductive organs than those of earlier ages. Vascular plant diversity, as measured by number of genera, peaks in the Emsian (Knoll et al. 1984; Niklas et al. 1983, 1985). This paper documents and considers certain aspects of overall growth form, architecture, and size of shoot and root systems among Lower Devonian (Pragian–Emsian) plants. Aerial (shoot) systems are reviewed to provide a context in which to evaluate existing and newly discovered information about underground (rooting) structures, thereby demonstrating the relative level of morphological and evolutionary complexity of plants achieved by the end of the Early Devonian. While examples are best provided from Laurussia, some cosmopolitan taxa are included, as well as lesser known plants from other paleocontinents that are similar in general aspects of growth form and complexity. Notable exceptions are plants with novel features apparently endemic to the Early Devonian of southern China, which are discussed in chapter 6.

Understanding of the array of both above- and below-ground organs present in the Pragian and Emsian will allow for more precise inferences about interactions between plants and soil, including the role of plants in soil development and composition and erosion. This is critical to the interpretation of possible environmental and biotic selective pressures acting on early land plant genomes, and to deductions about possible factors involved in the density and distribution of populations of early land plants, including the type(s) of physiological processes present in early land plants.

GROWTH FORM AND SIZE OF ABOVEGROUND STRUCTURES

Because fossilized remains of many early land plants are fragmentary, growth form and size may be difficult to ascertain. These features usually are inferred from general morphology or anatomy, especially overall branching pattern and sporangial location in aerial regions and, where preserved, construction of basal regions.

Size

Pragian–Emsian plants range from small in stature and simple in organization, recalling rhyniophytoids and early zosterophylls of the Silurian–earliest Devonian, to substantial plants several meters tall with considerable complexity of aerial systems. As examples of smaller plants, *Eogaspesiea gracilis* Daber (1960) and *Renalia hueberi* Gensel (1976) had slender axes less than 2 mm wide and at most 11 cm tall, varying in amount of branching in aerial regions. *Urpicalis steemansii* Gerrienne (1992a) from the Lower Devonian (Emsian) of Belgium also is diminutive and is anisotomously branched, with fusiform sporangia borne along the length of, and terminating, some lateral branches. Nevertheless, basal regions are unknown for *Renalia* and *Urpicalis*, and axial remains at best provide equivocal evidence of growth habit. The zosterophylls *Oricilla* Gensel (1982), *Ensivalia* Gerrienne (1996b), *Odonax* Gerrienne (1996c), and some as yet undescribed forms from New Brunswick, Gaspé, Arctic Canada, Germany, and China also apparently were small in size. Evidence from vegetative regions in these plants is sparse but suggestive of infrequent and mostly iso- to slightly anisotomous branching.

Larger plants, referable to either zosterophylls or trimerophytes (basal members), are estimated to have been 0.25 to 0.50 m tall based on size of fragments and axial taper (although this may be misleading since some axes are known that do not taper until nearly at the apex). They exhibit mostly isotomous to anisotomous, more or less synchronous branching, some terminating in sporangia. *Pertica*-like trimerophytes and plants such as *Chaleuria* Andrews, Gensel et Forbes and *Oocampsa* Andrews, Gensel et Kasper are strongly pseudomonopodial, with synchronous dichotomies in distal regions. The estimated height of *Pertica*-like plants exceeds 1 m (Gensel and Andrews 1984). Specimens of *Chaleuria* are at least 26 cm long, and some incomplete individuals of *Oocampsa* observed in outcrop are more than 45 cm long, indicating that these plants also would have exceeded 0.5 m in height, and perhaps were much taller.

Growth Form

Two broad and somewhat variable categories of growth habit can be recognized in early land plants: rhizomatous (or creeping/recumbent) and tufted. Examples of these habits are illus-

trated by discoveries of extensively preserved remains, such as the recently collected specimens of the zosterophyll *Bathurstia denticulata* Hueber with a rhizomatous habit (figures 5.1 and 5.2) (Kotyk 1998; Kotyk and Basinger 2000). Rhizomatous growth also can be inferred for *Sawdonia ornata* Dawson, *Drepanophycus* spp. (Schweitzer 1980b; Schweitzer and Giesen 1980; Hueber 1992), *Nothia aphylla* Lyon (chapter 4), and several other taxa. Plants that appear to exhibit a tufted habit include *Distichophytum* (*Rebuchia*) *ovatum* (Dorf emend Hueber) Schweitzer, *Hicklingia edwardii* Kidston et Lang, and some new species of *Zosterophyllum* Penhallow from Bathurst Island (see figures 5.11 and 5.12B) (Basinger et al. 1996; Kotyk 1998). The habit of other taxa, such as *Thrinkophyton* Kenrick et Edwards, *Gosslingia* Heard, *Crenaticaulis* Banks et Davis, and several species of *Psilophyton* Dawson, *Pertica* Kasper et Andrews, and others, all with pseudomonopodially branched aerial regions, has been more difficult to determine. Critical information concerning basal regions is lacking for many of these taxa.

Gametophyte–Sporophyte Connections

Schweitzer (1983), Remy and Hass (1991b), and Rothwell (1995) have postulated that many early

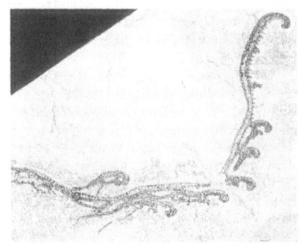

Figure 5.1.
A possible juvenile plant of *Bathurstia denticulata* from the Pragian Bathurst Island beds, Canadian Arctic Archipelago. K-branches occur at regular intervals along the axis. For each K-branch, one axis is a circinately tipped stem and the other is a noncircinate axis that appears rootlike (see figure 5.2). Some of the K-branches bear a similarly organized secondary K-branch. US680-6375. × 0.43.

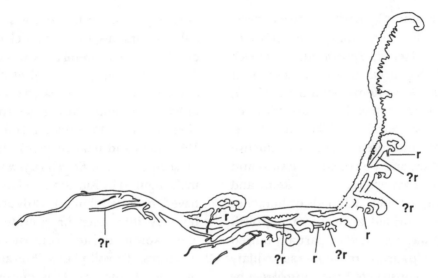

Figure 5.2.
Tracing of the *Bathurstia* specimen shown in figure 5.1. Rooting structures are indicated by *r* or, if interpretation of axes as rooting is less certain, as *?r*.

land plants developed from gametophytes organized the way *Sciadophyton* Steinmann is, with sporophytic shoot systems arising from gametophytic "cups," either as clusters (buschel) of axes or in a radiating manner (sternformig). Although fossils illustrating such an attachment have not yet been conclusively documented, Remy et al. (1992) noted diversity in size, branching pattern, and density of axes among remains assigned to *Sciadophyton* and suggested that more than one taxon may be represented, and Schweitzer (1983) had earlier associated sporophytes of *Taeniocrada langii* Stockmans (= *Stockmansella* Fairon-Demaret) and *Zosterophyllum rhenanum* Kräusel et Weyland with larger and smaller forms of *Sciadophyton*, respectively.

Documented here is a single specimen of *Crenaticaulis verruculosus* Banks et Davis from Gaspé that may support this concept. It consists of a carbonaceous region with a polygonal outline from which emanate several axes (figure 5.3). Two are clearly attached, and one exhibits the distinctive epidermal pattern of the genus (papillae, two opposite rows of teeth). This thalloid structure, which possibly represents either the gametangiophore cup or a sporophytic "corm," is associated with isolated axes of *Crenaticaulis* Banks et Davis. If such a basal region was common among early land plants, it is possible to envisage aerial systems exhibiting either

tufted or more creeping habits developing from such a structure, differing in the distance between branches or the tendency for axes to grow immediately upward rather than prostrate.

Selected Aspects of Aerial Shoot Architecture

Branching patterns in early land plants have been reviewed in detail by Edwards (1994), with additional aspects discussed by Rothwell (1995). Specific features of early land plant branching and architecture are discussed here in relation to growth form.

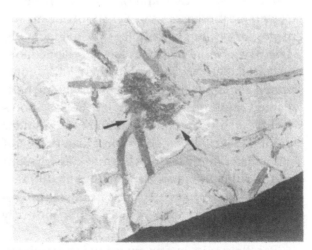

Figure 5.3.
Possible basal region of *Crenaticaulis verruculosus* from the Emsian of Gaspé. Attached axes are indicated by an *arrow*. NCUPC collections. × 0.8.

Among the zosterophylls with more extensive pseudomonopodial branching (i.e., *Crenaticaulis, Gosslingia* Heard, *Thrinkophyton* Kenrick et Edwards, *Deheubarthia* Edwards, Kenrick et Carluccio, *Anisophyton* Remy, Schultka et Hass), several produced subaxillary tubercles or branches, or scattered budlike structures. Edwards (1994) interprets these as distinctive types of branches, and we concur. Edwards and Kenrick (1986), Rayner (1983), and Remy and Hass (1996) suggest that such buds or branches also may have provided a means of regrowth after the plant was uprooted or partially buried by sediment. Upwardly trending subaxillary branching, as reconstructed for *Anisophyton* by Remy et al. (1986), might have facilitated interception of light, or provided a means of support by leaning on neighbors, or even served to propagate the plant vegetatively.

The occurrence of such buds or arrested (dormant) apices along axes is not restricted to zosterophylls but is found also in *Renalia hueberi* (affinity uncertain, but interpreted as a stem-group lycophyte by Kenrick and Crane 1997a), the rhyniacean (*sensu* Hass and Remy 1991) *Aglaophyton major* (Kidston et Lang) D. S. Edwards (Remy and Hass 1996), *Huvenia kleui* Hass and Remy (1991), and plants of uncertain affinities such as *Nothia aphylla* Lyon (chapter 4) and *Bitelaria dubjanskii* T. et A. Ischenko (Johnson and Gensel 1992). In *Aglaophyton, Nothia,* and *Bitelaria,* buds commonly occur in dense clusters at the bases of aerial shoots or on rhizomes. Remy and Hass (1996) postulate that in *Aglaophyton* some buds may have developed into new aerial shoots, while others may have functioned as gemmae. In *Nothia aphylla,* lateral buds located along rhizomatous or aerial shoots developed into plantlets (chapter 4). Slender shootlike axes developed from at least some of the buds in *Bitelaria* (Johnson and Gensel 1992) and possibly in *Huvenia.* Branching in all these plants is notably irregular; for example, in *Bitelaria,* four to five closely spaced branches may depart in all directions.

Parallels exist in few extant plants, although where the apical initial(s) is "lost," multiple new meristems may become established. Nevertheless, subterranean axes of extant *Psilotum nudum*

normally produce numerous separate pyramidal apical initials, some of which give rise to randomly arranged lateral axes, while others do not develop further (Takiguchi et al. 1997). Root apical meristems are initiated on the rhizophore of *Selaginella* below the existing rhizophore apex after it ceases to function (Lu and Jernstadt 1991; Kato and Imaichi 1994). An anatomically preserved apex of *Rhynia gwynne-vaughanii* originally figured by Kidston and Lang (1920b) may have functioned similarly (Edwards 1993, 1994), but very little other direct evidence is available about apical meristem construction and activity in shoots of fossil plants (Edwards 1993, 1997; but see chapter 4). It is important, then, to attempt to characterize apex anatomy and development of Pragian–Emsian (or older) plants as part of the interpretation of branching patterns.

Growth habit in prelycopsids or lycopsids, including *Asteroxylon* Kidston et Lang, *Drepanophycus* Göppert, *Baragwanathia* Lang et Cookson, *Leclercqia* Banks, Bonamo et Grierson, and some Middle Devonian taxa, has been interpreted as rhizomatous or, for *Drepanophycus,* recumbent (Hueber 1992). Some had rhizomes far larger than any known in zosterophylls; for example, some *Drepanophycus* species are up to 4 cm wide (figure 5.4), and it has been suggested that these may have been succulent (Gensel 1992; Hueber 1992). While rhizomes may divide isotomously, lateral branches much smaller than the major axes are more common and have been interpreted as aerial shoots (Schweitzer 1980b; Raynor 1984). Li and Edwards (1995) discuss problems in the interpretation of such branches, including ascertaining whether axes were recumbent, rhizomatous, or aerial. A conspicuous feature of some zosterophylls and prelycopsids is the formation of lateral, H- or K-pattern branches (hereafter called K-branches); while for some plants this may be a consequence of fossilization and the flattening into one plane of rhizomes with departing aerial shoots (Gensel et al. 1969; Hao 1989a; Gerrienne 1988), in many others the K-pattern appears to be a real and developmental phenomenon. Apparently creeping axes of *Drepanophycus spinaeformis* Göppert from Bathurst Island

Figure 5.4.
A bedding surface covered with several apparently rhizomatous axes of *Drepanophycus gaspianus* in an outcrop west of Dalhousie Junction, New Brunswick. Axis width is approximately 3 cm. Note possible bifurcation to right. × 0.3.

(Basinger et al. 1996; Kotyk 1998) have been found with several fully developed K-branches grading distally to buds, all regularly spaced in an apparent developmental sequence (see figures 5.7C and 5.8).

Within the trimerophytes (a paraphyletic group according to several authors—e.g., Kenrick and Crane 1997a), simpler forms (*Psilophyton* spp.) exhibit isotomous to anisotomous branching (figure 5.5A–D), with sporangia terminating

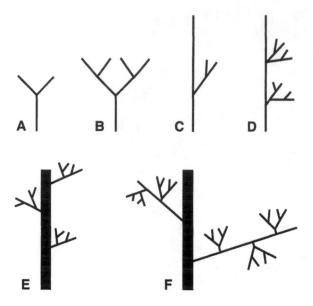

Figure 5.5.
Diagrams of major branching patterns in early Devonian plants. A few appear to be isotomous as at *A* or *B*. Anisotomous axes with isotomous laterals are shown at *C* and *D*. Clearly pseudomonopodial branching is shown at *E*. At *F*, both main axis and first-order lateral branches are pseudomonopodial, with dichotomizing ultimate branches along first-order axes.

some lateral branch systems. *Psilophyton crenulatum* Doran (1980) has been depicted as consisting of rhizomatous regions and upwardly extended aerial shoots, but basal regions are unknown or not recognized for other *Psilophyton* species. Some species of *Psilophyton* (*P. dawsonii* Banks, Leclercq et Hueber, *P. forbesii* Gensel, *P. crenulatum*, *P. coniculum* Trant and Gensel) reveal one or more of the following consequences of variable apical meristem activity: regions of dense branching alternating with more distantly spaced lateral branches, the presence of both isotomous and anisotomous branching, and the presence of aborted branches in a sequence of two closely spaced divisions. There is increasing evidence, therefore, that major innovation in branching architecture, with accompanying changes in anatomy and a new level of indeterminacy in main and lateral axes, occurred among nonlycophytic lineages during the Emsian. Many of the more complex trimerophytes, such as *Pertica* and the possibly related forms *Gothanophyton* Remy et Hass (1986) and a new genus of *Pertica*-like plant from Gaspé (Gensel 1984), exhibit pronounced pseudomonopodial branching, with a distinct

main axis and subordinate lateral branches (figure 5.5E–F). The developmental implication is strong anisotomy of apical division, or even derivation of lateral branches from distinct lateral branch primordia. Furthermore, lateral branches are arranged helically, decussately, tetrastichously, and so on, suggesting sophisticated, regular positioning of lateral/subordinate primordia within apices not seen in the simpler forms.

The anatomy of *Pertica quadrifaria* Kasper et Andrews and *P. varia* Granoff, Gensel et Andrews is unknown, but in *P. dalhousii* Doran, Gensel et Andrews, the apparently similar *Gothanophyton*, and the new genus from Gaspé (Gensel 1984), the vascular strand is deeply lobed, containing only primary tissues (Remy and Hass 1986; Gensel 1984; Gensel, pers. obs.). Mechanically, this seems insufficient to support a meter or more of stem only by turgor (Speck and Vogellehner 1988), although thick-walled hypodermal tissue, such as is known in axes of some *Psilophyton* species and in the new genus from Gaspé, could have contributed some support. Intertwining and mutual support may also have been a factor in gaining height, for in remains of both *P. varia* and the new genus from Gaspé, upright axes are observed to lie about 10 to 15 cm apart and parallel to one another, indicating that plants may have grown in dense stands of individuals or clones (basal regions are unknown).

The pattern of branching of lateral branches is known to vary among trimerophytes. In *Pertica quadrifaria* (Kasper and Andrews 1972) and possibly *P. dalhousii* (Doran et al. 1978), the lateral branches are synchronously dichotomous and of limited length, terminating in sterile tips or sporangia after four to six bifurcations. All lateral branches known for *Trimerophyton* Hopping (1956) exhibit apparent initial trichotomy (i.e., two closely successive dichotomies) followed by dichotomies, terminating after a short distance in sporangia. *Pertica varia* (Granoff et al. 1976), as presently known, is more variable, with distinct fertile and vegetative branches: the former branching synchronously and isotomously and terminating in sporangia after about six bifurcations over a 5 to 7 cm distance; and the latter more pseudomonopodial and at least 17 cm long. Both vegetative and fertile lateral branches of the new genus from Gaspé (Gensel 1984, and pers. obs.) and possibly *Gothanophyton* (Remy and Hass 1986) are pseudomonopodial and have second-order branches departing along their length. These second-order branches divide synchronously and isotomously and terminate in recurved vegetative tips or upright sporangia. Thus, both first-order lateral branches as well as main axes of such plants may be interpreted as more indeterminate than those of, for example, *Psilophyton* or *Pertica quadrifaria.*

Further elaboration of the pseudomonopodial pattern is exhibited by the mid to late Emsian *Chaleuria* Andrews, Gensel et Forbes (1974), *Oocampsa* Andrews, Gensel et Kasper (1975), and *Loganophyton* Kräusel et Weyland (Gensel, pers. obs.). The distinct main axes of these plants produce lateral branches at least 6 cm long, and often much longer, in a dense helix. These first-order branches in turn bear helically arranged higher-order branches, either in the form of dichotomizing ultimate sterile or fertile appendages (*Chaleuria, Oocampsa*), or second-order branches that in turn bear ultimate appendages (*Loganophyton*). Thus, apical anisotomy and indeterminacy had become established in all but the most distal order of branching.

Many of these taxa occur in large numbers, with axes aligned parallel to each other, suggesting that remains are autochthonous or parautochthonous, and that they grew in dense clumps or monospecific stands. This has resulted in the hypothesis that these plants "turfed in" an area (Tiffney and Niklas 1985). In some Early Devonian sedimentary deposits, biomass of axial fragments may be large and potential for litter production high.

POSITION OF SPORANGIA ON BRANCHING SYSTEMS

Placement and organization of fertile regions is highly variable among Pragian–Emsian plants, although two general patterns are recognized: Sporangia are found scattered along shoots or aggregated into spikes in many lycophytes, while they are terminal or restricted to terminal regions of the plant in other groups (e.g., the specialized fertile zone of Kenrick and Crane 1997a). Plants bearing sporangia in dense spikes occur as early as the late Silurian (Basinger et al. 1996). In both the zosterophyll-lycopsid lineage and in trimerophytes and their allies, determinate apices are increasingly confined to specific regions of the plant body—namely, in distal regions, or within lateral branch systems. From the Silurian to the Emsian, then, there is an increase in the number of branching orders produced prior to cessation of growth or termination of apical growth in sporangium production.

In its simplest and most archaic form, the plant body is interpreted to have been an unbranched axis that terminated in a sporangium (e.g., some rhyniophytoids, although few remains are actually well enough known from the fossil record to demonstrate absence of branching). Complexity increases first with dichotomous branching of the axis, each ultimate division terminating in a sporangium (several rhyniophytoids), then by anisotomous to pseudomonopodial branching, where subdivided lateral branches bear scattered or densely clustered sporangia. Fertile regions of trimerophytes and derivatives especially reflect this trend, as noted in the previous section. In *Psilophyton*, sporangia are borne in terminal clusters located at the ends of much-divided aerial systems. Fertile trusses terminate appendages attached to main axes in *Pertica quadrifaria, P. dalhousii,* and *P. varia,* although equivalent trusses are attached along first-order axes in the new genus from Gaspé and possibly in *Gothanophyton.* Sporangia are borne in ultimate dichotomizing units on both main axes and first-order lateral branches in *Ibyka* Skog et Banks and *Anapaulia* (chapter 7). In aneurophytes, there may be up to four orders of branching before fertile units are encountered.

While in nearly all cases, sporangia are borne in apical regions of aerial stems, seemingly the most advantageous placement for spore dispersal, some reports place sporangia at or near ground level. *Drepanophycus devonicus* (Weyland et Berendt) Schweitzer et Giesen (1980) has been reported to bear round to ovoid bodies, possibly representing sporangia, on creeping axes with attached rooting structures. If these bodies truly are sporangia, and the roots did not arise later in the life of the plant, then their location raises interesting implications concerning spore dispersal—perhaps via decay or by feeding arthropods, as has on occasion been invoked for other early Devonian plants (Kevan et al. 1975; Edwards and Selden 1993; chapter 3). It also is possible, however, that these bodies represent some type of vegetative propagative structure. *Stockmansella remyi* Schultka et Hass (1997) has been reconstructed with similar habit, but with some aerial shoots of limited height and sporangia borne on both creeping and aerial shoots. Sporangia found on creeping axes were dehisced, prompting Schultka and Hass (1997) to interpret these regions as originally erect, and becoming procumbent as the plant grew.

ROOTING STRUCTURES

The existence of roots, and their origin, remains an enigma. It commonly is held that very early land plants lacked any kind of true rooting structure; rhizomes with rhizoids, such as occur in some of the Rhynie Chert plants, would be the closest analogue. Since many Pragian to Emsian plants had reasonably large aboveground axial systems, the question of anchorage and water uptake becomes critical. For many, rhizoids alone would seem insufficient. In living plants, a substantial portion of the total biomass exists as below-ground root mass to support aerial structures, although the amount is influenced by mycorrhizal symbiosis with the roots. The proportion of root mass to shoot mass is less clear for nonseed plants than for woody trees, and very poorly understood for early plants. The presence of vesicular–arbuscular mycorrhizae in rhizomes of the Rhynie Chert plant *Aglaophyton major* (Taylor, Remy, et al. 1994, 1995; Remy and Hass 1996) indicates an early origin of plant–fungus interdependence.

By definition, a root is an endogenously formed organ that functions in anchoring and absorbing water and nutrients from the soil (after Bell 1991). Roots typically are characterized by a root cap and root hairs (at some stage of development), an exarch protostelic vascular strand, an endogenous origin of lateral roots, a gravitropic response, and an absence of leaves. Primary roots arise from a particular region of the developing embryo; in seed plants, this is opposite the site of shoot apical organization, while in all pteridophytes except the extant Psilotaceae, which lack roots entirely, it is lateral to the shoot apex. Carboniferous lycopsids appear to have lacked true roots, the corm or stigmarian base apparently formed by dichotomy of the shoot during embryogeny, and the "rootlets" being homologous with leaves. In most pteridophytes and in some seed plants, the primary root does not persist, but instead shoot-borne roots arise, endogenously in most plants, that function in anchorage and absorption. Thus, there are developmental, functional, and anatomical features that can be used to recognize roots, although some structures such as the rhizophore of *Selaginella* remain difficult to categorize (for current interpretations, see Lu and Jernstedt 1996; Kato and Imaichi 1994).

Obviously, recognition of rooting structures and mycorrhizal associations is difficult in impression or compression remains of early land plants. The terms *rootlike* and *rooting structures/organs* have been used for structures in fossil plants for which characteristics of roots are absent or unknown, but which resemble roots and are positioned such that they may have functioned to anchor the plant to a substrate. It is less obvious whether these organs developed as do true roots, or even functioned to absorb water and minerals. Absence of a root cap and root hairs from these organs may be real or just the result of poor preservation of such delicate structures. Anatomy may not be definitive in early land plants with exarch aerial axes.

Putative rooting structures have been reported attached to a small number of early land plants (table 5.1), and new discoveries of rooting structures indicate that such structures may have been diverse by the Pragian. The presence of rooting structures, some much larger than those reported here but usually isolated,

Table 5.1.   Reports of Rooting Structures in Early Devonian Plants

Rhizoids

Aglaophyton major	Remy and Hass 1996
Rhynia gwynne-vaughanii	Kidston and Lang 1917
Horneophyton lignieri	Kidston and Lang 1920b
Nothia aphylla	chapter 4
Serrulacaulis furcatus	Hueber and Banks 1979
Trichopherophyton teuchsanii	Lyon and Edwards 1991

Rootlike structures, possibly adventitious roots

Asteroxylon mackiei	Kidston and Lang 1920b
Drepanophycus spinaeformis	Rayner 1984; Schweitzer 1980b
Drepanophycus qujingensis	Li and Edwards 1995
Drepanophycus devonicus	Schweitzer and Giesen 1980
Hsua robusta	Li 1992
"Taeniocrada" langii (= *Stockmansella*)	Schweitzer 1980a
Crenaticaulis verruculosus	this chapter
Bathurstia denticulata	this chapter
Zosterophyllum sp. nov.	this chapter

has also been inferred from fossil soils, as discussed by Elick et al. (1998a) and in chapters 11 and 13 and references cited therein.

Previous reports of early rooting structures and rhizoids were recently summarized by Li (1992), Edwards (1993, 1994), and Li and Edwards (1995). Li (1992) describes the type of rooting structures present in several genera, comparing them to the putative rooting structures he found in *Hsua robusta* Li. In *Hsua*, the presumed rootlike structures consist of small, dichotomously divided, laterally positioned entities that resemble an ultimate branch system, but they are supplied by a vascular strand as robust as that of the stem and much more robust than the strands supplying the lateral branches. They are borne in regions not far from sporangium-bearing lateral branch systems, although it remains to be demonstrated whether these branch systems represent aerial shoots or are rhizomes with very short erect portions (the lateral branch systems), as depicted by Li (1992). Alternatively, these rootlike structures may have developed after aerial systems fell to the ground. Similar structures, equally difficult to interpret, are the synchronously dichotomizing structures borne on presumed aerial shoots in *Psilophyton dawsonii* and interpreted as rootlike by those authors (Banks et al. 1975). Li and Edwards (1995) discuss the much-

divided rooting structures found in several *Drepanophycus* species. Of particular interest is the zosterophyll *Trichopherophyton teuchsanii* Lyon et Edwards (1991), in which some rhizomes (or roots?) bear structures that could represent either rhizoids or root hairs (Edwards 1994).

New Discoveries of Rooting Structures

Renewed collecting and study of the Early Devonian flora of Bathurst Island, Arctic Canada, and recent collections of fossils from New Brunswick and Gaspé, Canada, have provided new opportunities for interpretation of rootlike structures, including lineages for which they have not been previously documented. The plants are referable to zosterophylls and prelycopsids or lycopsids of some workers. For those taxa whose aerial axes bear emergences or leaves, the rooting structures are distinguishable from stems because they (1) attach to aerial ones but bear no emergences or microphylls, (2) are generally much narrower in diameter and shorter in length than stems, (3) are either unbranched or irregularly branched, in contrast to a more predictable branching pattern of the stem, (4) may have a sinuous or delicate appearance, and (5) proceed in a direction opposite from that of the aerial axis. Also, in plants with naked axes, criteria 2 to 5 are generally applicable. The rooting structures are interesting in their location and morphology, in that some are more abundant than previously demonstrated and in some plants they have an intimate association with lateral branches. There still is no evidence of root caps and undoubted root hairs, and anatomy is not yet known.

Provenance of the Fossils

Lower Devonian land plants preserved as compressions and impressions have been recovered from fine-grained calcareous sandstones of the Bathurst Island and Stuart Bay Formations (Basinger et al. 1996; Kotyk 1998). These flysch-dominated units are the subject of reinvestigation by de Freitas and others (de Freitas et al. 1993; de Freitas, pers. comm. 1998). The sediments range from Ludlow to Emsian, dated by de Freitas et al. (1993) on the basis of the distinctive assemblage of graptolites and other invertebrates found in many instances on the same bedding planes as the fossil plants. The majority of plants, and all those reported here, were recovered from beds of Pragian age. Sedimentology and associated marine fauna indicate deposition in deep marine environments, with plants swept out to sea from the emergent Boothia Uplift and deposited on the ocean floor as they became waterlogged. Estimated distance from shore is about 10 to 20 km. As a result, relatively undamaged but isolated plant fossils have been found. Given the high latitude, only blocks that have weathered from the original beds are obtainable; permafrost occurs within centimeters of the ground surface and precludes excavation.

The plants were collected mainly along major streams or rivers on east-central Bathurst Island. Preparation of the root-bearing fossils included uncovering carbonaceous remains obscured by sediment using steel needles. Some rock surfaces were treated with dilute hydrochloric (10%) and hydrofluoric acids to increase the contrast or visibility of specimens. Attempts to isolate cuticles by maceration were unsuccessful.

Remains of rooting structures associated with, and attached to, the zosterophyll *Crenaticaulis verruculosus* were obtained from the D'Aiguillon section of the Battery Point Formation near Seal Rock, north shore of Gaspé Bay, Quebec, at locality U of Gensel and Andrews (1984). Rootlike structures are also found perpendicular to bedding surfaces containing *Sawdonia ornata* at Cap-aux-Os, north shore of Gaspé Bay, Quebec (locality S of Gensel and Andrews 1984; see also chapter 11), although connection to *S. ornata* has not been established. Beds of the Campbellton Formation, located near Dalhousie Junction, New Brunswick, contain similar, much-branched rootlike axes on bedding planes. These were obtained from a horizon below and just to the east of beds bearing *Chaleuria cirrosa* Andrews, Gensel et Forbes (locality A of Gensel and Andrews 1984). The Gaspé sequences are assigned to the *annulatus-lindlarensis* spore zone by Richardson and McGregor (1986), considered to be mid to late Emsian (the New Brunswick sequences) in age. Specimens are typically preserved as impressions or compressions in gray to buff-colored, fine-grained sandstone, the New Brunswick specimens being finer grained than those at Gaspé. Specimens were prepared by uncovering with fine needles. Again, maceration was not possible.

Description: Bathurst Island Plants

Bathurst Island remains with rooting structures fall into two categories: (1) shoot-borne rooting structures either scattered along apparently trailing or rhizomatous axes, or forming part of K-branching structures (*Drepanophycus, Bathurstia*), and (2) tufted rooting structures occurring as descending naked axes from plants with dense aggregations of branched ascending shoot systems, both arising from a common central region (*Zosterophyllum* sp. nov.).

Drepanophycus spinaeformis

Numerous axes of this taxon were recovered, ranging from 2.2 to 17 (ave., 7.3) mm wide and bearing relatively long spinelike leaves (figures 5.6A and 5.7D). Axes show a dark central line interpreted as a vascular strand. Leaves are 1.1 to 13 mm (ave., 5.3 mm) long, 0.8 to 2.4 mm wide at the base, becoming a little wider in the middle and tapering distally. Each is supplied by a dark strand interpreted as a

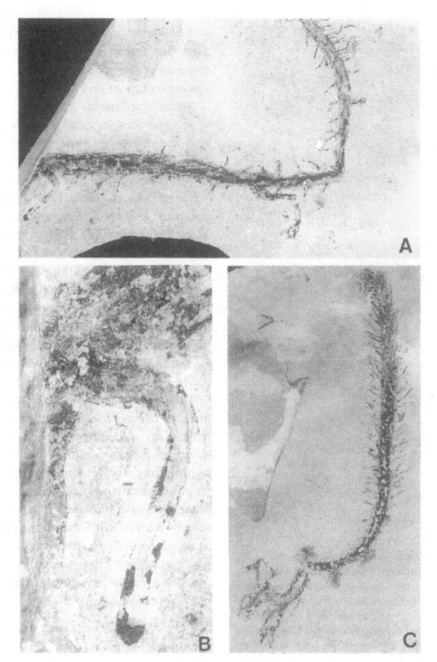

Figure 5.6.
Drepanophycus spinaeformis from the Pragian Bathurst Island beds, Bathurst Island, Canadian Arctic Archipelago. **A:** Fragment of one axis with at least three unbranched rooting structures attached. US664-7969. × 1.1. **B:** Detail of specimen in (A), showing one of the attached rooting structures. Note distal end appears intact (*bottom of photo*). US664-7969. × 6.0. **C:** Another axial segment with several attached, intertwined rooting structures. US 628-8307. × 1.0.

vascular trace. Some axes show K-branching, and while numerous, these often are incomplete (figure 5.7B,D). Buds occur on some axes in the same position as the branches (figure 5.7C). Sporangia are attached by short stalks along the axes in between the leaves on several specimens. These specimens compare well with axes identified as *D. spinaeformis* by Hueber (1972).

On many of the *Drepanophycus* specimens, slender naked axes depart at nearly right angles from both presumed creeping and aerial axes and extend, often sinuously, for up to 2 cm (see figures 5.6 and 5.7A). These naked axes are 1.2

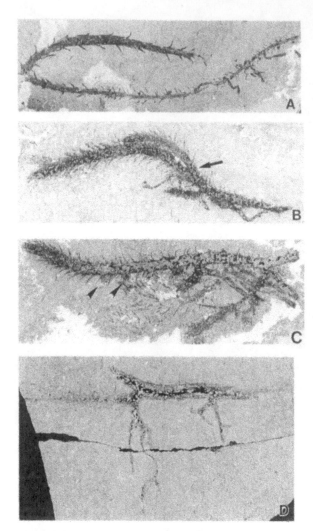

Figure 5.7.

Drepanophycus spinaeformis from the Pragian Bathurst Island beds, Bathurst Island, Canadian Arctic Archipelago. **A:** Fairly long, bent leafy axis with several unbranched attached rooting structures that arch over the axis surface, probably as result of mode of deposition. US604-6380. × 0.45. **B:** Leafy axis bears possibly two K-branches; one member of K-branch is leafy, the other smooth and rootlike. Note axis bifurcates at *arrow*. US636-8212. × .75. **C:** Leafy axis with apex to left, numerous K-branches to right. *Arrows* indicate buds that occur in similar position to K-branches. Each K-branch consists of an apically directed leafy component and a proximally directed nonleafy axis that is interpreted as rootlike. (See diagram in figure 5.8.) US636-8206. × 0.7. **D:** Fragment of axis with two lateral K-branches. See text for description. Proximally directed member branches of each K-branch appear smooth and rootlike. US678-8078. × .75.

to 1.8 mm wide proximally and narrow medially, and they sometimes widen distally. Such undivided rootlike structures occur in about 90 percent of the specimens in the collection. One shows a possible apex (see figures 5.6A, left; 5.6B). Rarely, these rootlike axes divide, and on

one specimen they are seen to curve over the axes (see figure 5.7A) before extending away from them, perhaps as a result of repositioning during transport and burial.

Some specimens produce K-branches in which one daughter axis is smooth and rootlike and the other leafy and shootlike (see figure 5.7B–D). In USPC 678-8078, the first dichotomy of a K-type branch (shown at the left of figure 5.7D) produces to the left a wide and leafy daughter axis that extends several centimeters along the rock surface. The right-hand daughter axis, however, is more slender and lacking in leaves; it bifurcates at least twice and extends for about 3.7 cm before preservation ends. The K-branch located to the right on this specimen (probably distally) also exhibits an initial bifurcation and differentiation of daughter branches.

Specimen USPC 636-8206 shows a similar, but more extensively developed, pattern (figures 5.7C and 5.8). Six K-branches depart at 1.1 cm intervals along the rhizome, which appears to terminate in an apex to the left. Distal to the last K-branch are two apparent buds, possibly representing immature or incipient lateral branches (figure 5.8, arrows). For each K-branch, one daughter axis, directed apically, is leafy and clearly a shoot, while the distally directed axis is naked and slender, extending as much as 2 cm but incomplete. These branching systems are so densely arranged as to overlap one another. The naked axes are interpreted as rooting organs.

Bathurstia denticulata

Rooting structures in the zosterophyll *Bathurstia denticulata* are similar to those of *Drepanophycus* in consisting of either slender naked axes departing from the creeping rhizomes or proxi-

Figure 5.8.

Tracing of US636-8206, shown in figure 5.7C. Note the leafy, apically directed members of each of the K-branches in contrast to the smooth, basally directed members. As in figure 5.7C, buds occur at *arrows*.

mal regions of aerial axes, or as one constituent of K-branches. In the former, they arise from the ventral surface of the rhizome, are smooth, and are either unbranched or, rarely, isotomously divided. They are 1.0 to 2.3 mm wide and up to 5 cm long, widening distally. Some exhibit a central vascular strand (figure 5.9A).

In specimen USPC 493-6576 (figures 5.9A and 5.10), two axes apparently depart at nearly right angles from the main stem and extend parallel to one another for about 0.85 cm. They are 2.2 mm wide near their departure, tapering to 1.8 mm distally. The branch located to the right (see figure 5.9A, arrow) bifurcates, with one half extending as a naked unbranched rootlike structure and the other forming a rounded apex probably representing a crosier. Another forking may have occurred, evidenced by the slender naked axis, possibly a root, that departs near the base of the crosier. The branch located on the left is 2 cm long and incomplete. Other similar-appearing axes overlie this system (see figure 5.10, o) and may represent unrelated plant remains. A K-branch, with a crosier and part of a second axis, also departs from the rhizome near the point where it turns upward. The specimen in figure 5.9B shows small, apparently truncated lateral K-branches; these may have been newly formed or their more distal regions broken off.

Specimen USPC 680-6375 (see figures 5.1 and 5.2) consists of an extensively preserved

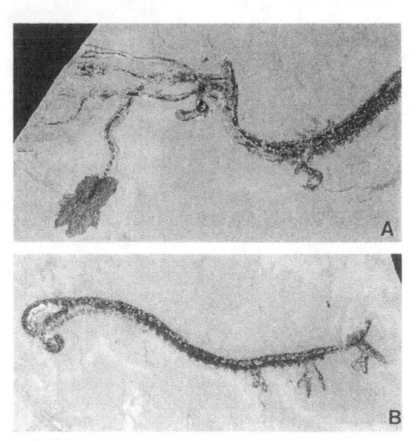

Figure 5.9.
Bathurstia denticulata from the Pragian Bathurst Island beds, Bathurst Island, Canadian Arctic Archipelago. **A:** Axis with characteristic emergences with two lateral branches. The left-hand region of the left-most K-branch consists of smooth axes that appear rootlike. The apically directed portion terminates in a circinate tip. Branch to the right is less extensive (younger?) and only the circinately coiled component is determinable. US493-6576. × 0.8. **B:** Axis with two circinately coiled apices to the left. At right, three lateral K-branches depart; the proximally directed segments are interpreted as rootlike. US636-8217. × 0.65.

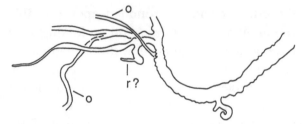

Figure 5.10.
Tracing of specimen shown in figure 5.9A. *r?* = rooting structure. Note that a rooting structure is attached near a circinate tip of the lateral branch. *o* = overlying, probably not attached, axes. US 493-6576.

rhizome/aerial shoot ending in a crosier (upper right). Five and possibly six K-branches occur on its ventral surface, ranging from small crosiers in the region nearest the apex, to several-times-repeated lateral systems of K-branches farthest from the apex. In several of the K-branches, one daughter branch is naked and downwardly directed, which we interpret as a rooting structure. The other daughter branch forms a small, circinate shoot or divides again. In the case of the K-branches bearing K-branches themselves, the partitioning into a rooting branch and a shoot branch occurs most clearly in divisions within the more distal (older) lateral branches (see figures 5.1 and 5.2).

Zosterophyllum sp. nov.

Several zosterophyll specimens were found that exhibit a tufted appearance—i.e., there are both ascending, branched, 1-mm-wide axes, many of which bear sporangia, and shorter, descending, unbranched axes along which no sporangia are visible. The aerial axes all appear to arise more or less parallel to one another from a central region (figure 5.11).

The putative root tuft in USPC 667-6374 (see figure 5.11) is 3.0 to 5.2 cm long, making up one sixth to one third of the length of the entire plant. On this and other specimens, poor preservation and dense overlapping of axes make details of their axial morphology difficult to discern. Branching in the presumed rooting structures was not observed. The width of individual rooting axes is roughly equal (1.0 mm wide) to that of the aboveground axes on specimen USPC 609-6376.

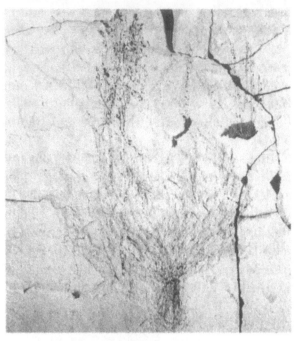

Figure 5.11.
Zosterophyllum sp.nov. from the Pragian Bathurst Island beds, Bathurst Island, Canadian Arctic Archipelago. Note the upwardly directed aerial axes, some of which are fertile, and the tuft of downwardly directed axial structures. The latter are interpreted as rooting structures. US667-6374. × 0.3.

There appears to be a specific region from which the ascending and descending axes depart in some of the specimens. For example, in USPC 667-6374, a line can be traced between the upper and lower portions (arrows). In another specimen, not figured, a region of coaly residue a few millimeters wide occurs, from which axes emanate. This organization could correspond to the horizontal, frequently divided rhizome area figured by Walton (1964), or even to a cormose base.

In these specimens, it appears that the plant settled into the sediment on its side and was fossilized without reorientation of its shoots and roots. In others, some reorientation has occurred, as for example, in figure 5.12B, where part of the basal shoot region (composed of many axes) curves to the side and the other part extends downward. The specimen is very weathered, so it is difficult to discern individual axes easily in the whole basal region, but enough detail is present to indicate that they were numerous. Curving may be the result of current

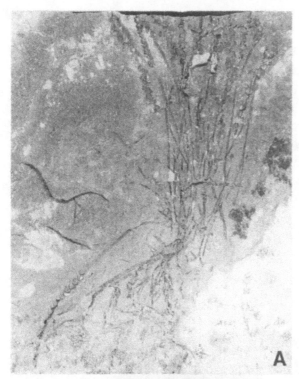

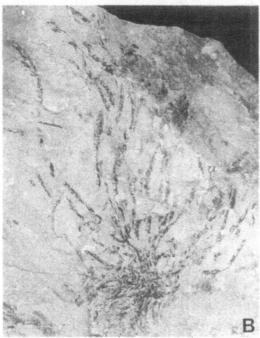

Figure 5.12.
Zosterophyllum sp.nov. from the Pragian Bathurst Island beds, Bathurst Island, Canadian Arctic Archipelago. **A:** Specimen with numerous erect aerial fertile shoots and one bent aerial shoot (*lower left*) and part of tuft of rooting structures. US609-6376. × 0.7. **B:** A similar specimen in which aerial axes trend upward (note fertile ones to *left*) and a tuft of axes extend downward and to the side from a central region. US643-8089. × 1.2.

action. In another specimen (figure 5.12A), one of the aerial shoots, bearing sporangia, is bent downward; adjacent to this fertile shoot are more sinuous downward-trending axes that are similar to rooting structures of other specimens.

In some specimens, coarse grains of sand are preserved bound within the tangle of rootlike structures. These most likely were trapped by and transported with the plant, as the sedimentary matrix is uniformly very fine-grained. This is strong evidence that the structures we interpret as rooting organs were actually growing underground.

Description of Quebec and New Brunswick Fossils

Crenaticaulis verruculosus

Specimens from near Seal Rock on the north shore of Gaspé Bay preserved in a buff, medium-grained sandstone include rootlike structures attached to axes of *Crenaticaulis verruculosus* (Banks and Davis 1969), identifiable on the basis of its distinctive papillate epidermal pattern and two opposite rows of toothlike emergences. The rootlike structures are smooth, sinuous, narrow, and with less coalified residue than aerial shoots. In one specimen, several slender (0.5 mm wide), smooth axes depart at right angles to the presumed rhizomatous or aerial axis (figure 5.13A). These occur in regions of the axis where otherwise no branching occurs. The smooth axes do not branch, but many are incompletely preserved and it is unknown whether branching occurred in more distal regions. A second axis shows similar smooth rootlike structures departing at a more acute angle, extending from the axes for about 2 mm, and while some remain undivided, others bifurcate once or twice (figure 5.13B). In another axis, a larger, laterally positioned smooth axis departs, bends backwards, and then ends abruptly (figure 5.13C). Determining whether these structures are actually attached or just overlie stems of *Crenaticaulis* may be difficult, but of those illustrated we can be reasonably certain of attachment. Interestingly, fine hairlike extensions protrude from some of these small rootlike structures

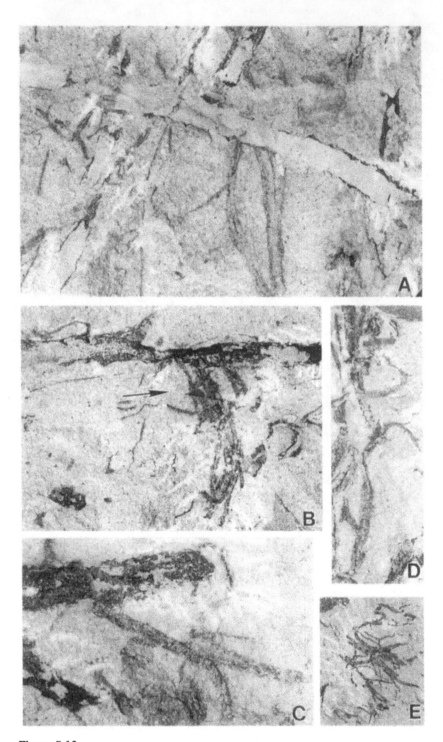

Figure 5.13.
Crenaticaulis verruculosus from the Emsian of Gaspé. **A:** Axis showing characteristic toothlike emergences along margins and papillate surface, with slender axes extending from it at nearly right angles. NCUPC collections. × 3.3. **B:** Another typical axis with several slender rootlike structures departing at *arrow*. NCUPC collections. × 2.6. **C:** Axis with one bent rootlike structure. NCUPC collections. × 4.5. **D:** Associated rootlike structures similar to those attached to *Crenaticaulis*, showing fine hairs along sides. NCUPC collections. × 27. **E:** Associated, much-branched rootlike structures, resembling those documented for *Drepanophycus spinaeformis* by Schweitzer (1980b) and Rayner (1984). × 1.3.

(figure 5.13D, arrows), but preservation does not permit confirmation that these represent root hairs. In both position and morphology, the rootlike structures differ from the subaxillary branches that frequently are found on *Crenaticaulis*. The subaxillary branches, clearly located near a branching point, are undivided and papillate, and they lack the toothlike enations characteristic of stems of the genus (Banks and Davis 1969).

Of the many isolated rooting structures that are associated with, but not attributable to, *Crenaticaulis*, some cross bedding planes while others lie on bedding surfaces. These are up to 3 mm wide and vary considerably in length. All are incomplete (figure 5.13E), but they divide isotomously several times, becoming less than 1 mm wide distally, with some divisions occurring at very wide angles. These isolated structures strongly resemble the rootlike structures described by Rayner (1984) for *Drepanophycus spinaeformis* from the Emsian of Scotland, and given that rare short lengths of rhizomes with falcate leaves similar to *D. spinaeformis* occur in these beds, they might represent rooting structures of that taxon.

Sawdonia ornata

At the type locality for *Sawdonia ornata*, near Cap aux Os, Gaspé, one bed preserves numerous smooth, much-branched, slender axes extending across bedding planes. They are 0.8 to 1 mm wide, and they branch up to three times, sometimes at nearly a 90° angle (figure 5.14). They are similar in appearance to those found associated with *Crenaticaulis* and, while axes of *S. ornata* are found in close association, none have been found attached. We suspect that they represent rooting structures of *S. ornata*, because at that level in the bedding sequence, it is virtually the only species present.

Unidentified Rooting Structures from New Brunswick

One type of isolated rootlike structure obtained from the Dalhousie Junction sequences is similar in branching and size to those from Gaspé. Aerial portions of several plant species occur in adjacent layers but the affinities of the rooting structures are unknown. Similar and larger rootlike structures were discovered recently in a presumed Lower (or at most, lower Middle) Devonian deposit in northern Maine.

Comparisons

This study documents for the first time the probable rooting structures of the zosterophylls *Bathurstia* and *Crenaticaulis*, as well as the tufted rooting organs for a species of *Zosterophyllum*. It also adds new information about rooting structures for a basal lycophyte. The differences in size and complexity of rooting organs among early land plants suggest that evolutionary change leading to these structures was initiated earlier than the Pragian.

The downwardly trending rooting structures of *Zosterophyllum* sp. nov. appear to arise in the same region as aerial axes, but how this happens is unclear, although the possibility of an early-formed cormose region was discussed earlier. Many of the structures illustrated here for *Drepanophycus*, *Bathurstia*, and *Crenaticaulis* either depart from apparently rhizomatous axes or form parts of K-branches. They are not as highly branched as those reported previously. Schweitzer (1980b) and Schweitzer and Giesen (1980) figure structures borne on axes of *D. spinaeformis* and *D. devonicus* that appear to arise in the position of, or from the basal region of, lateral K-branches, and that divide several times. Those described for *D. spinaeformis* from the upper Pragian of Germany arise from a creeping or vertical rhizome, are delicate and vascularized, divide two to three times close to the rhizome, and extend only a few centimeters from the rhizome before preservation ends. Their position near to or as part of a K-branch system is similar to that found in Bathurst Island specimens. Possibly, differences in rooting organ size and complexity from the Pragian to Emsian time are a consequence of evolutionary change, but environmental factors also might affect size (Li and Edwards 1995).

Rooting structures described for *D. spinaeformis* from the Emsian of Scotland by Rayner

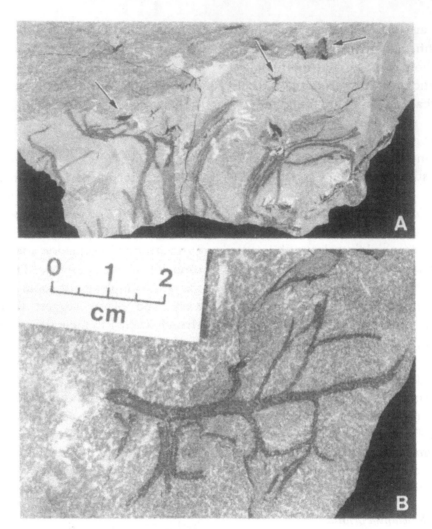

Figure 5.14.
Rootlike structures associated with *Sawdonia ornata*, Emsian of Gaspé. **A:** Root-like structures at right angles to bedding surface. *Arrows* indicate edge-on views of *Sawdonia ornata* axes with spinelike emergences. NCUPC collections. × 1.2. **B:** A much-branched rootlike structure from same beds as (A). NCUPC collections. × 1.5.

(1984) are similarly slender and much divided. They extend laterally for several cm., appear sinuous, and sometimes cross the bedding plane at a steep angle. Considerably smaller, divided axes depart from some of these. Rayner notes that they occur on both leafless and leafy axes, thus possibly being either subterranean or subaerial.

Rooting structures were described by Li and Edwards (1995) for *D. qujingensis* from the Emsian Xujiachong Formation, eastern Yunnan, China. They are randomly arranged on apparently unbranched leafy stems and are up to 3 cm long, dividing up to five times and spreading horizontally through the sediment.

Vascular strands similar in size to those of lateral branches and sporangia are present. Differences in relative frequency/concentration of rooting structures among the Yunnan specimens were hypothesized to be environmentally controlled.

There is some similarity between the rooting structures of *Drepanophycus* and *Bathurstia* from Bathurst Island and those described as belonging to *Taeniocrada* (= *Stockmansella*) *langii* by Schweitzer (1980a) from the Early Devonian of Germany. Several closely spaced, sinuous axes, 0.3 to 1 mm wide and delicate in appearance, depart from the lower surface of the rhizomes.

They divide up to three times and seem to penetrate the rock. Many appear to have a dark central strand interpreted as a vascular bundle. If, as suspected, *Taeniocrada* represents axes of taxa allied with the Rhyniaceae *sensu* Hass and Remy (1991), then similar types of rooting structures are known from at least two, and possibly three, distinct lineages.

Unlike these rooting structures, the rarely branched rooting structures of *Crenaticaulis* resemble most closely the so-called adventitious roots prevalent on rhizomes of the unrelated prefern *Rhacophyton* Crépin (Cornet et al. 1976). Rooting structures in other zosterophylls differ, consisting either of rhizomes bearing rhizoids (*Serrulacaulis* Hueber et Banks, *Trichopherophyton*) or of numerous downwardly trending smooth axes emanating from a central region (*Zosterophyllum*).

The rooting structures reported by Elick et al. (1998a) and in chapter 13 from the Lower Devonian of Gaspé in some cases are much larger than those found attached to plants in these studies. This may be due in part to their being observed *in situ* rather than on plants that have mostly been transported at least a short distance from their site of growth.

Origins of Rooting Systems, Particularly in Lycophytes

As they appear to be derived by division of aerial or rhizomatous axes, many of the rooting structures just described appear little differentiated from shoot systems except in their probable function. No information is available about the origin or evolution of root caps or endogenous lateral rooting.

Of particular significance among the plants described here is the first division of a K-branch, resulting in a shoot–root pair, a pattern identical to that occurring in the embryo of the Upper Carboniferous lepidodendrid *Lepidocarpon* (Phillips 1975; Phillips et al. 1979), where the rooting organ arises as a result of dichotomy of the embryonic shoot. In both instances (i.e., rooting structures on rhizomatous regions of some zosterophylls and pre-lycopsids and the stigmarian

root system of lepidodendrids), the rooting structure is derived by a dichotomy of a shoot apical meristem, and the root meristem is therefore modified (transformed) from a shoot apical meristem. Root and shoot apical meristems in these plants thus are homologous.

These data suggest a way roots may have arisen in some nonseed plants. With the advent of extended or indeterminate growth of the sporophyte, rooting structures were produced by dichotomy of the shoot apex and subsequent differentiation (by transformation) of a root apex (as may occur in the formation of rhizophores in *Selaginella*). K-branches found in *Bathurstia* and *Drepanophycus* indicate that, among some members of the zosterophyll/lycopsid lineage, rooting organs had become intimately associated with branch initiation, so that a branch primordium immediately underwent dichotomy to produce divergent shoot and root apices. K-branches in these taxa therefore include a proximate dichotomy into a shoot–root unit that appears to be a reiteration of the plant body. Such units represent vegetative propagation, and, as they appear to have possessed only an ephemeral dependence on the parent axis, they are interpreted here as vegetative propagules or plantlets. Association of root and shoot in this manner would have been a highly advantageous innovation, ensuring early independence of a branching system and contributing to an aggressively spreading habit.

Discrimination between a plantlet and an embryo may be largely a matter of timing, and their functions as well as their developmental stages are likely to be homologous. Therefore, precocious occurrence of the shoot–root dichotomy would introduce the root into embryogeny. The embryonic "root" in the Carboniferous lycopsid *Lepidocarpon* appears to be derived from dichotomy of the embryonic shoot in this manner (Phillips et al. 1975; Phillips 1979), and it is consistent with the interpretation of the subterranean organs of lycopsids as shoot homologues.

Development of the earliest land plants is believed to have involved only growth and branching of a shoot system. Evolution of rooting organs by transformation of the apex would therefore be associated with postembryonic development and in response to the "need" of large sporophytes for anchorage and absorption; it would not in the first instance be predicted to be associated with embryogeny. The differentiation of homologous subaerial and aerial organ systems of the zosterophyll–lycopsid lineage appears to have been well established in the Pragian and may have occurred much earlier if, as suggested by Schweitzer (1983), *Drepanophycus* is demonstrated to have existed as early as the Lochkovian.

A similar origin of the root in seed plants may be hypothesized, but it is also possible that the evolution of the primary root in embryos of seed plants may never have involved dichotomy of the embryonic shoot. Rather, the embryonic root of seed plants, and indeed the endogenous rooting organs that characterize this lineage, may be of a *de novo* origin. If shoots and roots of seed plants are nonhomologous, they would have an independent origin in development and evolution. Molecular genetic research in *Arabidopsis* and other seed plants thus far indicates that, for the most part, apparently different genes are involved in some aspects of shoot and root apical meristem differentiation (Benfy 1999), although the possible occurrence of homologues has not been tested.

Evolutionary origins of the roots of seed plants are likely to lie within the precursors of the progymnosperms, as substantial underground systems would be predicted to be a requirement for relatively large and/or woody plants. Stephen Scheckler's study of progymnospermous rooting organs (pers. comm. 1999) would appear to support this. As noted, unfortunately there is little known of the belowground organs of trimerophytic plants, espe-

cially whether rooting organs, as opposed to prostrate stems or rhizomes, were common within this group. Nor is it known if the rooting organs of this group were produced by dichotomy of the shoot apical meristem as in zosterophylls and early lycopsids, or if they originated by some other means.

There is no evidence of rooting organs in plants of rhyniophytoid grade. Given that progymnosperms and early ferns (e.g., *Rhacophyton*) had rooting organs, then it seems that rooting organs evolved within the trimerophytes, but it is also possible that the rooting organs of the various trimerophyte-derived lineages need not be homologous. An understanding of the initiation and development of trimerophyte rooting system(s) will be critical to the resolution of questions regarding not only the origin of the seed plant root, but also the origin of roots in the nonseed plant lineages derived from trimerophytic stock, including the sphenopsids and ferns.

ACKNOWLEDGMENTS

This research has been supported by the Natural Sciences and Engineering Research Council of Canada (RGPIN 1334 to JFB; PGS-A to MEK), the Polar Continental Shelf Project of Natural Resources Canada (PCSP/ÉPCP Publ. No. 01399), the Northern Scientific Training Program of the Department of Indian Affairs and Northern Development Canada (to MEK), and the University of North Carolina Faculty Research grant program and George Cooley Fund (to PGG). The authors recognize the contributions to field work by the following individuals: E. E. McIver, S. A. Hill, D. L. Postnikoff, L. Postnikoff, S. Kojima, and Y. Nobori.

6

The Posongchong Floral Assemblages of Southeastern Yunnan, China—Diversity and Disparity in Early Devonian Plant Assemblages

Hao Shou-Gang and Patricia G. Gensel

Most knowledge about the morphology, comparative level of complexity, evolutionary trends, interrelationships, and diversity of early land plants has been derived from fossil plants of the Laurussian paleocontinents (Laurentia and Baltica—i.e., present-day central and northern Europe, Greenland, and North America) because of the long history of research in these areas (Banks 1968; Chaloner and Sheerin 1979; Edwards 1980, 1990; Gensel and Andrews 1984).

A considerable amount of new information about the morphology and diversity of early land plants has accrued since Banks (1968, 1975c) divided the "psilophytes" into three new subdivisions: the Rhyniophytina, the Zosterophyllophytina, and the Trimerophytina, mainly based on this material. For example, it is now known that land plants existed earlier in time than previously supposed, dating from the Ordovician (Gray 1985, 1993; and see chapters 2 and 8), that abundant remains of early land plants exist in some geographic regions previously not considered productive, and that some entities do not fit into existing classification schemes. Kenrick and Crane (1997a) reevaluate the affinities of many of these early land plants, using phylogenetic analysis via cladistic methodology, thus advancing our understanding of some relationships among nonseed plants. Their work also demonstrates plants or charac-

ter complexes where more study is needed and proposes or tests some hypotheses about evolutionary change among them.

Even though some work on fossils outside Laurussia dates from the beginning of the 20th century, it is within the last few decades that plant fossils of Silurian and Devonian age from other paleocontinents have been investigated using modern techniques. This is especially true for Australia and China. Reinvestigation of Australian early land plants by Tims and Chambers (1984) and of the stratigraphy of the deposits by Garratt (1978) and Garratt and Rickards (1984) has resulted in an interpretation of a Lower Plant Assemblage of Silurian (Ludlow) age and an Upper Plant Assemblage of Lower Devonian (Pragian to Emsian) age, with some of the same taxa occurring in both horizons, in particular the comparatively large, complex lycopsid *Baragwanathia*. Questions remain concerning the age of the Lower Plant Assemblage, partly because of the presence of identical plant types in sediments of such disparate ages. If the Lower Plant Assemblage is correctly dated, such a complex form as *Baragwanathia* differs considerably from the types of fossils being found in Silurian sediments in other parts of the world. It further implies that some plant types would have existed unchanged for more than 20 million years. Other taxa recorded in these floras by Tims and

Chambers (1984) include zosterophylls, lycopsids, possible rhyniophytes (*Salopella*), and possible trimerophytes (*Dawsonites subarcuatus*, *Hedeia* spp.).

Records from southeastern Asia are sparse, with the exception of from China. The remainder of this account centers on early Devonian plants from China, particularly the Pragian assemblage from southeastern Yunnan, because of the intensive study of the plants and the distinctive nature of some taxa found in those deposits.

In China, the first detailed description of Lower to Middle Devonian fossil plants was made by T. G. Halle (1927). Only a few reports appeared before the 1970s, including those from Yunnan by Halle (1936), Sze (1941), Hsü (1946, 1966), and Li and Cai (1977, 1978, 1979). During the past two decades, many early land plants, mostly of Devonian age, have been collected and described from China with increased emphasis on interpreting their morphology and relative complexity. The major Devonian plant localities are summarized by Li and Wu (1996) and the major plant types by Cai and Wang (1995). China currently is believed to consist of amalgamated parts of up to seven microcontinents or terranes—that is, the Ta Li-Mu, North China, South China, Tibet, Shan-Thai, Kazachstan and Siberian microcontinents (Scotese and McKerrow 1990; Nie et al. 1990; Li and Wu 1996), all of which were separate entities during the Paleozoic. For the Lower Devonian, some of the richest plant localities occur in eastern and southeastern Yunnan Province, which during the Devonian was part of the South China (Yangzi Platform and South China Fold System) paleocontinent. Other early Devonian plant-bearing localities are in Sichuan, Guizhou, and Guangxi provinces, all part of the South China paleocontinent, and in western Yunnan, which was part of the Shan-Thai paleocontinent. A Silurian plant assemblage has been described from the Junggar region of northwestern Xinjiang in northwestern China, probably originally part of the Kazachstania paleocontinent (Cai et al.1993; Li and Wu 1996). No irrefutable Lower Devonian plant localities are yet known from the North China paleocontinent.

Plants described from the Pragian-age Posongchong Formation of southeastern Yunnan Province, China, are of considerable interest in that they are abundant and well preserved. Further, while some of them are similar to plants reported from Laurussia, others are quite distinctive, either in the combination of characters possessed by a single taxon or in the level of complexity manifested by vegetative and fertile plant organs. The goal of this account is to present a summary of current knowledge on the assemblages of the Posongchong Formation, South China, and to consider the following questions: To what extent are these floral assemblages different from coeval ones from other regions? How does knowledge of the structure and morphology of these plants affect existing concepts about early land plants? Might the pattern of diversification and evolution of early vascular plants in that region be different because of geographical location and paleoenvironmental differences?

THE POSONGCHONG FORMATION: DISTRIBUTION AND GEOLOGY

The Lower Devonian (Pragian) Posongchong Formation of southeastern Yunnan was named by Liao et al. (1978), based on a suite of strata exposed along the Posongchong Valley of Guangnan district. These strata, rich in plant megafossils, are widely distributed in Wenshan, Guangnan, Mengzi, Xichou, and Yanshan districts (figure 6.1), and they extend from southeastern Yunnan into Viet Nam. Among them, the fossil materials of Wenshan and Guangnan are well preserved, while those of other localities are less well preserved because of metamorphism.

Deposits of the Posongchong Formation consist of shallow-water terrestrial-derived sediments. A lower fluvial part consists of gray to yellow-gray, coarse and fine-grained sandstones, while an upper part, consisting largely of dark shale or silty shale with lenticular bodies, is interpreted as mainly a near-shore deposit of a lake or lagoon. The age of these sediments is considered to be Pragian (Siegenian) based on an early report of plants (Liao et al. 1978), dis-

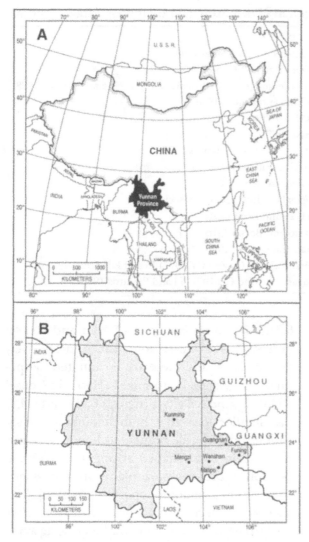

Figure 6.1.
Maps showing location of Yunnan province (**A**) and of Posongchong Formation outcrops in China (**B**). The dots in (**B**) indicate districts in southeastern Yunnan province where outcrops of the Posongchong Formation occur.

After map in *Atlas of the People's Republic of China*. pp. 1–2. Beijing: Foreign Languages Press and China Cartographic Publishing House.

persed spore data (Wang 1994) that place the Posongchong Formation in the PE-*Verrucosisporites polygonalis–Dictyotriletes emsiensis* spore zone of Richardson and McGregor (1986), some fish data (pers. comm., Pan Jiang 1993), and its position below the better-dated early Emsian Pojiao Formation. It also is correlative to the Naogaoling Formation in adjacent Guangxi dated as Pragian (Siegenian) using conodonts (see Hao 1989a; Hao and Gensel 1998). Gerrienne (1996a) discusses limitations

of the correlations, but his biostratigraphic coefficient analysis also indicates a late Pragian age. Many of the plants described to date are from the upper part of the formation.

The first recent investigation of megafossils was made by Li and Cai (1977) from the Zhichang section of Wenshan and the Posongchong section of Guangnan, followed by studies to the present, mainly concentrated on the Zhichang section and more recently the Changputang section of Wenshan and the Gegu section of Mengzi (Geng 1983, 1985; Hao 1988, 1989a, 1989b; Hao and Beck 1991a, 1991b, 1993; Hao and Gensel 1995, 1998; Li and Edwards 1992, 1996, 1997). Based on the papers published and in press, the Posongchong flora includes four more or less broadly ranging genera (*Zosterophyllum, Psilophyton, Hedeia,* and *Baragwanathia*—the first two are cosmopolitan and the latter two are found in Australia and China) and 11 endemic genera (*Stachyophyton, Huia, Eophyllophyton, Discalis, Gumuia, Catenalis, Yunia, Celatheca, Adoketophyton, Demersatheca,* and *Halleophyton*). Proposed higher taxonomic affinities of these plants are shown in table 6.1. Clearly, the affinities of many of these taxa are currently unknown, but others fit nicely into established lineages. The range of variation in overall form, histological construction, and the morphology of lateral appendages are discussed next, followed by consideration of affinities. Line drawings of the Posongchong plants published thus far are shown in figures 6.2, 6.6, 6.7, and 6.9. Since research on newly collected fossils from these deposits is ongoing, new taxa will undoubtedly be described.

GENERAL FEATURES OF THE POSONGCHONG PLANTS

Size

All the plants are herbaceous with relatively slender axes. Most have an axis diameter of less than 5.0 mm (range, 0.8 to 4.0 mm, but some are up to 20 mm). *Catenalis digitata* (affinity uncertain) (figure 6.4A) and the trimerophyte *Psilophyton primitivum* (figure 6.3A) possess the narrowest axes known from this flora, being from 0.8 to 2.5

Table 6.1. Taxa described to date from the Posongchong Formation and their postulated affinities

Taeniocradaceae *sensu* Geng (1985)	*Huia recurvata* Geng 1985
Trimerophytes	*Psilophyton primitivum* Hao and Gensel 1998
	Possibly *Hedeia sinica* Hao and Gensel 1998
Zosterophyllophyta	*Zosterophyllum australianum* (Lang and Cookson 1930) Hao and Gensel 1998
	Zosterophyllum sp. Hao and Gensel 1998
	Discalis longistipa Hao 1989a
	Gumuia zyzzata Hao 1989b
Lycophyta	*Halleophyton zhichangense* Li and Edwards 1997
	Baragwanathia sp. Hao and Gensel 1998
Incertae sedis	*Stachyophyton yunnanense* Geng 1983
	Eophyllophyton bellum Hao 1988, Hao and Beck 1993
	Catenalis digitata Hao and Beck 1991a
	Yunia dichotoma Hao and Beck 1991b
	Celatheca beckii Hao and Gensel, 1995
	Adoketophyton subverticillatum (Li and Cai 1977) Li and Edwards 1996
	Demersatheca contiguum (Li and Cai 1977) Li and Edwards 1996

mm. The widest axes seen in the flora belong to the lycopsid *Baragwanathia* (figures 6.2L and 6.3C), in which axes are up to 32 mm wide, possibly representing a creeping rhizome. Axes of *Stachyophyton* (figure 6.2J) can be up to 20 mm wide; these have been suggested to represent planated branch systems (Geng 1983; Wang and Cai 1996) and if so, they may have been borne on larger main axes. The lycopsid *Halleophyton* has 15-mm-wide rhizomes (figure 6.2K).

Growth Form and Branching Pattern

In the more basal regions of axes, some plants have a dichotomous pattern that consists of trailing (K- or H-shaped branching) and upwardly growing branches, as seen in *Discalis, Gumuia,* and *Zosterophyllum.* This morphology, common among early land plants, may reflect an adaptation to living in a moderately disturbed environment such as stream or lake margins or for maintaining clonal growth. For other plants, however, there is no clear evidence of their basal parts, and their overall growth form is unknown, most fossils apparently representing aerial shoots.

In the Posongchong plants, three-dimensional isotomous branching is common in aerial axes, as shown by *Gumuia zyzzata* (figure 6.2B), *Zosterophyllum australianum* (figures 6.2A and 6.3A,B), and *Psilophyton primitivum* (figures 6.2E and

6.3A). *Zosterophyllum* sp. 1 of Hao and Gensel (1998) has at least four times successively divided axes, and these are terminated by sporangia aggregated into spikes (Hao and Gensel 1998). *Catenalis digitata* (figures 6.2H and 6.4A) has slender axes that branch dichotomously, with terminal fertile branchlets being more planar and in a fan-shaped arrangement. Other, possibly more complex plants in the assemblages show pseudomonopodial branching (i.e., successive alternate branches result from obviously unequal dichotomies of the growing axis). Such a growth pattern, seen in *Huia* (figure 6.2G), *Eophyllophyton* (figure 6.7), and *Celatheca* (figures 6.8A and 6.9), apparently allows more effective trapping of sunlight for photosynthesis (Niklas and Kerchner 1984). Of interest is that plants with such pseudomonopodial branches still retain lateral or apical isotomous fertile or sterile branches as seen in *Psilophyton, Celatheca,* and *Eophyllophyton.* In these types of branching, the Chinese plants are similar in evolutionary complexity to Pragian and Emsian plants from other regions (Raymond 1987; Edwards 1990; Edwards and Davies 1990; Raymond and Metz 1995).

Eophyllophyton is interesting in that basal axes produce lateral branch systems that terminate in recurved, dichotomous, anchor-like structures (Hao and Beck 1993). Similar appendages, although less rigid in appearance,

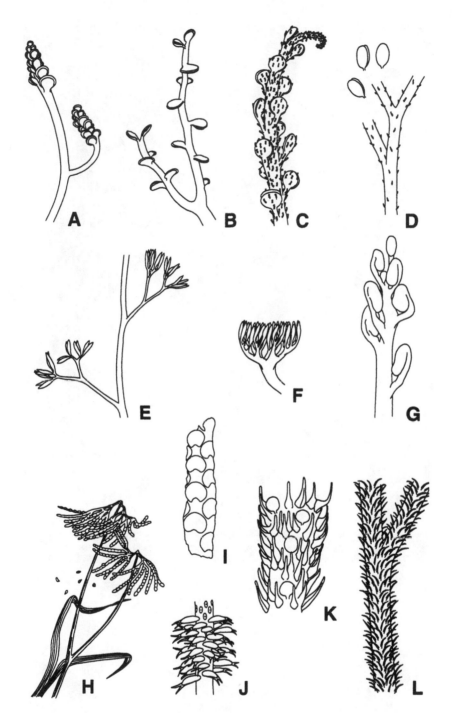

Figure 6.2.
Diagrams of plants described from the Posongchong Formation. All drawings are based on Hao's collections except where otherwise indicated. **A:** *Zosterophyllum australianum*, × 0.6. **B:** *Gumuia zyzzata*, × 1.0. **C:** *Discalis longistipa*, × 1.2. **D:** *Yunia dichotoma*, × 1.2. **E:** *Psilophyton primitivum*, × 1.2. **F:** *Hedeia sinica*, × 1.5. **G:** *Huia recurvata*, × 1.2. **H:** *Catenalis digitata*, × 0.8. **I:** *Demersatheca contigua*, × 1.7, after Li and Edwards (1996). **J:** *Stachyophyton yunnanense*, × 1.5, after Geng (1983). **K:** *Halleophyton zhichangense*, × 1.1, after Li and Edwards (1997). **L:** *Baragwanathia* sp., × 1.1.

Figure 6.3.
Representative plants from the Posongchong Formation. **A:** Part of shoot systems of *Zosterophyllum australianum* (*right*) and *Psilophyton primitivum* (*left*). PUH-G. 301. × 0.8. **B:** Several strobili of *Zosterophyllum australianum*. PUH Z01. × 1.3. **C:** *Baragwanathia* sp. PUH-G. B1. × 1.4. **D:** Fertile region of *Hedeia sinica*. New collections, more extensively preserved, indicate the fertile structures are pendulous. PUH-G 201. × 3.

Figure 6.4.
General view of *Catenalis digitata* and details of fertile regions. **A:** A terminally borne, dichotomous fertile system. *Arrow* (at L) indicates a specimen of *Lingula* sp. PUH-Ch 803. × 1.8. **B–D:** Parts of fertile branchlets. *Arrows* indicate impression of lateral margin of the sporangium formerly present in those areas (B,D) and the crescent-shaped dehiscence slits of sporangia (C). PUH–Ch 818, 817, 814. × 9.

occur in *Psilophyton* and some species of *Pertica* as part of the aerial lateral branch system.

Appendages

The initial adaptations for early vascular plants to a land environment are not only to develop morphological structures and physiological functions for efficient photosynthesis, but to transport and retain water, and to effectively reproduce. In regard to features that may be adapted for these functions, some Posongchong plants appear unusually advanced for Early Devonian times. In addition to plants with smooth axes (e.g., *Gumuia*, *Zosterophyllum*, *Psilophyton*, *Huia*, *Stachyophyton*, *Hedeia*), a number of plants of the Posongchong flora show several different forms of lateral appendages, ranging from emergences to leaflike appendages to true leaves (microphylls,

perhaps megaphylls). *Discalis longistipa* (Hao 1989a) axes (figure 6.2C) are completely covered with small, straight-sided, nonvascularized appendages with expanded apices reminiscent of the emergences of *Psilophyton princeps* (Hueber and Banks 1967) and *Anisophyton* (Remy et al.1986). Axes of *Yunia dichotoma* (Hao and Beck 1991b) have sparsely distributed, tapered, spine-like emergences (figure 6.2D). Small conical appendages with rounded ends occur on axes of *Eophyllophyton* (figures 6.5A and 6.7). A depression on the top of these appendages may be the site of a stoma (Hao 1988), but this requires further confirmation.

In the Posongchong assemblages, morphologic differentiation of leaflike organs, including types possibly referable to micro- and megaphylls, appears quite advanced relative to coeval plants, possibly indicating a longer time in which evolutionary change could have occurred (i.e., earlier origin) or perhaps representing the effect of different selective forces during isolation of the area from Laurussia or other continents. The leaflike structures might have served to increase photosynthetic area and/or they may have served a protective function—the latter is a particularly attractive possibility since in many cases the leaflike structures cover sporangia and possible apices. Essentially two types of appendages resembling leaves are known from these plants: large fan-shaped structures and both large and small, slightly lobed, entities. In both cases, anatomy, which might confirm that they truly represent leaves instead of modified stems, is lacking.

Each stalked, fan-shaped, leaflike appendage of *Adoketophyton subverticillatum* (Li and Edwards 1992) has a sporangium attached to the adaxial surface of the appendage stalk (figures 6.5 and 6.6C). These structures consist of a parallel-sided stalk that expands into a triangular lamina with straight sides and rounded or undulate distal regions. The stalks are short, narrow, and decurrent, and they appear rigid. They are up to 1.3 mm wide and about 2.5 mm long. The blade is thin, very regular in shape, equal in width to the stalk at the base, and 10 mm wide distally, and it appears rather rigid. The sporangium is

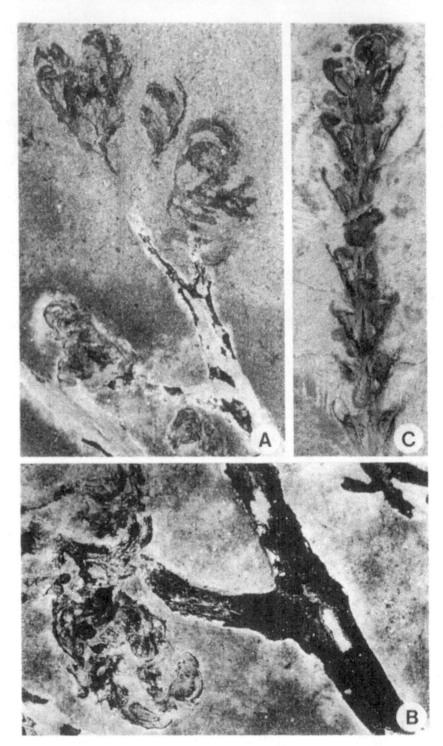

Figure 6.5.
Selected specimens of Posongchong plants. **A:** Leaves of *Eophyllophyton bellum* attached along axes. BUP-6. 137. × 9. **B:.** Detail of fertile region of *Eophyllophyton bellum*, showing leaves in mostly side view; dark areas represent parts of sporangia. BUP-6.110. × 12.6. **C:** Fertile region of *Adoketophyton subverticillatum*, showing appendages and sporangia. PUH AO1. × 1.6.

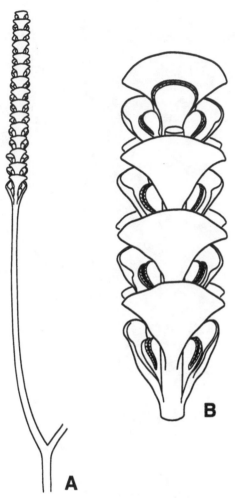

Figure 6.6.
Partial reconstruction of *Adoketophyton subverticillatum.* Axis and strobilus (**A**) and a detailed view of strobilus (**B**), showing sporangia occurring in axils of leaflike structures.

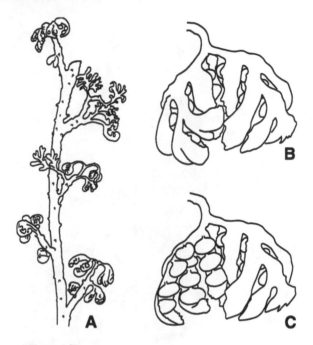

Figure 6.7.
Eophyllophyton bellum. **A:** Reconstruction of a part of the plant. **B:** Line drawing of an entire cluster of fertile leaves. **C:** Line drawing of leaf cluster with some leaf segments removed to show rows of sporangia attached to inner (upper?) surface.

inserted just distal to where the stalk begins to widen, about 1.8 mm from the axis. Internal histology of the leaflike structure is inconclusive in regard to whether it truly represents a sporophyll. These structures may have functioned for both photosynthesis and protection.

Less rigid appearing, lobed, and inwardly folded leaflike structures are known in *Eophyllophyton bellum* (figures 6.5A,B and 6.7) and *Celatheca beckii* (figures 6.8 and 6.9). Those of *Celatheca* surround sporangia only, while leaflike structures are both vegetative and sporangium bearing in *Eophyllophyton.*

The aerial regions of the shoot system of *Eophyllophyton bellum* bear laminate, highly divided leaflike structures (figures 6.5A and

6.7), interpreted by Hao and Beck (1993) as being true leaves, possibly representing an early megaphyll. The leaves appear irregularly to pinnately divided and fan shaped. They are simple structures in several ways. First, the leaves are very small, ranging in width from 1.4 to 4.0 mm, in length from 2.0 to 4.5 mm, and in thickness from 40 to 200 μm. Second, the leaves are deeply divided (lobed?), and each segment of the leaf is vascularized by a single vein consisting of very few tracheids, to produce an overall branched venation pattern. Third, epidermal and mesophyll cells lack distinct differentiation, the mesophyll cells still being longitudinally extended so as to parallel the epidermal cells and the vascular strand. This lack of anatomical differentiation could be interpreted either as representing a modified stem or as a primitive leaf that is not yet very differentiated from an axial branching system. If these truly represent leaves and were derived from branch systems, they would represent megaphylls according to some definitions. They differ from the undoubted axial structures in the plant in their discrete, fan-shaped form, in

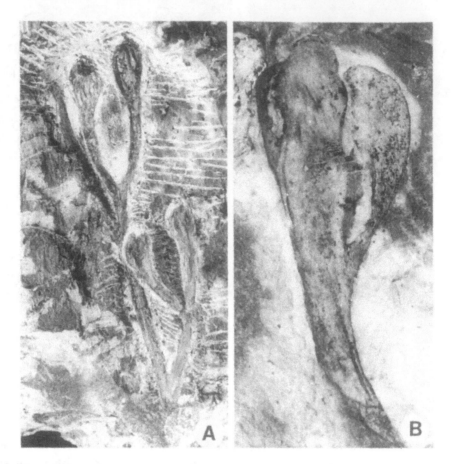

Figure 6.8.
Celatheca beckii. **A:** Axes and sporangia-bearing regions. PUH-G.110. × 3.0. **B:** Leaflike structures (three of the total four) covering sporangia. PUH-G.126. × 9.6.

their attachment as appendages, and in the absence of emergences on the leaflike structures.

Some of the leaves in *Eophyllophyton* bear circular to reniform sporangium-like bodies. The sporangia frequently are arranged in rows on their adaxial surfaces next to the veins, are attached by a short stalk, and may dehisce along their distal margin. Sporangium-bearing leaves typically curve inwards (figures 6.5B and 6.7).

In *Celatheca beckii* (Hao and Gensel 1995), each of the four terminally located, obovate sporangia is covered by a leaflike structure that may show slight lobing and wavy margins. The sporangia are borne close together at the end of an axis, and in surface view only the leaflike structures can be seen because each one folds around a sporangium (figures 6.8B and 6.9). It is reminiscent of the manner in which cupules enclose ovules in Carboniferous plants, but there is no evidence in favor of these entities representing

ovules. The leaflike appendages appear faintly striated, but vascular strands have not been found in them. Axes lack evidence of vegetative laminar structures.

Other occurrences of leaves include the lanceolate vascularized microphylls of *Baragwanathia* sp. (Hao and Gensel 1998), up to 24 mm in length and flattened–oval in basal outline, becoming ovate in outline distally. *Halleophyton* (figure 6.2K) exhibits broader leaves (Li and Edwards 1997). Spirally arranged, flattened sporophyll-like structures with bifid tips (figure 6.2J) occur in strobili of *Stachyophyton yunnanense* Geng (1983), each with an adaxially borne sporangium. The details of the attachment and relationship between sporophyll-like structure and sporangium require further clarification. Basal branches are dichotomous and were interpreted as planated by Geng (1983). According to a recent report of new collections of

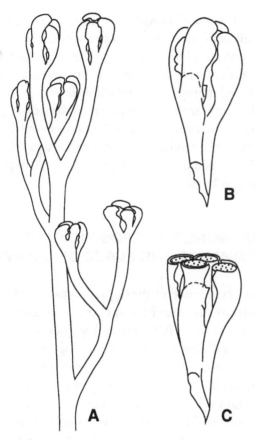

Figure 6.9.
Celatheca beckii. Diagrams of parts of the plant.
A: Axes with terminal fertile regions. **B:** Detail
of an entire fertile structure; leaflike structures arrayed in fours. **C:** Cut-away of distal
region to illustrate the four sporangia located
inside the leaflike structures.

Stachyophyton, the vegetative lateral branches are
borne spirally, extend undivided for a short distance, and then divide distally into about eight
segments forming cuneate leaflike structures
(Wang and Cai 1996). Relative positioning and
organization of fertile and sterile regions still is
unclear although Wang and Cai state that distal
regions of some lateral branches are fertile, and
they present a reconstruction.

Reproductive Structures

Various sporangial types are present among the
known plants from this region, including globose to reniform ones with transverse distal
dehiscence (*Zosterophyllum, Adoketophyton*), more
fusiform ones such as *Psilophyton* and *Hedeia* (fig-

ures 6.2F, 6.3A, and 6.3D), and the elongate-oval
ones of *Celatheca* and *Stachyophyton* or ovate ones
of *Huia* (figure 6.2G). Sporangia attributed (but
not found attached) to *Yunia* (Hao and Beck
1991b) appear similar to zosterophyll sporangia
but are longer than wide and more spatulate
(figure 6.2D), thus differing in shape from those
of most zosterophylls. They also are very different from trimerophyte sporangia, which are
fusiform and split open along one side only.

As noted, the reniform sporangia of *Adoketophyton* and elongate-oval sporangia of *Celatheca*
are covered by leaflike structures (if leaves,
these would represent sporophylls), and those
of *Stachyophyton* are associated with a laminar
structure (figure 6.2J). Two additional plants
exhibit unusual sporangial features, which
might reflect protection from exposure. In *Catenalis,* sporangia appear little differentiated and
sunk into the axis (figure 6.4B–D), occurring in
a single row along the distal ends of some axes.
They apparently were globose and dehisced
near the distal margin. Hao and Beck (1991a)
noted that they fall off the axes (possibly as a
result of formation of an abscission layer). Sporangia of *Demersatheca* (Li and Edwards 1996)
appear similar to those of zosterophylls and
Catenalis but are borne in up to four rows in a
strobilus. These sporangia are sunken into the
stem so that only the abaxial valve is visible in
surface view (figure 6.2I). A region of thickening and splitting indicates the distal dehiscence
zone, and Li and Edwards demonstrated that a
discrete adaxial valve is present. Less definite is
the exact sporangial shape, the presence of a
sporangial stalk, and how they are borne on the
stem (i.e., is the sunken appearance of sporangia an original feature present throughout
development or a result of taphonomy?). The
plant parts available for study were fragmentary;
vegetative regions of this plant are unknown.

Anatomy

Evidence for preserved internal structure exists
thus far for only four plants in the flora: *Huia
recurvata* (Geng 1985), *Yunia dichotoma* (Hao
and Beck 1991b), *Eophyllophyton bellum* (Hao

and Beck 1993), and *Stachyophyton yunnanense* (Wang and Cai 1996). Preservation is by limonite and carbonized remains, revealing the structure of the vascular strand and details of water-conducting cells and some limited preservation of cortical cells. In the first three genera, axes possess a columnar protostele and centrarch protoxylem. The diameter of the xylem is about a fifth (*Eophyllophyton bellum*) to a quarter (*Yunia dichotoma, Huia recurvata*) of the total axis diameter. The metaxylem tracheids are characterized by helical, spiral, and even reticulate secondary wall thickenings, between which are regions of presumed primary wall containing perforations. As such, they are comparable to G-type cells of Kenrick and Edwards (1988) and Kenrick and Crane (1991). Xylem maturation in the protostele of *Stachyophyton* possibly is exarch (preservation is incomplete), and metaxylem cells are described as having scalariform to possibly reticulate secondary wall thickenings.

Another interesting feature is found in permineralized axes of *Yunia dichotoma*. Axes exhibit a circular protostele in which there occur two lunate to circular regions of parenchyma (indi-

vidual cells are elongate in longitudinal section). Small cells interpreted as protoxylem occur to the outside of these regions. After the axis bifurcates, each stele contains only one parenchyma/protoxylem region, so the whole is interpreted as being centrarch, with precocious division of protoxylem before branching. Metaxylem wall thickenings are annular to scalariform, with a thinner wall between thickenings, showing small perforations (=G type).

TAPHONOMY, DEPOSITIONAL ENVIRONMENTS, AND PLACE OF GROWTH

The occurrence of these plant types in specific horizons in the Posongchong Formation is shown in table 6.2. Selected sequences will be discussed next in terms of lithological features, presence or absence of taxa, and their relative abundance. It should be realized that detailed stratigraphic correlation between the different sections is difficult to establish and in need of refinement, and that these observations may be modified by future collecting.

Table 6.2. Occurrences of plants thus far described (published) from selected sections of the Posongchong Formation

Zhichang Section: Seven lithological units	
Lithological unit 2	*Adoketophyton subverticillatum, Stachyophyton yunnanense, Huia recurvata, Demersatheca contiguum, Eophyllophyton bellum*
Lithological unit 3	*Eophyllophyton bellum, Gumuia zyzzata*
Lithological unit 4	*Eophyllophyton bellum, Gumuia zyzzata, Psilophyton primitivum, Zosterophyllum* sp. 1, *Celatheca beckii*
Lithological unit 5	*Catenalis digitata, Discalis longistipa, Eophyllophyton bellum, Yunia dichotoma, Huia recurvata, Halleophyton zhichangense*
Lithological unit 7	*Zosterophyllum australianum*
Changputang Section: Nine lithological units	
Lithological unit 3	*Hedeia sinica, Zosterophyllum* sp. 1
Lithological unit 6	*Catenalis digitata*
Lithological unit 7	*Zosterophyllum australianum, Psilophyton primitivum, Gumuia zyzzata*
Lithological unit 8	*Celatheca beckii, Huia recurvata, Eophyllophyton bellum, Hedeia sinica*
Gegu-Daigu Section: Seven lithological units	
Lithological unit 2	*Baragwanathia* sp.
Lithological unit 4	*Zosterophyllum australianum, Adoketophyton subverticillatum*
Lithological unit 5	*Huia recurvata*
Lithological unit 6	*Eophyllophyton bellum*

Plants occur in lithological units other than the ones listed, but these have not yet been formally described. Details of lithology can be found in the text and in Hao (1989a) and Hao and Gensel (1998). For lithological units, lower numbers are older.

Zhichang Section

The lower part of this section consists of mostly higher-energy deposits with more transported plant remains. The upper part indicates a relatively quiet environment of deposition, with greater abundance of finer-grained, darker sediments and more *in situ* plant remains. The transition appears to occur near the top of the third lithological unit.

In the second lithological unit of the Zhichang section, some vertical separation of plant types exists, in that *Eophyllophyton* occurs in the upper layers and *Stachyophyton* and *Adoketophyton* occur in the lower layers. In the fifth lithological unit of the Zhichang section, two major lithologies occur, lenticular arenaceous mudstone and a thin argillaceous sandstone. Numerous relatively undisturbed, apparently basal rhizomes of *Discalis longistipa* occur in the lenticular mudstone levels, suggesting nearly *in situ* deposition of this taxon. *Eophyllophyton* stems lie across some of the *Discalis* specimens. Since vegetative and fertile stems and leaves are still connected, *Eophyllophyton* may have been transported, probably only a short distance, to this depositional site. Remains of *Yunia* and *Huia* are intermixed. Fragments of *Catenalis digitata* form a monospecific assemblage in the overlying argillaceous sandstone.

Eophyllophyton bellum occurs in the second through fifth lithological units of the Zhichang section. In fact, it is known from a total of nine beds from four different sections—six at Zhichang, and one each from the Changputang, Gegu, and Daliantang sections. Four of these occurrences are monospecific. Fairly large portions of the plant are preserved in the fourth and fifth lithological units of the Zhichang section in an arenaceous mudstone. The third layer of the Zhichang section is a gray to white, sandy mudstone and sandstone in which broken-apart stems and leaves occur. The latter may be a high-energy channel deposit.

Changputang Section

Large, fairly intact portions of *Catenalis digitata* complete with distal, fanlike fertile branches are found in the sixth lithological unit of the Changputang section. The lithology of sediments here indicates deposition in a quiet water environment. Some aspects of the morphology of the plant suggest it may have been entirely or partly aquatic, although definitive evidence is not available. The morphology of the plant suggests it may have been entirely or partly aquatic. The co-occurrence here of a single brachiopod species, *Lingula* sp. (figure 6.4B, arrow), with *Catenalis* may indicate a marginal marine depositional environment, and if so, the plants may have grown along a shoreline, such as of a lagoon.

In the seventh horizon of Changputang section, *Zosterophyllum australianum*, *Z.* sp., *Gumuia zyzzata*, and *Psilophyton primitivum* are found intermixed. Preserved in a dark arenaceous mudstone, plant remains often are large, fairly intact, and orientated in the same direction. Immature fertile spikes and extensive branching systems of *Z. australianum* (figure 6.3A,B) are frequently found here. Some of the best-preserved specimens of *Gumuia* occur intermixed with *Z. australianum*, and even more abundantly beneath it. *Psilophyton* remains are rare. It appears that several of the plants were transported only a short distance prior to preservation. They may have formed an association in that region during that time (Hao and Gensel 1998).

The eighth lithological unit of this section is a poorly bedded siltstone; the upper region contains intermixed *Celatheca*, *Huia*, and *Eophyllophyton* (only a few fragments), underlain by *Hedeia sinica* and some as yet undescribed zosterophylls. Overall, *Eophyllophyton bellum* is a dominant element of the Zhichang sequence but is very sparse in the Changputang section, where, despite numerous collecting visits, only a few isolated leaves were obtained from the eighth lithological unit. Similarly, *Huia recurvata* is found fairly abundantly in the second and fifth units of the Zhichang section, but only a single short fertile shoot has been obtained in the Changputang section thus far. *Adoketophyton* and *Stachyophyton* have not yet been found in the Changputang section, although they occur in the second lithological unit of the Zhichang section. In contrast, *Zosterophyllum australianum* and other as yet unidentified zosterophylls are very abundant at all levels of the Changputang section but are relatively rare at Zhichang. If we assume these plants have not been transported far from where they lived, then the plant assemblages just described may reflect communities (or parts of communities) and it can be concluded that distinct plant associations or communities apparently existed in the Pragian of China as in the Pragian and Emsian of Laurussia.

AFFINITIES, DIVERSITY, AND DISPARITY

As presently known, the Posongchong flora is characterized by having abundant and diversified zosterophylls, including the three genera *Discalis*, *Gumuia*, and *Zosterophyllum*, the last being represented by at least three species. Several plants that appear related to *Zosterophyllum* but bear long-stalked sporangia either more laxly or more closely than in *Zosterophyllum* are currently being studied. Lycopsids are represented by the pre-lycopsid *Halleophyton* and the lycopsid *Baragwanathia*. Only one very simple representative of trimerophytes (i.e., *Psilophyton primitivum*) is known. *Hedeia* possibly may represent another, although ongoing research indicates it is more complex in organization than most psilophytons.

Some combinations of structural and histological characters of the Chinese plants differ from those found in contemporaneous vascular plants from Laurussia so that determining their affinities is problematic. First, in Laurussian plants, centrarch xylem usually is closely correlated with terminal fusiform sporangia as seen in some rhyniophytes and trimerophytes and exarch xylem with lateral, round to reniform sporangia as in zosterophyllophytes, barinophytes, or lycopsids (Banks 1968; 1975). Yet *Huia recurvata* in the Posongchong flora has lateral ovate sporangia (with a long stalk, although they also could be interpreted as being borne at the ends of lateral branches in a close arrangement) and centrarch xylem. In evaluating its

affinities, Geng (1983) recognized a family of possible rhyniophytes, the Taeniocradaceae, in which he included *Taeniocrada decheniana* and *Huia recurvata*. Although these plants traditionally have been considered rhyniophytes, the redefinition of Rhyniaceae by Hass and Remy (1991) and Kenrick and Crane (1991, 1997a) to encompass *R. gwynne-vaughanii* and *Stockmansella* type plants, leaves the familial or suprafamilial status of *Huia* and *T. decheniana* unclear. All of these taxa conform to the level of morphological evolution established for Pragian–Emsian plants by Banks (1980), Edwards and Davies (1990), and Gerrienne (1996a). The number of genera and species recognized in this deposit is consistent with diversity in other well-studied Lower Devonian deposits.

Another taxon exhibiting a mixture of features is *Eophyllophyton*, whose axes exhibit a centrarch protostele, but small circular to reniform structures interpreted as sporangia, resembling those of zosterophylls, are borne on the adaxial surfaces along the veins of laminate leaves. Tracheid pitting is G-type, commonly found in zosterophylls and some lycopsids. It is of interest that despite the globose sporangial morphology and G-type wall pattern, *Eophyllophyton* is sister to the eutracheophyte clade in Kenrick and Crane's (1997a) analysis, not the Lycophytina (perhaps a result of cladistic "long branches"?). Other plants included in their analyses, with globose to reniform sporangia and either centrarch haplosteles or unknown anatomy, are positioned in a polytomy with zosterophylls and lycopsids and considered by them as part of the Lycophytina clade. Thus the possibility exists that stem group lycophytes may have had centrarch protosteles; to pursue this, new data and analyses are needed.

The presence of parenchyma bands in axes of *Yunia* and how to interpret them is unclear, but Hao and Beck (1991b) suggest this organization might represent a primitive pith similar to that of zygopterid and filicalean genera of the early Carboniferous primitive ferns. Despite uncertainties as to affinity, this is a type of stelar histology generally found in younger, more advanced taxa in Laurussia. The sporangia associated with *Yunia* axes are very different in organization from those of trimerophytes and we suggest that further study is needed to determine if they truly belong with the anatomically preserved axes. If so, relating the taxon *Yunia* to the trimerophytes should be reexamined. More anatomical studies of these and other Posongchong plants are needed to understand better the possible cause or significance of the anatomical discrepancies and questions concerning affinity. The plants discussed here, plus certain Laurussian ones, may be representatives of as yet unrecognized lineages.

The truly interesting and at present not well understood feature of the Posongchong flora is the existence of the several disparate types of plants that thus far are endemic to that region and of uncertain affinity. As demonstrated previously, they show considerable diversity in vegetative and fertile structures (sporangium morphology, attachment pattern, and, in some, presence of vegetative and fertile leaflike structures) and are quite different from those commonly attributed to contemporaneous Laurussian early vascular plants. For example, the appressed sporangia of *Catenalis digitata* (Hao and Beck 1991a) and *Demersatheca* (Li and Edwards 1996) are quite different in position relative to the axis than most other contemporaneous plants. Four other plants, *Stachyophyton*, *Eophyllophyton*, *Adoketophyton*, and *Celatheca*, are disparate in some combination of sporangial morphology such as shape, presence of leaflike structures, and location or orientation of leaflike structures. In all but one taxon, the leaflike structures are associated with sporangia. The bifurcated sporophyll and adaxial elliptical sporangium of *Stachyophyton yunnanense* closely resemble the condition in *Protolepidodendron* from the Middle Devonian, but vegetative leaf morphology of the two differs. The inward curving of leaf-pairs making up a fertile unit in *Eophyllophyton bellum* is reminiscent of sterile and fertile organs of some Middle to Upper Devonian Aneurophytales such as *Aneurophyton* or *Tetraxylopteris* (although no relationship is inferred here). However, its anatomy, a centrarch protostele, has remained at the evolu-

tionary level of the Lower Devonian and its sporangial morphology is reminiscent of zosterophylls. The strobilar construction of fertile regions in *Adoketophyton subverticillatum*, coupled with the stalked, fan-shaped sporangium-bearing appendages, led Li and Edwards (1992) to suggest that it is perhaps closely related to the Barinophytales. Recent collections suggest that this plant has a possible centrarch xylem (also suggested by Li and Edwards), thus differing from the exarch condition in barinophytes. This again raises the issue, treated in part by Kenrick and Crane (1997a), of the importance of xylem maturation sequence as a character distinguishing major plant groups. Although *Celatheca* is similar to *Psilophyton* and *Pertica* in having terminal sporangia, it differs obviously in sporangial outline and surrounding leafy appendages. The putative affinities of *Catenalis digitata* with algae as postulated by Hao and Beck (1991a) was based on its strong superficial resemblance to the latter. This requires further consideration, given the presence of a central conducting strand and resistant walled spores, but its putative aquatic habitat remains a possibility.

How does the Posongchong flora compare to coeval floras in China? Cai and Wang (1995) record deposits with fossil plants of probable Lower Devonian age in the following: the Guijiatun Formation of the Cuifengshan Group in eastern Yunnan, a suite of clastic rocks in the Changning area of western Yunnan (a part of the Shan-Thai paleocontinent), the lower part of the Dangchai Formation of Dushan in southern Guizhou, the lower part of the Cangwu Group in eastern Guangxi, and the main part of the Pingyipu Formation in Yanmenha, Sichuan. For several of these outcrops, plants identified as *Zosterophyllum myretonianum* or *Z.* sp., *Taeniocrada* sp., and *Drepanophycus spinaeformis* are reported. Many of these require further study to establish correct identity, since some identifications are based on fragmentary remains or sterile axes.

A more diverse assemblage of plants is known from the Pingyipu Formation in Sichuan, where 12 genera and 18 species are recognized (Li and Cai 1978; Cai and Li 1982; Geng 1992a,b), including *Zosterophyllum myretonianum* Penhallow, *Z. yunnanicum* Hsü, *Z. sichuanensis* Geng, *Z. ?* sp. Li and Cai, *Sporogonites sichuanensis* Li and Cai, *Psilophytites* sp. Li and Cai (possibly = *Psilophyton* sp. Geng), *Hostinella* sp. Cai and Li, *Uskiella* sp. Geng, *Sciadocillus cuneifidus* Geng, *Eogaspesiea gracilis* Daber, *Hicklingia cf. edwardii* Kidston and Lang, *Oricilla unilateralis* Geng, *Leclercqia complexa* Banks et al., *Amplectosporangium jiangyouensis* Geng, *Drepanophycus spinaeformis* Goeppert, *D. spinosus* Kraüsel and Weyland, and *D.* sp. Geng. The age of the Pingyipu Formation is uncertain; although it was originally considered as Lochovian or mid to late Lochovian by Li and Cai (1978) and Cai and Li (1982), Geng (1992a) suggested it ranges from late Lochovian to earliest Emsian, mostly being Pragian. Cai and Wang (1995) discussed the age and, based on plant types and depositional environment, tentatively regarded it as ranging from Pragian to early Emsian. Plant taxa recognized here are largely Laurussian forms, and the distinctive components of the Posongchong Formation not known in Laurussia are absent (even if one considers only structures such as leaves or distinctive anatomy in lieu of taxa). Some taxa are not well known, so that further study is needed. For example, more data would confirm the identification of taxa such as *Leclercqia* and *Oricilla*. Remains recognized as *Psilophyton* by Geng consist of vegetative axes only, and plants assigned to *Drepanophycus* spp. may be revised when more data about them are available. *Amplectosporangium*, according to the description, consists of a terminal fertile branch system in which oval sporangia are attached to the inner sides of dichotomizing axes in a single row but details are not well documented. Undoubtedly, further collecting and study of these plants, as well as lithological and stratigraphic determination of the various sequences comprising the Pingyipu Formation, are needed. Further exploration for additional fossils also would improve floristic determinations from other localities.

Emsian plants are known from several sites in various provinces (see Cai and Wang 1995) but floras are not yet known to be very diverse. The Xujiachong Formation of Yunnan contains *Zos-terophyllum* spp., *Hsüa robusta*, *Drepanophycus qujingensis*, *Psilophytites* sp., and a new taxon *Emplecophycus yunnanensis*, again mostly similar to Laurussian forms. Few plants other than fragmentary remains are known from earlier sediments, the most interesting one being the presumed Lower Silurian *Pinnatiramosus*. Middle Devonian floras are dominated by lycopsids, many being quite different from the Pragian and Emsian plants of southern China. From this brief comparison, it is evident that while studies of early land plants in China have progressed considerably in the past few decades, many challenges and questions remain concerning evolutionary change through time within any given region. It is equally clear that to best unravel what is happening biologically and geographically, more consideration must be given to the sedimentary and tectonic/paleogeographic context, as well as to detailed study and reconstruction of the plants.

Thus, as presently known, the Posongchong flora is as diverse as many Laurussian ones, but it exhibits disparity and complexity of fertile and sterile organs and some aspects of vascular anatomy among some component taxa. This exists within a background of less complex, cosmopolitan taxa (zosterophyllophytes and trimerophytes). Possible causes for this floral disparity may include the following. Hao and Beck (1991a) suggested that a long isolation of the South China block might account for the endemism present in the Posongchong flora. This also was considered by Li and Edwards (1992), Edwards (1997), and Hao and Gensel (1998). Eastern Yunnan is a part of the South China Block, a separate paleocontinent positioned closer to Gondwana continents than Laurussian ones during the early Paleozoic (Fang 1989). Paleomagnetic data (Liu and Liang 1984; Fang et al. 1989; Bai and Bai 1990) indicate that by Early Devonian time, eastern Yunnan occupied an equatorial zone (near 0° or a little more south), under tropical conditions of temperature and photoperiod. Coeval Laurussian localities were more southerly (10° to 30° S) (Cocks and Scotese 1991), as were most Gondwanan paleocontinents with the exception of Australian, Shan-Thai, and Ta Li Mu ones, which were located relatively close to one another (Scotese and McKerrow 1990; Fang et al. 1989; Bai and Bai 1990).

Sediments of the Posongchong Formation were produced from a plane of denudation that had existed from the Middle Ordovician through the Silurian to the Lochovian of the Early Devonian. This setting thus differs geologically from the emerging land (transition from deep marine to terrestrial) history of Laurussia. Deposition of Posongchong sediments is predominantly terrestrial, and both floodplain/stream and near-shore lake or lagoonal environments can be recognized (Li and Wu 1996), which may have provided various habitats for diversification of early land vascular plants. Because of these factors and the influence of the ocean current (Hao and Beck 1991a), the climate is inferred as hot and humid (supported by absence of red beds). The numerous black, fine-grained sediments are rich in organics, whose source is not yet known. While many layers indicate a reducing environment, some oxidation (?secondarily) has occurred, based on the pale gray to yellow color of some layers. Thus we suggest, on the basis of present evidence, that the geographical position of the land mass, its geological history, and perhaps some environmental parameters explain, at least in part, the distinctive flora present in what is now part of southern China.

With regard to phytogeography, because of the recent discovery of *Baragwanathia* sp. and *Hedeia* from the Posongchong flora (Hao and Gensel 1998) and the great similarity of the cosmopolitan members at the generic and species level (i.e., *Zosterophyllum* sp. 1 and *Psilophyton primitivum*) to coeval Australian plants, we have suggested that the Posongchong flora has a close relationship with the upper *Baragwanathia* floral assemblage (Tims and Chambers 1984) of Australia and that both of them may belong to the same phytogeographical unit during the Early Devonian (especially the Pragian). This is provisionally named the northeastern Gondwana unit. On the basis of current data, this unit (province) obviously differs from Laurussian and other paleogeographical units. Because of the presence of several endemics in Yunnan and other taxonomic differences, the southeastern Chinese and Australian floras could be considered as separate subunits (Hao and Gensel 1995, 1998). Even as we propose this, we recognize the gaps in knowledge about floras and taxa that exist (as do Edwards 1990 and Edwards and Wellman in chapter 2), in particular the absence of Silurian and Devonian floras from northern Gondwana paleocontinents other than China and Australia. Clearly, these hypotheses require further testing, in part by addition of new data on coeval as well as younger and older floras in these regions.

ACKNOWLEDGMENTS

The work was supported by a grant from NSFC (#49742004), the State Education Commission of China to S. G. Hao, and also by NSF of U.S.A. for Hao's attending the IOPC-5 conference and conducting collaborative research with P. G. Gensel in the U.S.A. Support from the National Science Foundation grant # 92-08362 and a University of North Carolina Faculty Research Grant to PGG is also gratefully acknowledged. The line drawings were done by Ms Susan Whitfield, Biology Department, UNC-Chapel Hill, North Carolina.

7

The Middle Devonian Flora Revisited

Christopher M. Berry and Muriel Fairon-Demaret

The Middle Devonian (Eifelian, Givetian) flora usually receives little attention in syntheses of Devonian plants when compared with the recent intensity of studies in the Early and Late. In this paper, we shall try to give an overview of the floras of this age, presenting modern concepts of the plants involved and their ecological associations.

We include in this study reports from the Eifelian up to the lowermost beds of the Frasnian (earliest Late Devonian). In North America, whence probably the most complete succession of Middle Devonian floras is presently known, many elements of the Middle Devonian flora continue into the Early Frasnian before being lost from the record (Scheckler 1986c; Edwards et al. 2000). It is therefore impossible to date megafossil assemblages accurately with regard to the Middle–Late Devonian boundary if we rely on identification of the plants themselves. It is similarly impossible at present to recognize the Eifelian–Givetian stage boundary.

In discussions of the Middle–Late Devonian boundary, it is often stated (e.g., Chaloner and Sheerin 1979; Meyen 1987) that the appearance of *Archaeopteris* marks the Late Devonian; however, the presence of *Svalbardia* in Middle Devonian sediments, a plant indistinguishable from *Archaeopteris* in many ways, does little to convince us that a clearly recognizable boundary exists. It seems a "transition flora" existed before the Late Devonian flora became firmly established.

Most previous accounts of the Middle Devonian flora have concentrated on the supercontinent of Laurussia, as does this one, because it is from there that the most accessible, relatively well dated, and clearly illustrated fossils are known. This chapter is influenced by our studies (both individually and jointly) on Middle Devonian assemblages from Europe (most noticeably Belgium), New York State, and northwest Venezuela. We include Venezuela because of the similarity of its extensive flora with that of New York State and the current speculation as to the provenance of the host terrane, which could perhaps derive from Laurussia rather than Gondwana (Restropo and Toussaint 1988). However, with the easier availability of detailed reports from China (Cai and Wang 1995) and the former Soviet Union (Yurina 1988), it is now possible to extend our generalizations further afield.

Previously, in most reports, the Middle Devonian flora was thought to be from the Givetian (Gensel and Andrews 1984; Meyen 1987; Allen and Dineley 1988). However, it is increasingly clear that there are a number of good middle and late Eifelian assemblages in Europe (e.g., Lindlar, in Streel et al. 1987; Goé, in Hance et al. 1996) that contrast with the predominantly Givetian age given to the most diverse and important compression assemblages of New York State. Palynological revision might increase the ages of some of these localities, too (e.g., the age of the Blenheim–Gilboa site is given as "no older than late Eifelian" by Richardson et al. 1993).

One of the most important advances in recent years has been the increasing emphasis placed on relating anatomical to morphological studies of vascular plants. We incorporate some of this information in our review. Also, because the study of plant morphology is presently considered unfashionable, textbooks are full of plant reconstructions that have gone unchallenged since the early years of this century. We try to provide more modern concepts of these plants here.

In our description of taxa, we start from the threefold division of Early Devonian vascular land plants conceived by Banks (1968), the Rhyniophytopsida [some members now included in the redefined Rhyniaceae by Remy and Hass (1991)], Trimerophytopsida, and Zosterophyllopsida, and we discuss the forms that may have survived or evolved from each of these groups. One of the features of the Early Devonian Flora, emphasized by recent studies on Chinese specimens, is the large number of land plants that may be termed aberrant (Banks 1992) in that they defy classification in Banks's scheme (see chapter 6). In the Middle Devonian, plants representing Banks's three subdivisions, and plants that apparently evolved from them (in particular, the trimerophytes and zosterophylls), are the only presently recognized lineages. Rhyniopsida are increasingly recognized as persisting into the Middle Devonian, but they then became extinct. Primitive Zosterophyllopsida also formed a minor component of Middle Devonian assemblages, and the more derived descendants (lycopsids) were more dominant and evolutionarily more vigorous. However, it is the plants that are believed to have evolved from Trimerophytopsida that demonstrate the most explosive taxonomic and architectural radiations.

RHYNIOPSIDA AND DERIVATIVES

In recent years, a completely new, more restricted concept of Rhyniopsida and its deriva-

tives has emerged. The Rhyniaceae, *sensu* Hass and Remy (1991), are recognized by the presence of S-type water-conducting cells [in which the helical thickenings have a spongy filling, the direction of the helix reverses frequently, and the lumen wall is covered by a microporate layer (Kenrick et al. 1991)] and by characteristic sporangia with a well-defined, thick abscission layer. Sporangia are both terminally and laterally positioned. Compression fossils of plants included in this group usually are ribbon-like with a narrow but prominent vascular strand (see figure 7.2C); they have in the past been sometimes attributed to the genus *Taeniocrada*.

Although Rhyniaceae have often been considered confined to the Early Devonian (e.g., Kenrick and Crane 1991), some of these plants can be recognized from Middle Devonian sediments. For example, fertile specimens of *Stockmansella (Taeniocrada) langii* were reported by Ishchenko (1965) from the Eifelian of the Donbass. A stele apparently composed of S-type cells was recognized by Berry (1993) from flattened sterile axes found in Venezuela (see figure 7.2C) in a horizon of Middle–Late Devonian age also containing *Leclercqia* cf. *complexa* and *Haskinsia sagittata. Stockmansella remyi* Schultka and Hass (1997) from the Upper Eifelian of Germany has narrow, 3 mm axes, with a central strand made up of S-type cells, and lateral sporangia. These sporangia, as is typical for Rhyniaceae, are ovoid with a number of longitudinal dehiscence slits, and are irregularly attached to the axis on thick pads. Schultka and Hass (1997) reconstructed the plant as having creeping axes with short aerial branches. Another plant that demonstrates irregularly arranged, apparently lateral abscission scars (of sporangia?), similar to those of Rhyniaceae, is *Palaeostigma sewardi* (e.g., Plumstead 1967) from the Upper Devonian of South Africa. Although considered a lycopsid by some (Anderson and Anderson 1985), its true affinities have yet to be determined. An attempt to detect diagnostic cell structure in the illustrated vascular strand (Seward 1932) by scanning electron microscopy (SEM) proved unsuccessful (N. Hiller, pers. comm. 1996).

Future investigations of "*Taeniocrada*"-like

axes with prominent narrow vascular strands may well demonstrate the presence of more Rhyniaceae in Middle and even Late Devonian sediments, although many are probably axes of zosterophylls, barinophytes, and other similar plants. Rhyniaceae may prove to be restricted to certain environments. Some Rhyniaceae are believed to have had a life cycle with a prominent gametophyte generation (Remy et al. 1993) and might have been confined to moist terrestrial habitats. *Stockmansella remyi*, preserved *in situ*, was believed by Schultka and Hass (1997) to have been a pioneering plant on newly formed alluvial plains in a deltaic environment. The presumed fleshy stems of most sporophyte genera, and their narrow terete strand of water-conducting cells that contribute little to flexural stiffness (Speck and Vogellehner 1994), suggest that these plants were unable to support themselves to a considerable height and required moisture to maintain turgor pressure and structural strength.

ZOSTEROPHYLLOPSIDA AND ALLIES

Primitive zosterophylls are recognized by lateral sporangia dividing into two valves, and by an elliptical vascular strand with exarch maturation and G-type water-conducting cells (Kenrick et al. 1991). The latter were formed with a two-layered wall; the inner, more decay-resistant wall was continuous over both the predominantly annular thickenings and the areas between these thickenings, where it formed around numerous simple pits. A large number of genera with various sporangial and emergence morphologies are recognized from the Early Devonian (Gerrienne 1996c).

There is a general consensus that zosterophylls and lycopsids form part of a sister group to most other land plants (Gensel 1992; Kenrick and Crane 1997a). Other possible members of this group are Barinophytales and Sphenophyllales. The latter are considered by some authors (e.g., Stewart and Rothwell 1993:196) more likely to be related to lycopsids than to horsetails, a view with which we have some sympathy but have yet to explore fully.

Zosterophyllopsida

The best-documented zosterophyll from the Middle and early Late Devonian is *Serrulacaulis* (Hueber and Banks 1979; see also figure 7.1B). This is composed of more or less terete axes with two opposing rows of triangular-prism shaped emergences arranged transversely and contiguously (Berry and Edwards 1994). Water-conducting cells are of the G-type. Tips are circinate and sporangia occur alternately in two rows along one side of the axis. Branching is dichotomous, and rhizoids are present on some stems. The reconstruction (Hueber and Banks 1979) suggests a creeping plant with some aerial shoots bearing sporangia. It is known from several Laurussian localities in Belgium and North America, as well as Venezuela.

Spiny axes superficially similar to *Sawdonia ornata* Hueber are known from New York State and Venezuela. However, the New York specimens are sterile (Hueber and Grierson 1961) and the Venezuelan ones, similar to *S. ornata* in habit and branching, apparently have more complex sporangia than any yet known from the Early Devonian (figure 7.1A). Sterile axes attributed to *S. ornata* have been described from the upper Eifelian of Germany (Köster quarry) by Mustafa (1978a); fertile examples from the same quarry noted by Schultka and Hass (1997) are not yet described. The determination of a single specimen of ?*Zosterophyllum bohemicum* (Obrhel 1961) from the Givetian of Czechoslovakia is inconclusive (Kräusel and Weyland 1933: pl. VII, figs. 6–8).

For Middle Devonian zosterophylls, fertile material is essential to be confident of identification unless the plant has a unique emergence morphology (e.g., *Serrulacaulis*). Diversity was apparently low, especially compared with the many disparate genera known from the Early Devonian. The occurrence of some fossils showing typical zosterophyll anatomy (e.g., *Euthursophyton*, in Mustafa 1978b; cf. *Stolbergia*, in Kasper et al. 1988) but without preserved fertile structures suggest that more diversity among Middle Devonian zosterophylls remains to be recognized.

The *Sawdonia*-like plants were probably taller than *Serrulacaulis* and may have reached a

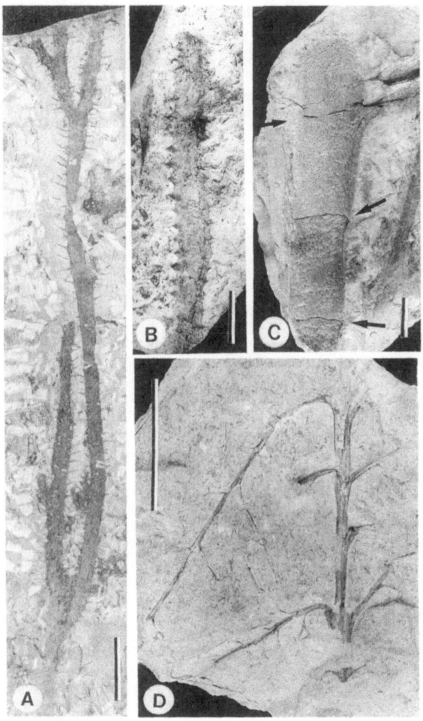

Figure 7.1.

Selected Middle Devonian plants. **A:** cf. *Sawdonia* sp. Zosterophyllopsida. Note
planated erect branching. Scale bar = 20 mm. Venezuela horizon 3. Berry collection, Cardiff. **B:** *Serrulacaulis* sp. Zosterophyllopsida. Characteristic triangular-
prism-shaped emergences visible. Scale bar = 10 mm. Goé, Belgium. Liège collection. **C:** *Anapaulia moodyi*. Iridopteridales. Note whorled insertion of lateral
branches and appendages marked by groove across stem. Scale bar = 10 mm.
Venezuela, float. Berry collection, Cardiff. **D:** cf. *Tetraxylopteris*. Progymnospermopsida. Note opposite and decussate arrangement of branches and ribbed
stem. Scale bar = 50 mm. Venezuela horizon 11. Berry collection, Cardiff.

height of 60 to 80 cm standing alone, but they are more likely to have grown in partially self-supporting stands where they may have reached a greater height.

Barinophytales

Protobarinophyton and *Barinophyton* are best known from strobili bearing two ranks of sporangia that are sessile or on a short stalk (Brauer 1981). Like Zosterophyllopsida, they have exarch primary haplosteles composed of G-type tracheids. The sporangia are arranged in spikes that are either terminally (*Protobarinophyton*) or terminally and laterally (*Barinophyton*) arranged on the main stems. Both genera are known from the late Early Devonian to Upper Devonian. *Protobarinophyton* also is recorded in the early Carboniferous (Scheckler 1984). Another plant that may be part of this complex is *Pectinophyton* (Høeg 1935), which has only a single row of sporangia.

No definitive "whole plant" concept of any member of this group has been established. However, Brauer (1981) illustrated an axis of *Protobarinophyton* that was 29 cm long and terminated by a strobilus. Given that the only supporting tissues we know of were the water-conducting cells, it is unlikely that these plants were tall, although sometimes branching was of a complexity comparable to some of the more-branched zosterophylls (Høeg 1935; Stockmans 1948). No axes larger than 10 to 15 mm are known to occur with G-type water-conducting cells, and none is known to have developed secondary growth or to have had more than a relatively narrow zone of supporting tissues, so it is unlikely that their fertile parts are simply distal parts of a large arborescent form. These plants might therefore be envisaged as low plants similar in habit to the zosterophylls, again perhaps gaining extra height by growing in dense stands.

Lycopsida

Three categories of early lycopsids were recognized by Gensel and Andrews (1984), and we use a modified version of their system here. Pre-lycopsids include those taxa that have the basic features of the lycopsid lineage (microphylls or spine-like leaves, globose lateral sporangia) but do not have all characters expected of true lycopsids (e.g., sporangia adaxial on microphylls, organised phyllotaxy). These plants are most common in Early Devonian deposits and are often included in Drepanophycales. Protolepidodendrales have complex divided or lanceolate microphylls with sporangia positioned adaxially on unmodified leaves. They apparently were herbaceous. Early arborescent lycopsids had developed a pseudobipolar growth, but some examples retained complex leaf morphologies and had not developed stigmarian rootlets. (*Arborescent* is used in this context to infer treelike form rather than stature.)

Pre-lycopsids

Although no fertile Middle or Late Devonian *Drepanophycus spinaeformis* material is known, Banks and Grierson (1968) reported cuticular and morphological preservation of sterile specimens attributed to this species in the early Late Devonian of New York. The genus, with dichotomous, robust, ground-running, and short aerial axes, adventitious rooting structures, and irregularly inserted thornlike leaves, survived from its Early Devonian cosmopolitan abundance. Hueber (1992) suggested that the fleshy stems were held rigid by turgor pressure and therefore required a moist habitat. They do not seem to have been a major part of typical Middle Devonian floras in Laurussia at least. *Drepanophycus spinosus* is a more questionable member of the genus; it is known from Belgium, the Eifel region of Germany, and Kazakhstan. Axes with densely arranged, short spinelike leaves have been attributed to genera such as *Thursophyton* and *Asteroxylon* (see Høeg 1967), yet fertile parts and true affinities remain unknown. Early Devonian *Drepanophycus* had G-type water-conducting cells (Kenrick and Edwards 1988) and was in this respect similar to the zosterophylls.

Protolepidodendrales

More abundant at many localities are remains of Protolepidodendrales (Berry 1997). These plants have stems that branch mostly isotomously and have pseudowhorled and helically arranged leaves. Sporangia are attached to leaves adaxially. Two well-delimited families are recognized, Protolepidodendraceae and Haskinsiaceae (e.g., Grierson and Banks 1983). In the Protolepidodendraceae, the leaves are narrow and divide into three (*Colpodexylon*, see figure 7.2B; *Minarodendron*), five (majority of *Leclercqia* leaves), or more segments (Banks 1944; Li 1990; Banks et al. 1972; Bonamo et al. 1988). In some, the leaves are very densely arranged (e.g., *Colpodexylon deatsii*—up to 20 leaves per whorl). Only *Leclercqia* has been shown to possess a ligule (Grierson and Bonamo 1979). In Haskinsiaceae, the leaves are broader, have a lanceolate, sagittate, or hastate blade, and are attached to the stem by a narrow petiole (Grierson and Banks 1983; Berry and Edwards 1996a). In both families, single sporangia are borne on unmodified leaves in the adaxial position. Protolepidodendrales had evolved circular, oval, and scalariform bordered pits in their tracheids (Grierson 1976). All xylem was primary, and maturation exarch.

Fertile specimens attributed to Archaeosigillariaceae (Kräusel and Weyland 1949) have not yet been demonstrated, but the plants attributed to the family may have been very similar to Protolepidodendrales. Recently, the genus *Archaeosigillaria* has been restricted more or less to the type specimen (Berry and Edwards 1997). The two species of *Gilboaphyton* Arnold have laminate leaves with two prominent lateral teeth, and a marked pattern of hexagonal, swollen leaf bases on the stem surface (Berry and Edwards 1997).

It is unlikely that any of these plants grew to a height of more than 1 meter. They are sometimes found in monospecific mats, in deposits where more than one species or genus of herbaceous lycopsid are present, or in horizons where many other types of plant are found. These plants may have lived on riverbanks and floodplains, quickly pioneering unstable and fresh ground surfaces, then being buried *in situ* on occasional flooding (Banks et al. 1985) as well as slumping as a mat of vegetation into the river channels. They may also have formed part of the understory component of more complex ecological systems.

Early Arborescent Lycopsids

The most complete records of early arborescent lycopsids are the accounts of *Longostachys latisporophyllus* (Cai and Chen 1996) and *Chamaedendron multisporangiatum* (Schweitzer and Li 1996). These plants, from the late Givetian or early Frasnian of China, are very similar in almost every way and will be considered together. They have main axes 7 to 35 mm in diameter, which branch dichotomously at the base, forming a conical rooting system, and at the top to form a crown of branches. The roots and lower trunk of *Longostachys* have a protostele with secondary xylem arranged in radial arms. The slender leaves of this plant have many spines arranged along the margins. Sporophylls have enlarged bases but do not form true cones. In *Chamaedendron*, multiple sporangia were reported on the sporophylls (although based on size, these structures might also be interpreted as megaspores within a single sporangium); the plant is heterosporous. No stigmarian rootlets were observed. These plants were estimated to range from 0.5 to 1.5 m in height, so, although they were similar in gross morphology to Carboniferous arborescent lycopsids and had secondary xylem, they did not attain the considerable vertical growth of later forms.

The most famous example of an early arborescent lycopsid was the Naples Tree, *Lepidosigillaria whitei* Kräusel and Weyland (early Late Devonian; Banks 1966). It exhibits a swollen base with roots and linearly and spirally arranged leaf scars, and it was reported to be unbranched to a height of 5 m (White 1907; Kräusel and Weyland 1949). *Sigillaria? gilboense* (Goldring 1926; Grierson and Banks 1963) from the Middle Devonian of Gilboa, New York State, had long, narrow, lax leaves up to 28 cm in length, probably borne on a stout upright trunk. *Protolepidodendropsis*, from the Givetian of Spitzbergen, was reconstructed by Schweitzer (1965) as a small tree with a dichotomous crown.

Anatomically preserved Middle Devonian

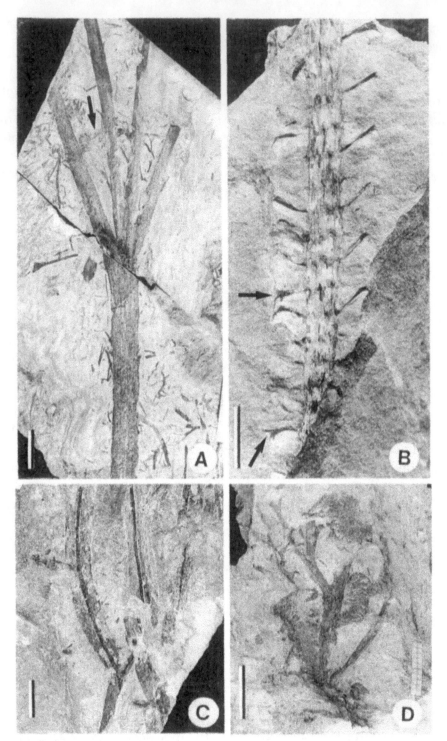

Figure 7.2.
A: *Pseudosporochnus nodosus.* Cladoxylopsida. Distinctive digitate branching of first-order branch, and laterally attached appendages (*arrow*). Scale bar = 20 mm. Goé, Belgium. Liège collection. **B:** *Colpodexylon* sp. Lycopsida. Note leaves divided into three segments distally (*arrows*). Scale bar = 10 mm. Venezuela horizon 3. Berry collection, Cardiff. **C:** "*Taeniocrada*" sp. Rhyniopsida. Note broad axes with distinct prominent narrow vascular strand. Scale bar = 10 mm. Venezuela horizon 4. Berry collection, Cardiff. **D:** cf. *Rellimia* sp. Progymnospermopsida. Showing recurved arms bearing dense clusters of elongate sporangia. Scale bar = 10 mm. Venezuela horizon 8. Berry collection, Cardiff.

(Givetian) lycopsids from Kazakhstan such as *Atasudendron* demonstrate a higher level of differentiation of stem tissues than is found in the Chinese lycopsids of the same age (Senkevich et al. 1993). Compact strobili containing both mega- and microsporangia were also reported from Kazakhstan (Senkevich et al. 1993). *Atasudendron* was said to be a small tree, 2 to 3 m high (Yurina 1988).

TRIMEROPHYTE DERIVATIVES

Trimerophytina was established by Banks (1968) to accommodate *Trimerophyton robustius* (Dawson) Hopping, *Dawsonites* Halle, and (in addendum) *Psilophyton princeps* Dawson emend. Hueber. Banks (1975c) later characterized the trimerophytes as follows: "Main axis branched spirally and dichotomously, lateral branches forked either into three units or dichotomously; fertile branches, much forked, terminated in a mass of paired fusiform-ellipsoid sporangia; axes smooth, punctate or spiny; xylem strand large, solid, centrarch, composed of scalariform tracheids."

Today, numerous species of the Early Devonian genus *Psilophyton* are known (Gerrienne 1995, 1997), characterized by mainly dichotomous branching, paired twisted sporangia, and a terete centrarch haplostele (where known). However, it has also become apparent that the more complex trimerophytes, with pseudo-monopodial main axes, are anatomically far advanced over the condition presumed by Banks (which was based on the model of *Psilophyton*). For example, *Pertica*-like plants (Gensel 1984) and anatomically preserved *Gothanophyton* from Germany (Remy and Hass 1986) have lobed xylem strands.

Kenrick and Crane's recent cladistic analysis (1997a) has confirmed that Banks's Trimerophytina (i.e., *Psilophyton* and *Pertica*) form a paraphyletic rather than monophyletic group. They chose the name Euphyllophytina to encompass the group of nonlycopsid trachaeophytes. We believe that at present the study of these early nonlycopsid land plants requires a break-

through in knowledge and understanding to arrive at a meaningful classification rather than simply a reinterpretation of the existing data and the surrounding mythology. It is clear that some of the synapomorphies used to define the Euphyllophytina are vaguely defined or only weakly argued. Because a term such as *basal euphyllophytes* can be interpreted at a number of levels, we still find the concept of trimerophytes as an acknowledged paraphyletic complex a useful one, at least until we have a better understanding of the diversification and evolution of the plants involved at the anatomical and morphological levels.

Although the definition of Trimerophytina is in need of modification or probable eventual rejection, it seems likely that most of the *Pertica* and *Psilophyton* plants are closely related. It is from within this complex that the most diverse radiation of a major group of land plants originated, including early "ferns," progymnosperms, iridopteridaleans, cladoxylopsids, and stenokolealeans. Documenting the patterns and processes of this radiation is one of the most exciting challenges of systematic paleobotany.

Wight (1987) proposed that the development of a ribbed stele from a terete haplostele was the result of changes in the spacing and position of lateral appendages and thereby occurred as nonadaptive change. His model indicated how members of the "radiate protoxylem group" (Beck and Stein 1993) within the trimerophyte derivatives may have evolved. In these plants, protoxylem bundles depart from a centrally located protoxylem strand and move outwards through the metaxylem and other stem tissues to enter lateral appendages with a one-to-one correspondence. Such plants include *Gothanophyton* and *Pertica*, aneurophytalean progymnosperms, stenokolealeans and some primitive seed-plants. A second group, the "permanent protoxylem group" (Beck and Stein 1993), has also been recognized within the trimerophyte derivatives. In these plants, appendage traces develop from the division of individual columns of protoxylem, which remain permanently toward the periphery of the xylem ribs. This group includes Iridopteridales and sphenopsids. The division between the two groups of trimerophyte derivatives, the "radiate protoxylem group" (Radiatopses of Kenrick and Crane 1997a) and the "permanent protoxylem group" (Moniliformopses) seems to be a fundamental one in terms of development and also phylogeny (table 7.1). First attempts at computerized modeling of hypothetical plant apices involving auxin concentration gradients promises a powerful method of investigating the evolution of stelar architecture in trimerophyte-derived plants (Stein 1993).

Psilophyton kräuseli Obrhel (1959) is the only species of the genus known from Givetian sediments, but it is known from compressions only. No Middle Devonian plant has yet been shown to have the anatomical characteristics of *Psilophyton*. Trimerophyte derivatives, though advanced in anatomical characteristics, are sometimes morphologically very similar to their ancestors. Basic structures, such as dichotomous trusses bearing paired sporangia, are common in trimerophyte derivatives, as are dichotomous sterile trusses. However, each morphologically recognized group seems to have at least one distinctive characteristic. Iridopteridales have a tendency toward predominantly whorled appendages and branches, progymnosperms have modified fertile units and cladoxylopsids have a tendency toward digitate branching and insertion of appendages without obvious geometrical regularity.

RADIATE PROTOXYLEM GROUP

Iridopteridales

The genus *Ibyka* Skog and Banks (1973) has a stout main (first-order) axis. Axes of the nth order carry both dichotomous appendages and branches of the n+1 order on them. Although Skog and Banks interpreted the arrangement of the lateral organs as helical, it has been suggested (Stein 1982; Berry and Edwards 1996b) that a predominantly whorled arrangement of leaves and branches was more probable, with branches replacing some appendages in the whorls, and this has been confirmed by recent work (Berry et al. 1997). Fertile organs are simple dichotomous trusses similar to the sterile appendages except terminating in paired sporangia. Plants sharing a similar pattern of morphology include *Anapaulia moodyi* (Berry and Edwards 1996b; see figure 7.1C), "*Hyenia*" *vogtii* (Høeg 1942) and "*H.*" *banksii* (Arnold 1941). These two species of "*Hyenia*" are represented by sterile compressions only, with "*H.*" *banksii* having much more densely arranged leaves.

Anatomically, *Ibyka* has a ribbed stele with traces departing to the lateral organs from permanent protoxylem strands in the tips of the ribs, but only a limited amount of badly preserved material was available for study. Better-preserved permineralized stems with such anatomy, yet with a typical nodal/whorled arrangement of trace departures, were considered by Stein (1982) to demonstrate sufficient distinct characters to erect the order Iridopteridales. Iridopteridaleans preserved as permineralizations only include *Arachnoxylon* (Stein 1981; Stein et al. 1993), *Asteropteris* (Dawson 1881; Bertrand 1913), and *Iridopteris* (Stein 1982).

A new plant, showing both iridopteridalean anatomy and well-preserved morphology, has recently been described from Venezuela (Berry and Stein 2000). Other examples of anatomically preserved questionable Iridopteridales are included in table 7.2. The leafless permineral-

Table 7.1. Some Middle Devonian trimerophyte derivatives and their possible descendent groups

	Middle Devonian Groups	Possible Descendent Groups
Permanent protoxylem group (Moniliformopses of Kenrick and Crane 1997a)	Iridopteridales and *Metacladophyton/ Protopteridophyton*	Equisetopsida (horsetails) and zygopterid ferns
Incertae sedis	Cladoxylopsida	? Other "fern" groups
Radiate protoxylem group (Radiatopses of Kenrick and Crane 1997a)	Progymnospermopsida Stenokoleales	? Seed plants ? Seed plants

Table 7.2. Summary of selected Eifelian/early Frasnian floras from Laurussia

	Lindlar (Mid Eifelian)	Goé North (Upper Eifelian)	Venezuela (?Givetian)	Riverside Quarry (Givetian)	Ronquières (Givetian)	Spitzbergen (Givetian)	Livingstonville (Lower Frasnian)
Zosterophyllopsida	Serrulacaulis[1]	Serrulacaulis[1]	Serrulacaulis[2], cf. Sawdonia[1]	Serrulacaulis[1]	Serrulacaulis[1]		Serrulacaulis[3], Sawdonia[4]
Lycopsida Herbaceous	Lycopodites[5], Protolepidodendron[6]	Leclercqia[1]	Haskinsia[7,8], Colpodexylon[7]	Gilboaphyton[9,10], Protolepidodendron[11]	Leclercqia[12], Lycopodites[13], Drepanophycus spinosus[13]		Colpodexylon[11]
Arborescent				Amphidoxodendron[11], ?Sigillaria gilboensis[11], Lepidosigillaria[11]		Protolepidodendropsis[14]	
Barinophytales					Lerichea?[13]		
Cladoxylopsida	Duisbergia[6], Calamophyton[15]	Calamophyton[16], Pseudosporochnus[17,18]	Wattieza[1]	Eospermatopteris[19], Pseudosporochnus[20]	Calamophyton[13], ?Pseudosporochnus[13]		?Pseudosporochnus[21], ?Cladoxylon[21]
Ibykales		Dixopodoxylon[22]	Compsocradus[23]	Ibyka[24]	Langoxylon[13]	'Hyenia' vogtii[25]	
Progymnospermopsida Aneurophytales	Rellimia[5]	Aneurophyton[1], Rellimia[26]	?Rellimia[1]	Aneurophyton[19]	Rellimia[13]	?Rellimia[14]	
Archaeopteridales					Svalbardia[13]	Svalbardia[25]	Archaeopteris[27]
Stenokoleales							Stenokoleos[28]
Others	Weylandia[5]			Prosseria[29]	Thamnocladus[13]	Enigmophyton[25]	

Question mark before genus indicates uncertain generic identification; question mark after genus indicates uncertain higher level taxonomic assignment.

1, This chapter; 2, Berry and Edwards (1994); 3, Hueber and Banks (1979); 4, Hueber and Grierson (1961); 5, Schweitzer (1974); 6, Schweitzer (1966); 7, Edwards and Benedetto (1985); 8, Berry and Edwards (1996a); 9, Berry and Edwards (1997); 10, Fairon-Demaret and Banks (1978); 11, Grierson and Banks (1963); 12, Fairon-Demaret (1981); 13, Stockmans (1968); 14, Schweitzer (1965); 15, Schweitzer (1972, 1973); 16, Leclercq and Andrews (1960); 17, Leclercq and Banks (1962); 18, Berry and Fairon-Demaret (1997); 19, Boyer et al. (1996); 20, Schuchman et al. (1969); 21, Banks et al. (1985); 22, Fairon-Demaret (1969); 23, Berry and Stein (2000); 24, Skog and Banks (1973); 25, Høeg (1942); 26, Leclercq and Bonamo (1971, 1973); 27, Carluccio et al. (1966); 28, Matten and Banks (1969); 29, Banks (1966).

ized axis *Dixopodoxylon* Fairon-Demaret (1969) from Goé, although badly preserved, has the anatomical characteristics of the "permanent protoxylem" group and is probably most closely allied to Iridopteridales. However, whorled appendage traces were not observed, as only one possible incipient trace was present. Sterile axes with many narrow dichotomous appendages, as seen in "*Hyenia*" *banksii*, are also present at Goé but remain to be studied. *Langoxylon*, although superficially iridopteridalean in anatomy, was described as having protoxylem points along the centers of ribs (Stockmans 1968), which suggests greater affinity with the "radiate protoxylem" group.

We have no definitive whole-plant concept of Iridopteridales, largely because we do not know for certain if the monopodial main axis is a small trunk or whether it is part of a much larger plant. No Iridopteridales have been found with secondary xylem, although some radially aligned metaxylem is present in one specimen of *Arachnoxylon minor* (Stein et al. 1983:1288), so we presently presume that the plants were monopodial in growth and reached a height of perhaps 1.5 to 2 m.

Metacladophyton/Protopteridophyton

Two Middle Devonian plants from South China, which are clearly trimerophyte derivatives, are presently of uncertain affinity. Both have a rhizome with a monopodial main axis arising from the rhizome apex and bearing lower orders of branching laterally.

Protopteridophyton Li and Hsü (1987), from the Givetian and Frasnian, has aerial branching systems that are more or less pinnate, but the smallest order of branching is terminated by simple dichotomizing sterile and fertile units. Part of the stem preserved as a permineralization demonstrates a V-shaped primary xylem strand with protoxylem strands arranged toward the periphery emitting lateral traces. The stele as preserved may be part of a larger system.

Metacladophyton tetraxylum Wang Zhong and Geng Bao-Ying (1997), from the Givetian, has main axes that contain four radially arranged V-

shaped xylem strands with peripherally located mesarch protoxylem. Before trace emission, these xylem strands join together centrally to produce an iridopteridalean-style actinostele. Traces are C-shaped and occur in whorls. Second-order axes have dichotomous units (fertile and sterile) arranged on them, reportedly in alternate, opposite, or subopposite patterns in two ranks.

The described anatomy of both plants demonstrates affinity with the "permanent protoxylem group" of Beck and Stein (1993) and close relationship to the iridopteridaleans. However, more detailed work is needed on the morphology and anatomy of these plants before more conclusive determinations can be made.

PERMANENT PROTOXYLEM GROUP

Progymnospermopsida

This group was founded by Beck (1960) for free-sporing plants that nevertheless had gymnospermous wood. The first plant demonstrated with these characteristics was *Archaeopteris*, which was widespread and had leafy branches borne on large trunks with the stature of modern trees (Beck 1962). These are common in Upper Devonian deposits and first appear in North America in the lowest beds of the Frasnian. However, wood of *Archaeopteris* (form genus *Callixylon*) and sterile and fertile foliage similar to *Archaeopteris* yet with more deeply divided leaves (often named *Svalbardia*) also occur in Middle Devonian deposits. For example *C. velinense* (Marcelle 1951), *S. avelinesiana*, and *S. boyii* (Stockmans 1968) were all described from the Belgian locality of Sart-Dame-Avelines, regarded as Givetian (Lacroix, in Bultynk et al. 1991).

Smaller, probably bushy plants with stout monopodial main axes also belonged to the earliest recognized group of progymnosperms, the Aneurophytales. These include *Aneurophyton* (Serlin and Banks 1978; Schweitzer and Matten 1982), *Rellimia* (Leclercq and Bonamo 1971), *Triloboxylon* (Scheckler 1975), and *Tetraxylopteris* (Bonamo and Banks 1967). *Rellimia* (*Protopteridium*) is especially common in Middle Devonian

(Eifelian) deposits and is closely related to *Tetraxylopteris*, which continues into the lowermost Late Devonian. Anatomically, these plants have three- or four-lobed protosteles often surrounded by secondary xylem with radially aligned tracheids and vascular rays (e.g., Stein and Beck 1983). This secondary xylem may even display growth layers (Dannenhoffer and Bonamo 1989).

Aneurophytales have recognizable branching patterns in the sterile part of the plants. *Rellimia* and *Aneurophyton* exhibit a one-third helical organotaxy. Opposite and decussate attachment of branches is present in *Tetraxylopteris* (figure 7.1D). Distal sterile branching systems are basically dichotomous units, yet in the fertile parts sporangia are arranged either laterally in simple (*Aneurophyton*) or in more complex pinnate arrangements [*Rellimia* (see figure 7.2D); *Tetraxylopteris*], this being the most recognizable morphological advance over their trimerophyte ancestors.

Stenokoleales

Stenokoleales are an enigmatic group of Middle Devonian to Early Carboniferous land plants recognized at present only from their permineralizations. They had three- to four-ribbed protosteles, and appendages often borne in pairs on alternate sides of the axis or in a helix from individual xylem ridges (Matten 1992). Secondary xylem was sometimes present (Matten 1992; Beck and Stein 1993). Beck and Stein (1993) suggested that this group represents a distinct order of plants within the "radiate protoxylem" group. Some authors have suggested close affinity with plants that were ancestral to seed plants (Matten 1992; Galtier and Meyer-Berthaud 1996).

Apart from this limited information, we have no morphological concept of this family. An intriguing possibility is the similarity of the branching patterns to those of morphologically preserved *Protocephalopteris praecox* (Ananiev 1960; Schweitzer 1968). Any inference between the anatomy and morphology of stenokolealeans and *Protocephalopteris* might be enhanced by the discovery of traces that might lead to aphlebia below the attachment of the second-order axes. At the moment, this remains speculation. However, when we do understand the morphology of stenokolealeans, they may be found to make up an important part of the Middle and early Late Devonian vegetation.

INCERTAE SEDIS

Cladoxylopsida

It is presently difficult to fit Middle Devonian Cladoxylopsida into either the radiate or the permanent protoxylem groups. This is because, although their xylem ribs and plates appear to have permanent mesarch, peripherally located protoxylem points that appear to divide to form traces, there are also a number of protoxylem points or clusters of smaller-diameter cells strung along the middle of each rib (Stein and Hueber 1989). The former feature is characteristic of the permanent protoxylem group, the latter of the radiate protoxylem group. Further three-dimensional mapping of xylem tissues in well-preserved material is necessary to establish the significance of these features.

Lorophyton (Fairon-Demaret and Li 1993) has provided us with the only specimen of a well-preserved whole-plant Devonian cladoxylopsid, albeit a juvenile. It had a short trunk with densely arranged, acutely inserted lateral branches that are especially closely packed toward the top. On these lateral branches were arranged small dichotomous appendages. *Lorophyton* had sturdy roots emerging from a swollen base. Other fossils of diverse genera belonging to this group probably had a similar overall morphology, perhaps reaching 3 to 6 m in height. In *Pseudosporochnus*, lateral branches had bulbous bases and branched in a digitate manner into three to five smaller branches some distance from the main trunk (Leclercq and Banks 1962; Berry and Fairon-Demaret 1997; see figure 7.2A). In *Calamophyton*, the division of lateral branches was more profusely dichotomous. The branches may have been shed during growth. The main difference between the genera is the morphology of lateral appendages on the branches, sterile examples ranging from simple dichotomously branching trusses (*Calamophyton*) to more complex structures with a central axis (*Pseudosporochnus, Wattieza*). Fertile lateral appendages, of similar morphology to the sterile examples (*Pseudosporochnus*) or modified in some way (*Calamophyton, Lorophyton*), bore paired sporangia at the end of each segment.

Anatomically, the cladoxylopsids were the most complex of the trimerophyte derivatives (Stein and Hueber 1989), having a number of variously interconnected and anastomosing strands and plates of primary xylem in their branches. Some demonstrate secondary growth (e.g., *Cladoxylon bakrii*, in Mustafa 1978b). Large permineralized axes, named *Xenocladia*, which may represent the trunks of Devonian cladoxylopsids, have substantial amounts of secondary growth or aligned metaxylem (Lemoigne and Yurina 1983). The traces entering each appendage of Devonian cladoxylopsids vary from single traces (*Calamophyton*) to up to three derived from adjacent xylem ribs (*Pseudosporochnus hueberi*, in Stein and Hueber 1989).

These plants form a very common part of the Middle Devonian assemblages worldwide. Cladoxylalean anatomy is known from the late Emsian (Stein and Hueber 1989), and potential compressions are known from the early Emsian (e.g., *Foozia*, in Gerrienne 1992b). Cladoxylopsids are likely to have formed one of the tallest parts of the vegetation in the early Middle Devonian.

ORIGINS OF MAJOR PLANT GROUPS

Although it is not pertinent to give an exhaustive account of the evolution of major plant groups, we shall briefly consider our own, and some current opinions. More detailed accounts of the early history of ferns and horsetails will appear elsewhere.

Lycopsids

Lycopsids are widely believed to be related to or descended from Late Silurian and Early Devonian zosterophylls. Cladistic analyses suggest that they form a monophyletic group nested within zosterophylls (Bateman 1996; Gensel 1992; Kenrick and Crane 1997a).

Equisetopsida (Horsetails)

The oldest confirmed horsetails (archaeocalamites) are from the late Tournasian (Early Carboniferous) of Scotland, although there are dubious records of *Archaeocalamites* from Upper Devonian deposits (see analysis in Mamay and Bateman 1991). Kräusel and Weyland (1926) established the class Protoarticulatae for the Middle Devonian plants *Hyenia* and *Calamophyton*, believing them to be primitive horsetails. Skog and Banks (1973) considered *Ibyka* to be intermediate between trimerophytes and Hyeniales. Stein et al. (1984) reviewed evidence for the origin of horsetails, considering the most likely candidates to be cladoxylopsids or iridopteridaleans, but concluded, using cladistic methodology, that neither stock was favored using the then-current data. Beck and Stein (1993) recognized that both iridopteridaleans and horsetails shared membership in the permanent protoxylem group. Kenrick and Crane (1997a) included *Ibyka* in Equisetopsida on the basis of protoxylem characteristics and alleged whorled branching. Berry et al. (1997 and in prep.) demonstrated conclusively the whorled nature of the branching in *Ibyka*. They were also able to repeat the analysis of Stein et al. (1984) using updated information on both iridopteridalean and cladoxylalean plants and show that iridopteridaleans are favoured as potential horsetail ancestors (in prep.).

Ferns

The phylogeny of ferns is currently the subject of much research, and the definition of the term *fern* is also a cause of debate. Hence, taxonomy of this group is in an extreme state of flux. The phylogeny of early ferns will be established only by using fossil data, as many of the groups involved are extinct. However, much can be accomplished by the restudy of classic material as well as the description of new taxa.

In recent years, the earliest examples of ferns

have usually been recognized as being in one of two groups. Some authors choose to see the earliest recognized ferns as being Late Devonian members of, or closely related to, the zygopterid ferns, including latest Famennian species of *Rhacophyton* (e.g., Galtier and Phillips 1996) or more recently mid Frasnian *Ellesmeris* (Hill et al. 1997). Other authors include Middle Devonian cladoxylopsids (e.g., *Pseudosporochnus, Calamophyton*) as being the earliest members of Filicales (e.g., Schweitzer 1973).

In our opinion, the case for cladoxylopsids being ancestral to, or closely related to, ferns has been overstated for the following reasons.

1. The systematic position of cladoxylopsids with regard to the permanent or radiate protoxylem groups has not been definitively shown.

2. The case for recognizing the progressive evolution of webbed megaphylls within the Cladoxylopsida (e.g., Stewart and Rothwell 1993: 222) has been comprehensively overturned (Berry and Fairon-Demaret 1997).

3. Recurvation of sterile appendages is not known in Devonian Cladoxylopsida. In fertile appendages, where present, it appears to be a derived characteristic.

4. Anatomical investigation of *Pseudosporochnus hueberi* by Stein and Hueber (1989) demonstrated the occurrence of a pair of distinctive opposed clepsydroid traces below the dichotomy of a branch. Branch anatomy is more complex, being the typical cladoxylopsid much-divided vascular system with traces of one, two, or three xylem strands entering appendages from adjacent ribs. Berry and Fairon-Demaret (1997) demonstrated the occurrence of large, three-dimensional dichotomous appendages in the position expected for the clepsydroid traces, suggesting they were simply the vascular supply of a relatively large dichotomizing appendage and in no way homologous to the clepsydroid vascular system of the axes of *Rhacophyton*. However, the cladistic analysis of Kenrick and Crane (1997a:130, 135, 242) groups *Rhacophyton* and *Pseudosporochnus* "as sister taxa [labelled ferns] based on the distinctive clepsydroid xylem shape." We do not concur.

New understanding of the morphology and anatomy of Iridopteridales, including new descriptions in progress, will allow better testing of the relationships between iridopteridaleans, cladoxylopsids, and *Rhacophyton*. Preliminary indications suggest *Rhacophyton* is more closely related to Iridopteridales than to Cladoxylopsida. It is among Iridopteridales and putative Iridopteridales that many of the characteristics of *Rhacophyton* are found (e.g., membership in "permanent protoxylem" group, characteristics of trace departure in *Metacladophyton*, alternate branching in the new Venezuelan genus).

The relationships between Devonian plants (including Cladoxylopsida and Iridopteridales) and other groups of primitive Carboniferous ferns are far from being resolved. A polyphyletic origin of Carboniferous ferns is a distinct possibility.

Seed Plants

Current analyses recognize seed plants as being a monophyletic group derived from within or close to the progymnosperms (e.g., Rothwell and Serbet 1994), although in an alternative hypothesis they are diphyletic (see Beck and Wight 1988 and papers therein; see also discussion in Kenrick and Crane 1997a) but still derived from within progymnosperms. However recent analyses of anatomical, vegetative characteristics (e.g., Matten 1992; Galtier and Meyer-Berthaud 1996) place seed plants closer to stenokolealeans than progymnosperms. This highlights the importance of developing a whole-plant rather than purely anatomical model for stenokolealeans to test this hypothesis further.

PLANT COMMUNITIES AND ASSOCIATIONS

It is common to find monospecific "assemblages" of Devonian plants, which implies that certain plants grew in monospecific stands in some habitats. In other localities, diversity may be very low and give no hint of a community structure. Rather than illustrate total diversity of plants from different continents, an approach taken before (Edwards and Berry 1991; Edwards et al. 2000), we give a table of plants from selected well-known localities where five or more well-preserved genera are present within a single horizon or within a vertical distance of a few meters of sediment. The localities and their flora, together with details of references, are shown in table 7.2.

Such accumulations of diverse taxa in deposited sediment may have arisen in different ways—for example, (1) the plants were growing physically close together on a single ancient landscape, or (2) coeval plants living in different environments were deposited after transport into the same place.

1. Lindlar, Germany

The Lindlar plants were recovered from a quarry in the Rhineland. The majority of the specimens derived from a single small fossiliferous lens (Schweitzer 1966). The plants include small, probably herbaceous lycopsids (*Protolepidodendron?* = *Leclercqia* and *Lycopodites*), progymnosperms (*Rellimia* or *Protopteridium*), and the cladoxylopsid *Calamophyton*. *Duisbergia* is interpreted as a lycopsid by Schweitzer (1966), but we believe it may represent a cladoxylopsid. Similarly, we currently suspect *Hyenia elegans* (Schweitzer 1972) to be conspecific with *Calamophyton primaevum* (Schweitzer 1973). *Weylandia* (Schweitzer 1974) is a wholly enigmatic plant that requires fuller consideration. The age of the deposit is middle Eifelian (ADMac spore zone of Streel et al. 1987).

2. Goé North Quarry, Belgium

A large collection was made by Leclercq from the Carrière Brandt nord near the village of Betâne in east Belgium. Seven plant-bearing horizons were discovered in a lens with a thickness of about 10 m (Leclercq and Banks 1962). These are now believed to be latest Eifelian in age (AD pre-Lem spore zone; Hance et al. 1996). Mostly compression material is present, although some specimens have patchy limonite permineralization, including the possible iridopteridalean *Dixopodoxylon*. The cladoxylopsids *Calamophyton* and *Pseudosporochnus* are among the most common fossils, whereas the

zosterophyll *Serrulacaulis* is rare. Progymnosperms (*Rellimia* and *Aneurophyton*) are also locally common. One badly preserved but characteristic fragment of the lycopsid *Leclercqia* is present. The depositional environment has been interpreted as near-shore marine (Streel 1964).

3. Sierra de Perijá, Venezuela

Eleven plant-bearing horizons are known in the Lower Member of the Campo Chico Formation in Western Venezuela (Berry et al. 1993). About 20 species of plants have so far been identified. The most productive horizon (3) is a 1-m-thick layer of fine green sandstones and shales. Here a diverse assemblage of plants is found, with many slabs recovered displaying a mixture of plant remains. *Haskinsia sagittata* is sometimes found in monospecific assemblages on individual slabs (Berry and Edwards 1996a) but also mixed with other species. Two zosterophyll genera (cf. *Sawdonia*, *Serrulacaulis*) are recognized, as well as Iridopteridales and at least one genus of aneurophytalean progymnosperm. The Belgian taxon *Wattieza* Stockmans also occurs; it is a genus of cladoxylopsid very similar to *Pseudosporochnus* but with more complex lateral branching systems. Age of the assemblage is at present best inferred from the plant remains themselves, which may be latest Eifelian to earliest Frasnian (Berry et al. 1993). Although now part of the South American continent, the Sierra de Perijá terrane has a complicated pre-Mesozoic geological history, and its provenance is uncertain (Restropo and Toussaint 1988). The plant assemblages, particularly the lycopsids, are very similar to those in New York State.

More detailed sedimentological studies have not yet been undertaken, but a general overbank or floodplain environment was interpreted for this unit (Berry et al. 1993) as part of a larger deltaic sequence.

4. Riverside Quarry, Gilboa, New York State

This quarry, opened in 1920, was noted for the famous *Eospermatopteris* stumps that were found *in situ* (Goldring 1924, 1927; Hernick 1997).

These large stumps may be of cladoxylalean origin (Boyer and Matten 1996), but other authors have attributed them to progymnosperms [Goldring (1924) made a reconstruction combining the stumps with branching systems of what is now known as *Aneurophyton*] or lycopsids (e.g., Banks, in Hernick 1997). They demonstrate the presence of a forest of trees of considerable stature. Other plants that have been reported from the quarry include possible aneurophytalean progymnosperm foliage and lycopsids (both herbaceous and arborescent). *Ibyka* (Iridopteridales) was recovered from a very nearby exposure. A single specimen of undoubted cladoxylalean origin attributed to *Pseudosporochnus* (Schuchman 1969) is also known from a loose block in this quarry.

Environments of deposition are interpreted to include a forest subject to short-term flooding, near-shore sedimentation, and soil horizons (Matten 1996). The strata are believed to be equivalent to the Kiskatom Formation, of Givetian age (Banks et al. 1985).

5. Tour du Plan Incliné, Ronquiéres, Belgium

In digging the foundations for the tower during construction of the Plan Incliné, rich fossiliferous horizons were encountered. Stockmans (1968) reported a long list of fossil species, but only a handful can be confidently identified to well-established groups. We do not agree with his identification of his illustrated specimens of *Serrulacaulis* at this locality, although other specimens from his collection can be confidently assigned to this genus. Progymnosperms, both aneurophytalean and archaeopteridalean (*Svalbardia*), are present. *Langoxylon* is a permineralized stem that resembles Iridopteridales *sensu* Stein, although it differs in that it appears to have protoxylem along the midline of the primary xylem ribs. *Lerichea* resembles a barinophyte. Undoubted cladoxylopsid material is present as is the herbaceous lycopsid *Leclercqia* and enigmatic *Drepanophycus spinosus*.

The strata belong to the Alva and Mazy Members (Legrand 1967), Bois de Bordeaux Formation, of Givetian age (Lacroix, in Bultynk et al. 1991). The plant remains were deposited in a floodplain environment cut by active channels. Immature alluvial paleosols were developed, and there were episodes of lacustrine deposition (Legrand 1967).

6. Planteryggen Sandstone 8e, Spitzbergen

The upper *Svalbardia* horizon (Oberen *Svalbardia* horizon in Schweitzer 1965) contains arborescent lycopsids as well as the archaeopteridalean progymnosperm *Svalbardia* and a questionable specimen attributed to *Rellimia*. The enigmatic leafy *Enigmophyton* is not yet attributable to a major group. We believe "*Hyenia*" *vogtii* to have morphology consistent with an iridopteridalean rather than cladoxylopsid affinity. This horizon is believed to be of Givetian age.

7. Livingstoneville, New York State

In the road metal quarry (Banks et al. 1985), well-preserved fossils have been recovered from a lens of black shaley mudstone. These include zosterophylls, herbaceous lycopsids, cladoxylopsids, and *Archaeopteris*. Permineralized taxa include *Stenokoleos* and *Callixylon*. The locality is thought to be of lower Frasnian age (Banks et al. 1985). The environment of deposition is stated as "wholly terrestrial."

DISCUSSION

Associations, Assemblages, and Floras

Depositional environments of Middle Devonian strata that yield well-preserved and diverse plant compression fossils are predominantly fluvial to deltaic, and near-shore to littoral in nature. As such, the plant fossils are usually allochthonous and therefore may have derived from a variety of source areas. Herbaceous plants and those relying on moist conditions may have been swept into fluvial systems from riverbanks, whereas parts of taller plants such as cladoxylopsids and *Archaeopteris* may have entered the same channels while growing in drier environments. Catastrophic flooding might have led to even less specificity in the environments sampled. The fossil associations previously tabulated may therefore reflect more general vegetation growing in a variety of habitats in lowland coastal, often deltaic, environments rather than specific ecological conditions. Probably both specific local environments and more generalized regional floras are represented in these fossil associations, yet current studies most often have not attempted to resolve the difference between these two possibilities.

The chosen examples demonstrate a diverse flora of stable composition that consistently occurs in the macrofossil record from the middle Eifelian and into the lowermost Frasnian across Laurussia (table 7.2). The main change in the megafossil record is the appearance of the progymnosperms *Svalbardia* and *Archaeopteris* in the Givetian and Frasnian.

In Laurussia, therefore, lowland plant communities were largely dominated by trimerophyte derivatives, although in some localities or horizons herbaceous lycopsids are more abundant. These trimerophyte derivatives have been recognized in the past to have been predominantly cladoxylopsid and aneurophytalean or archaeopteridalean. We are now also able to recognize from compression fossils that iridopteridalean plants formed a substantial part of the overall flora, something that could previously only be guessed at from structurally preserved fossils restricted to North American localities. We might expect, when we can develop a morphological model for Stenokoleales, that these too would form a significant part of the whole community structure. Trimerophyte derivatives dominated the taller vegetation (over 1 m). Zosterophylls are widely recognized but apparently lack the diversity of their Early Devonian relatives, and the co-occurrence of *Serrulacaulis* and spiny *Sawdonia*-like zosterophylls is marked. The ecological niche of the zosterophylls had perhaps been largely taken over by the diverse herbaceous lycopsids that are so prominent in New York State floras and in Venezuela but not so abundantly recorded in Europe. Zosterophylls and herbaceous lycopsids are the dominant members of the shorter (less than 1 m) vegetation. More exacting sedimentological studies

may reveal an ecological reason for the lack of diversity and abundance of herbaceous lycopsids in Europe. Rhyniopsida may also be found to play a significant part in the Middle Devonian Flora, but their habitat might well have been very restricted.

Absolute abundance of fossils is probably not a good indicator of the relative composition of floras. For example, our model of cladoxylopsid growth suggests that a large volume of branch material is shed during the life of the plant, whereas the growth of an iridopteridalean plant might be achieved without loss of branch systems. One might expect the same number of cladoxylopsids to produce a much higher number of potentially identifiable megafossils than their iridopteridalean cousins.

The typical Middle Devonian plant community was therefore composed of bushy aneurophytalean progymnosperms, herbaceous lycopsids, cladoxylopsids, iridopteridaleans, and zosterophylls. This association of plants seems to represent the mixture of plants found in relatively stable and favorable habitats on the coastal environment across Laurussia.

This is, of course, a very generalized view of the Middle Devonian vegetation. As detailed later, bed-by-bed collecting and analysis of plant-bearing deposits might demonstrate more structure in the data. In Venezuela, where 11 plant-bearing horizons have already been documented in the Lower Member of the Campo Chico Formation, and a new one more recently found in the Upper Member, there are dramatic changes in the composition of individual assemblages, which most probably reflects changes in the local environment controlling both the growing plants and the taphonomic processes leading to preservation.

Horizon 3, as described earlier, is a classic example of the "typical" flora. Some slabs contain diverse remains such as cladoxylopsid, iridopteridalean, lycopsid, and zosterophyll fragments, all relatively well preserved in the green or gray sandstones. Other slabs, however, contain only abundant *Haskinsia* in dense intertangled mats (Berry and Edwards 1996a: figure 18), suggestive of *in situ* or near *in situ* preservation.

Sedimentary structures do not allow determination of the exact sedimentary environment, although it is undoubtedly fluvial and terrestrial. In the next stratigraphically higher plant-bearing horizon (4), composed of dark fine-grained shales, lycopsids are very rare, represented by a single short axis of *Leclercqia* cf. *complexa* (Berry 1994) and three fragmentary specimens of *Haskinsia*. The majority of the rest of the fossils from this horizon are abundant sterile axes of "*Taeniocrada*" (with S-type water-conducting cells suggesting a rhyniophyte affinity), often covering entire bedding surfaces. Scolecodonts are present. This suggests an entirely different environment, possibly marine or lagoonal. It is even possible to speculate that the "*Taeniocrada*" plant, by far the most abundant, may have been aquatic. The lycopsids do not appear to be *in situ,* but washed in.

Immediately above this shale unit is a 10-m-thick band of red fluvial sandstones and muds, with plant fossils found in the lowest beds. Here the assemblage (5) is dominated by four species of herbaceous lycopsids: *Colpodexylon coloradense* (Berry and Edwards 1995), *Haskinsia sagittata*, *H. hastata* (Berry and Edwards 1996a), and *Lycopodites.* These are well preserved and densely packed on some bedding surfaces. In contrast, the only other fossils are cladoxylopsid stems, occasionally well-enough preserved to demonstrate small amounts of the typical punctate surface pattern or digitate branching, but usually showing only the remains of the vascular strands in a state of decay. No identifiable attached appendages were found. This contrasts with the preservation of the herbaceous lycopsids, whose delicate leaves and sporangia remained intact. It suggests stream-bank vegetation made up exclusively of a number of species of herbaceous lycopsids, preserved virtually *in situ,* with cladoxylopsid material having been transported some distance from its source before being deposited in a decayed condition.

In the Upper Member of the Campo Chico Formation, *Archaeopteris* occurs alone in sediments containing abundant acritarchs and chitinozoans, demonstrating a marine environment of deposition of rafted and transported vegetation.

Other assemblages from the Venezuela succession reveal further information, and the elucidation of the paleoecology of the Venezuelan plants is a long-term objective of one of us (CMB). However, it cannot be accomplished until the completion of the basic botanical and systematic treatment of all the plants involved.

At Goé there is a marked difference between the assemblages of the Brandt north and south quarries (Hance et al. 1996), which are believed to be more or less coeval and are only 300 m apart (although separated by a small fault). The north quarry contains the "typical" Middle Devonian flora, whereas the south quarry yields only the cladoxylopsid *Lorophyton* and numerous unidentifiable spiny axes. The reasons for this are not clear.

Other sources of information that challenge the "typical" model of the Middle Devonian flora are localities where permineralized axes are the only type of fossil present. The Millboro Shale (Upper Eifelian) of south-western Virginia contains large percentages of stenokolealeans, progymnosperms, iridopteridaleans, and a small number of cladoxylopsids (Stein et al. 1983). The depositional environment is marine, and sorting, which determined the type of plant that reached the site of preservation, is likely to have altered drastically the composition of the source vegetation. Therefore, herbaceous lycopsids and zosterophylls may have either been eliminated taphonomically or not been present in the source vegetation. The Cairo flora (Matten 1975, 1996) is thought to have been deposited in a lacustrine environment. It is dominated by progymnosperm taxa with cladoxylopsids, *Rhacophyton,* and *Stenokoleos* (Matten 1992). Again, herbaceous lycopsids and zosterophylls are absent from the permineralized flora, but lycopsids similar to *Colpodexylon* occur in nearby coarser sediments (Matten 1996).

Habit

It is hard to estimate accurately the height of Early Devonian plants that, although only possessing narrow stems and axes, may have gained extra support from growing in close stands. More substantial trunks are found in the Eifelian cladoxylopsids, which, judging from the bases of the first-order branches, may have reached diameters of 10 to 20 cm. Large Middle Devonian permineralized cladoxylopsids (*Xenocladia*) demonstrate substantial quantities of secondary xylem growth and may represent the trunks of these plants. We speculate a height of perhaps 3 to 6 m for some of the taller of these plants. Schweitzer (1992) attributes polystelic trunks from the Middle Devonian of Spitzbergen, which are up to 30 cm in thickness, to *Protocephalopteris*, and reconstructed the plant as being 5 to 8 m tall. The aneurophytalean progymnosperms could also probably reach heights of 2 to 3 m, and they produced secondary xylem. All of these plants are recorded in Eifelian sediments, although the exact times of their first appearance are not well known.

Such plants mark the beginnings of the development of a tall vegetation of vascular plants. They had to compete in a vertical sense for light, as well as horizontally for space. Arborescent lycopsids are first noted in the Givetian and were obviously capable of considerable vertical growth. However, they probably did not branch profusely at the top, allowing light to penetrate easily beneath their branches. Larger cladoxylopsids probably gave greater shade. The advent of a true canopy was in the Frasnian with the arrival of *Archaeopteris*, or possibly in the Givetian if the plants we know as *Svalbardia* were capable of growing in the same way. *Archaeopteris*, with abundant planated leaves, was capable of creating shade on a large scale, especially if growing in a forest situation.

At present, the lack of ground-covering small trimerophyte derivatives in the published record is a mystery. Small examples, perhaps up to 50 cm high, are known to us from Venezuela and Colombia, at least. One of the difficulties in interpreting such material is that often it can be perceived as part of larger plants, or it is attributed to meaningless form-genera such as *Psilophytites*. They appear to have been upright plants

rather than sprawling ones such as zosterophylls and herbaceous lycopsids. We are confident that better concepts of such plants will be presented in the future.

FINAL COMMENTS

Recent advances in the study of Middle Devonian plants derive predominantly from the monographic treatment of aspects of the anatomy and morphology of individual plants. The careful evaluation of characters and the development of whole-plant concepts is essential to pursue the goal of understanding the evolution and relationships of early land plants. Reevaluation of existing museum collections as well as new discoveries in increasingly widespread geographical localities are both valuable to this end. However, environmental studies of these plants have not advanced so dramatically and are poor compared with understanding of Carboniferous ecosystems, for example. Yet bed-by-bed collecting and enhanced sedimentological and paleoenvironmental analysis are likely to provide significant advances when this is pursued vigorously.

Two major land-plant radiations are recognized by DiMichele and Bateman (1996) as having occurred by the end of the Early Carboniferous. The first was the Siluro-Devonian radiation of vascular plants, during which the zosterophyll, rhyniophyte, and trimerophyte groups became established. The second was the Late Devonian–Early Carboniferous radiation, which included the establishment of selaginellalean and rhizomorphic lycopsids, sphenopsids, seed plants, and several groups of "ferns." We, however, recognize a major radiation of the trimerophyte lineage beginning at the Early and Middle Devonian boundary and lasting into the Eifelian. During this time, cladoxylopsids, progymnosperms, iridopteridaleans, stenokolealeans, and other as-yet-unclassified trimerophyte derivatives adopted a wide number of anatomical and architectural innovations. A less dramatic but nevertheless important radiation of herbaceous lycopsid taxa also occurred. These new plants lived alongside a few relicts of the Early Devonian flora. This led to the development of a stable vegetation that persisted over a wide area of Laurussia.

This Middle Devonian radiation, which has been overwhelmed in the literature by reports and reviews of the Siluro-Devonian radiation and the Late Devonian development of the seed, remains comparatively unstudied and certainly under emphasized. Much work remains to document this important period in the history of terrestrial vegetation.

ACKNOWLEDGMENTS

We would like to thank Professor Maurice Streel for advice on European stratigraphy and geology, and Pat Gensel and Bill Stein for their constructive comments on the first draft of this paper.

8

The Origin, Morphology, and Ecophysiology of Early Embryophytes: Neontological and Paleontological Perspectives

Linda E. Graham and Jane Gray

Most chapters in this book recount paleontological or paleochemical evidence bearing on tracheophyte diversification and concomitant changes in atmospheric chemistry and climate. This chapter is focused on the evolutionary transition from aquatic algal ancestors to pretracheophyte terrestrial embryophytes, the possible morphological and physiological characteristics of earliest land-adapted embryophytes, and the environment that may have led to terrestrialization. Conclusions or hypotheses are deduced from neontological evidence (including molecular phylogenetics and physiological information), and comparison between microfossil remains linked to earliest (Ordovician) pretracheophyte land plants and the cells and tissues of extant seedless plants and algae.

Neontological and paleontological evidence is consistent with the presence of embryophytic land plants in the Early Phanerozoic before the appearance of tracheophytes. The morphology and affinity of early embryophytes deduced from fossil spores and other microfossil evidence is consistent with neontological data that support a bryophytic relationship, with liverwort-like plants appearing first. However, the known megafossil record of bryophytes is younger than that of the earliest tracheophyte fossils. Neontological evidence is also consistent with a charophycean green-algal ancestry for Ordovician embryophytes, but the known fossils interpreted as charophyceans also come too late to provide support.

Characters of earliest embryophytes that may have conferred resistance to decay during dormancy or periods of drought avoidance (and perhaps also short-term desiccation), or promoted fecundity under low moisture conditions, included histogenetic meristems, resistant compounds in vegetative cell walls, placental transfer cells, and spores enclosed by a sporopollenin wall. Such features are linked to putatively pre-adaptive features of modern charophyceans. We also argue that desiccation tolerance, an ancestral protoplasmic attribute that is widespread among algae and embryophytic spores, was essential to plant life on land. Further, we suggest that the transition to land most likely occurred in ephemeral freshwater habitats characterized by unpredictable water availability. The low probability that such habitats would generate substantial sedimentary remains may explain the absence of known charophycean fossils of Ordovician age or earlier, and the absence of fossils representing transitional forms.

OVERVIEW

We begin with a review of the ultrastructural, biochemical, and molecular evidence that supports the concept that modern green algae of class Charophyceae (Mattox and Stewart 1984) are more closely related to the kingdom Plantae (embryophytes) than any other group of extant protists. We survey the evidence that extant embryophytes are monophyletic (derived from a single, common ancestor), bryophytes diverged from this lineage prior to the origin of tracheophytes, and liverworts (hepatics) are the earliest-divergent group of extant embryophytes (Mishler et al. 1994; Lewis et al. 1997; Qiu et al. 1998). Such phylogenetic relationships justify comparative study of charophyceans and bryophytes, particularly liverworts, with the goal of understanding the origin of critical innovations and identification of pre-adaptations. The discussion includes a brief survey of charophycean features hypothesized to represent pre-adaptations to autapomorphic bryophytic characters (histogenetic meristems, placental transfer tissues, and sporopollenin-enclosed spores), as well as plesiomorphic features (resistant compounds of vegetative cell walls, and cytoplasmic desiccation resistance) that are shared by charophyceans and bryophytes.

A number of modern charophyceans and bryophytes provide examples of short-term physiological dormancy in response to low temperatures and desiccation. Some are also known to be able to survive extended desiccation, a condition known as desiccation tolerance or anhydrobiosis—dormancy induced by low humidity or desiccation—sometimes as ordinary vegetative cells, but most commonly as zygospores, oospores, or meiospores produced as the result of sexual reproduction, or as asexual resting cells (akinetes). Such physiological tolerance is shared with other poikilohydric organisms (characterized by limited control over protoplasmic hydration), including prokaryotes, noncharophycean protists, fungi, lichens, and a few homoiohydric tracheophytes.

Both long- and short- term-dormancy are adaptive survival strategies for exploiting fluctu-ations in environmental favorableness, and we hypothesize that both were plesiomorphies essential for transmigration onto land and for early land life in habitats of inconstant water supply. We argue that at least some evolutionary pre-adaptations critical for success in the terrestrial habitat first appeared in nonmarine aquatic ancestors in response to selective pressures imposed by life in drought-prone, environmentally inconstant habitats, which are here termed *environments of unpredictable unfavorableness*. The physiological capacity for desiccation tolerance would have enhanced dispersal and survival of ancient charophyceans in such environments (as it does today), and of later-appearing early embryophytes, which most likely occupied terrestrial habitats of inconstant water supply. Vegetative adaptations specifically associated with the origin of the homoiohydric tracheophyte sporophyte—related to avoiding or escaping the effects of water deficits—developed after the poikilohydric gametophytes of embryophytes were established on land (Gray and Boucot 1977).

Because of the significance of the charophycean algae in the ancestry of embryophytes, we provide a summary of their fossil record. Our focus is primarily on Paleozoic taxa, although their relationship to modern charophyceans is obscure. This survey reveals that most such fossils are regarded as marine forms, contrasting sharply with the primarily freshwater habitats of modern charophyceans. Did the charophycean lineage undergo primary radiation in freshwaters (as proposed by Graham 1993), with some now-extinct branches having moved secondarily into marine habitats, or did charophyceans originate in marine habitats and move secondarily into freshwaters? Alternative hypotheses explaining this environmental discontinuity are offered.

We discuss previously controversial paleontological evidence that the earliest-known (mid to late Ordovician) spores were produced by bryophyte-like, liverwort-like plants (Gray 1984, 1985; Gray and Shear 1992). We also discuss the neontological evidence that previously enigmatic microfossils, cellular scraps ("dispersed cuticle") and banded tubes ("nematoclasts"), which first appear in the record prior to earliest tracheophytes, represent remains of the sporangial epidermis of organisms resembling modern liverworts and early-divergent mosses. Finally, we offer possible explanations for the absence of megafossil records of pretracheophyte embryophytes from the earliest Paleozoic.

RESPONSIBILITIES

This is a cooperative effort by a neontologist and a paleontologist who assume responsibilities for different sections of this paper. Although complete agreement on importance and interpretations of some kinds of evidence has not been achieved, consensus was reached on the major conclusions. Graham assumes responsibility for molecular systematic information and interpretations of structures in charophycean algae and their relationship to bryophyte structures; Gray is responsible for discussions of desiccation tolerance, the charophycean fossil record and environment, and environments most likely to have led to terrestrialization of aquatic algae in the Early Paleozoic. Other aspects represent a cooperative effort.

CHAROPHYCEAN ANCESTRY FOR EMBRYOPHYTES: NEONTOLOGICAL EVIDENCE

During the past quarter century, substantial ultrastructural, biochemical, and molecular evidence has accumulated in support of the concept that charophycean green algae (the class Charophyceae *sensu* Mattox and Stewart 1984) are the closest extant protistan relatives of the embryophytes. Such a relationship is no longer controversial. Modern taxa included in the Charophyceae include the biflagellate *Mesostigma* (Melkonian and Surek 1995), the morphologically simple (sarcinoid or filamentous) and putatively early-divergent *Chlorokybus* and *Klebsormidium*, the Zygnematales (which include familiar filamentous genera such as *Spirogyra* and *Zygnema* as well as numerous unicellular or pseudofilamentous desmids such as *Stauras-*

trum), and the more complex, later-divergent Coleochaetales (including *Coleochaete*) and Charales (such as *Chara* and *Nitella*). *Stichococcus* and *Raphidonema*, unicellular or pseudofilamentous forms once included by some authors, are not presently regarded as charophyceans. Variations in systematic treatment of this modern group and fossil representatives exist. We will informally refer to any or all fossil or extant algae thought to be associated with this group as charophyceans, and refer to the modern Charales and their acknowledged fossil relatives as the charaleans.

Compelling ultrastructural evidence for the close relationship of charophyceans to the ancestry of embryophytes includes the occurrence of multilayered structures in the flagellate cells of both groups. Such structures are rare or absent from other green algal groups (Pickett-Heaps and Marchant 1972; Graham 1993; McCourt 1995) (although similar structures occur in a variety of other flagellate protists, suggesting antiquity). Another link is provided by similar ultrastructure, enzymatic content, and cytokinetic behavior of peroxisomes (Frederick et al. 1973; Graham and Kaneko 1991; Brown and Lemmon 1997) of charophyceans and embryophytes. Acid hydrolysis– and decay-resistant, autofluorescent (phenolic-containing) polymers occur in at least some thallus cell walls of certain charophycean algae (*Staurastrum* and *Coleochaete*) and all bryophytes examined (Kroken et al. 1996). Complex phragmoplasts similar to those occurring at cytokinesis in all embryophytes are also found in *Coleochaete* (Marchant and Pickett-Heaps 1973; Brown et al. 1994) and *Chara* (Pickett-Heaps 1967a; Cook et al. 1998). Other aspects of cytokinesis, including formation of a cell plate and primary plasmodesmata, also suggest phylogenetic continuity between advanced charophyceans and embryophytes (Pickett-Heaps 1967b; Cook et al. 1997).

An impressive body of molecular evidence also indicates that earliest embryophytes are derived from ancestors that, if still extant, would be classified with modern charophycean algae. Ribosomal DNA, *rbcL*, and other sequences have been used to establish this relationship. A non-comprehensive list of references to such work includes Chapman and Buchheim (1991), Wilcox et al. (1993), Manhart (1994), Ragan et al. (1994), Surek et al. (1994), McCourt et al. (1995, 1996), and Bhattacharya and Medlin (1998). However, the nucleic acid sequence data have yielded topologically different phylogenies, and there is still significant controversy regarding branching patterns within class Charophyceae. Molecular architectural evidence, including intron insertion and gene transfer events, has proven useful in evaluating sequence and morphology-based phylogenies. Examples include transfer of *tufA* sequences from the chloroplast to the nuclear genome (Baldauf et al. 1990) early in the charophycean radiation, sequential acquisition of introns in two chloroplast tRNA genes (Manhart and Palmer 1990), and insertion of an intron into the nuclear-encoded V-ATPase subunit A gene (Starke and Gogarten 1993). Such evidence, consistent with cytokinesis information and some of the molecular data just cited, suggests that Coleochaetales and Charales diverged from the evolutionary line that ultimately led to embryophytes at a later time than did the Zygnematales (figure 8.1).

This evidence implies that we can deduce the uniquely shared, derived characters (autapomorphies) of embryophytes by identifying features present in extant but early-divergent embryophytes that are absent from modern Coleochaetales and Charales. This process relies on the assumption that modern land plants are monophyletic, and on identification of the earliest-divergent modern group of embryophytes, discussed in the next section.

Evidence for Embryophyte Monophyly

A substantial list of putative autapomorphies of the embryophyte clade has been defined (Graham 1996). Among these are conserved size and architecture of the plastid genome; preprophase microtubule bands; histogenetic apical meristem with cells having more than two cutting faces; expression of centrioles only at spermatogenesis, and some other aspects of sperm development; multicellular sporophyte; sporic meiosis; and sporopollenin-impregnated spore exines (Graham 1996). Independent derivation of more than one lineage of embryophytes from distinct charophycean ancestors would require that each lineage separately acquire all of these features. That this is unlikely provides a strong argument for monophyly of extant embryophytes (Graham et al. 1991; Graham 1993; Kroken et al. 1996). This conclusion is also supported by substantial molecular data and combined molecular and morphological

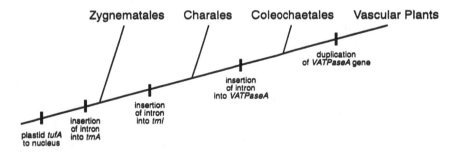

Figure 8.1.
Phylogeny of the more recently derived orders of Charophyceae. Molecular architectural changes, such as intron insertions and changes in gene location from one genome to another, are consistent with morphological evidence and some of the DNA sequence data in suggesting that Zygnematales diverged earlier than Charales and Coleochaetales. Although presence of the V-ATPase-A intron has not as yet been reported from Charales, such a finding would alter the relationship between Charales and embryophytes shown in this diagram.

analyses (e.g., Mishler et al., 1994; Waters and Chapman 1996).

Identification of Earliest-Divergent Embryophytes

Available molecular sequence information suggests that bryophytes diverged earlier than tracheophytes (Mishler et al. 1994; Lewis et al. 1997). Molecular architectural evidence for early derivation of bryophytes includes the following: (1) all tracheophytes except the three lycophytes examined exhibit a 30-kilobase inversion in the plastid genome that is absent from all examined bryophytes (Raubeson and Jansen 1992), and (2) in the liverwort *Marchantia*, the mitochondrial DNA consists of a single molecule that lacks the large, repeated sequences and homologous recombination mechanisms typical of tracheophytes (Ohyama et al. 1991). Further, RNA editing, as detected by differences between cDNA and genomic DNA sequences, occurs in mitochondria of all embryophytes examined except the three bryophytes (the mosses *Sphagnum palustre* and *Physcomitrella patens* and the liverwort *Marchantia polymorpha*) examined by Hiesel et al. (1994). RNA editing is also absent from the charophycean *Coleochaete*, which suggests that its absence in bryophytes is primitive rather than derived (Hiesel et al. 1994). Such evidence overturns the longstanding paradigm, indicated by the primarily vegetative megafossil record, that bryophyte divergence followed that of tracheophytes. The molecular evidence supports the fossil spore record that has long been recognized, beginning with Gray and Boucot in 1971 (Gray and Boucot 1977).

The relative branching order of modern bryophyte groups, liverworts, mosses (including *Takakia*), and hornworts is currently debated. Although some DNA sequence analyses (e.g., Hedderson et al. 1996) suggest that hornworts are the earliest-divergent bryophyte group, other such data place liverworts in this position (Mishler et al. 1994; Lewis et al. 1997). Significantly, an examination of 352 plant taxa revealed that three mitochondrial group II

introns occur (with occasional losses) in mosses, hornworts, and tracheophytes but are absent from liverworts, green algae, and all other eukaryotes (Qiu et al. 1998). Sztein et al. (1995) found that the single charophycean alga and four liverwort taxa examined were unable to conjugate IAA (auxin), in contrast to all mosses, hornworts, and tracheophytes tested. Comparative tubulin immunolocalization studies have documented the occurrence of localized spindle pole organizing regions with astral microtubules in dividing cells of certain liverworts. These resemble the microtubule organizing centers of charophycean algae (Brown and Lemmon 1990, 1992), but they are absent from other embryophytes. This body of data provides very substantial evidence that liverworts are the earliest-divergent bryophytes, and it justifies studies in which liverworts are compared to advanced charophyceans and the most ancient embryophyte fossils.

Some Critical Innovations: Histogenetic Meristems, Placental Transfer Cells/Tissues, Cell Wall Phenolic Polymers, Spore Wall Sporopollenin, and Desiccation Tolerance

Histogenetic meristems, efficient placental tissues at the sporophyte/gametophyte interface, decay-resistant sporangial walls, and sporopollenin-invested spores characterize all bryophytes, including liverworts. Such features, as well as desiccation tolerance (characteristic of at least some bryophytes), thus are likely to have imparted adaptive advantage to earliest terrestrial embryophytes.

Tissue-Producing Meristems

Tissue-producing meristems could generate thalloid morphologies several cells or more in thickness, thus providing substantially lower surface-area-to-volume ratios than those of filamentous or monostromatic ancestral algae. Thicker thalli are potentially more resistant to desiccation, and larger gametophytes are potentially capable of greater fecundity. Histogenetic meristems would also be essential for development of the earliest multicellular sporophytes.

Even the simplest bryophyte sporophytes consist of differentiated epidermal, sporogenic, and placental tissues. Larger, more complex gametophytes and sporophytes putatively represent adaptations that foster greater fecundity when water availability limits fertilization frequency.

The Bryophyte Placenta

Experiments have demonstrated the function of the bryophyte placenta, characterized by transfer cells having elaborate wall ingrowths, in increasing the flux of photosynthates across the apoplastic (lacking plasmodesmata) junction from the gametophyte to the (at least partially) nutritionally dependent sporophyte generation (Browning and Gunning 1979; Renault et al. 1992). Since this nutrient supply fuels spore production, increases in its size or flux would be expected to increase fecundity, allowing spore production to amplify the results of rare successful fertilizations (Graham 1996).

Acid Hydrolysis–Resistant, Autofluorescent Phenolic Cell Wall Polymers

Acid hydrolysis–resistant, phenolic cell wall polymers have recently been reported to occur in a variety of bryophyte tissues, including mature placentae and sporangial epidermis of liverworts and mosses, liverwort elaters, and moss leaves and rhizoids. These materials are highly autofluorescent in violet and ultraviolet excitation, suggesting the presence of phenolics. Their presence in sporangial epidermis, rhizoids, and leaves is suggested to confer resistance to microbial decomposition (and possibly also some desiccation and UV-resistance) (Kroken et al. 1996).

The Sporopollenin-Invested Spore

The sporopollenin-invested spore was a particularly critical feature acquired by earliest land plants or emergent aquatics prior to terrestrialization (Gray and Boucot 1977). Equipped with such propagules, earliest embryophytes, represented in the fossil record only by tiny Ordovician spore tetrads, were no doubt able to disperse atmospherically between habitats compatible with vegetative growth. Sporopollenin

walls may have provided short-term protection of spore protoplasm from desiccation. It seems likely, however, that a primary function of the sporopollenin in spore exines is to provide mechanical protection to spore protoplasm during seasonal dormancy prior to germination, and to resist fungal and microbial attack. This is also its probable function in modern and ancient algal resting cells. Sporopollenin, first identified in the pollen and spore exines of tracheophytes, is the single most chemically resistant biological polymer known (Shaw 1971; Brooks 1971). Although phenolic content is likely (Southworth 1990; Hemsley et al. 1992), the biochemical composition of sporopollenin and pathways leading to its synthesis are very poorly understood (Southworth 1974, 1990). Hence, it is difficult to determine the homology of sporopollenin-like polymers. Molecular approaches, such as comparison of sequences of genes involved in sporopollenin synthesis or the regulation of its deposition, may be helpful.

Desiccation Tolerance

Desiccation tolerance, or anhydrobiosis, is a common feature of continental aquatic organisms subject to water stress when their aquatic habitats experience extreme fluctuations in water level, are frozen, or experience salinity changes. Desiccation tolerance implies extended survival in the dormant state beyond seasonal dormancy that is a routine, commonly mandatory part of the maturation of resting or reproductive cells of many organisms. Although sometimes difficult to distinguish, seasonal or other required physiological dormancy periods may last a few days to several months, whereas desiccation tolerance may ensure survival for decades. Desiccation tolerance is also distinct from poikilohydry, which is the state of having an internal water content in equilibrium with that of the atmosphere (Raven 1986). Although the cells or tissues of various poikilohydrous algae and bryophytes may also be desiccation resistant, poikilohydry and desiccation resistance are not necessarily correlated (Raven 1986).

In the desiccated state, resistant cells can survive environmental conditions that would destroy vegetative cells in a few hours or days. Desiccation tolerance is associated with metabolic quiescence, a suspension of metabolic activity below detectable levels until the organism is rewetted (Hinton 1968; Crowe and Clegg 1973, 1978; Bewley 1979, 1995; Bewley and Krochko 1982; Bewley and Oliver 1992; Leopold 1986; Hochachka and Guppy 1987; Crowe et al. 1992; Somero 1992; Oliver 1991). The recent demonstration of drought stress response genes induced by dehydration (Oliver 1991; Bewley and Black 1994; Bewley 1995; Bartels et al. 1990, 1996) in plants, including moss, indicates a genetic basis for desiccation tolerance that can be switched off and on under appropriate environmental conditions.

For poikilohydrous organisms in general, and charophyceans and bryophytes in particular, desiccation tolerance can be argued to be an optimal adaptation to life in ephemeral waters and the inconstant hydration of terrestrial environments. It is characteristic of a variety of moss and liverwort gametophytes, and it is common in mosses of both xeric and mesic habitats (Proctor 1982, 1984; Bewley and Oliver 1992), including corticolous (bark-inhabiting) and epilithic (rock-surface) species, where the advantage is obvious. Desiccation resistance is also well documented to occur in moss spores (van Zanten 1976, 1978a,b), particularly nongreen ones. Spores of some mosses have been demonstrated to be capable of germination after desiccation for years (van Zanten and Pocs 1981; Sussman 1965a,b). For example, *Ceratodon* spores have germinated after having been dried for 16 years, and *Oedipodium* spores after 20 years' dormancy. Similarly, spores of some liverworts, namely nongreen ones such as *Riella*, can survive desiccation for many years (Miller 1982; Sussman 1965a,b). Likewise, in Early Paleozoic terrestrial environments, anhydrobiosis would have provided the capacity to survive intervals unfavorable to vegetative growth and reproduction.

Because of the importance attached to homoiohydry and the seed habit in classic paradigms of terrestrialization, it is often assumed that tracheophytes were the first land plants, and that they were, upon the evolution of the seed, the first land plants to leave moist lowland habitats

and colonize a variety of habitats. However, the evidence presented earlier indicates that characters typically associated with homoiohydry (such as waterproof waxy cuticle, stomata, lignified water conduction tubes, and, in higher tracheophytes, seeds) were not essential to terrestrialization, or for colonization of marginally mesic habitats. Early spore-producing, poikilohydric organisms with desiccation-tolerant propagules or vegetative structures could have colonized any but the most extreme of Ordovician terrestrial habitats.

Putatively Pre-adaptive Charophycean Attributes

A number of charophyceans are known to have occupied terrestrial habitats or aquatic environments that were vulnerable to unpredictable desiccation, implying the potential existence of characteristics—pre-adaptations—that were ancestral to the features critical for dealing with the moisture stress that was inevitable in terrestrial life (as previously discussed). Here, the term *pre-adaptation* is used in the sense proposed by Simpson (1953): the "possession of characteristics making possible a change in adaptation and on which the new adaptation is built." This may imply an intensification of adaptations found in ancestors, or a change in function and morphology of ancestral adaptations in a descendant's new environment to which the ancestor was already minimally adapted (Simpson 1953). Here, we briefly discuss charophycean habitats, then survey charophyceans for evidence of pre-adaptive features.

The fact that the earliest-divergent charophycean alga, the unicellular flagellate *Mesostigma*, as well as other early-divergent charophyceans (*Klebsormidium* and Zygnematales) are exclusively freshwater strongly suggests that the primary radiation of this clade occurred in freshwaters. Although *Mesostigma* is exclusively aquatic, *Chlorokybus*, *Klebsormidium*, and Zygnematales may occur in a variety of terrestrial habitats, including soils (Graham 1990b, 1993; Grant 1990; Hoshaw et al. 1990). However, the charophycean taxa most closely linked by molecular and morphological evidence to embryophytes, Coleochaetales and Charales, have shown limited inclination for occupation of terrestrial habitats. Some species of *Coleochaete* and *Chaetosphaeridium* may occupy extremely shallow freshwaters, where they are subjected to periods of desiccation, and *Coleochaete orbicularis* is capable of growing in moisture-saturated air (Graham 1993). The enigmatic charalean *Nitella terrestris* (Iyengar 1960; Chacko 1966), which superficially resembles a moss, has been found in India, in moist soil bathed by a thin layer of water or at the edges of freshwater paddy fields where it is submersed by flooding for not more than two or three days at a time. Such environments are hypothesized to have selected for a number of charophycean features that have been proposed as pre-adaptations for life on land.

Prehistogenetic Meristems

Histogenetic meristems of embryophytes are foreshadowed by apical growth and spirally arranged, asymmetrical cell divisions that occur in nonapical regions of *Coleochaete* and charaleans. Such features are absent from Zygnemataleans and other charophycean algae, but apical meristems capable of producing tissues by spiral division patterns are typical of seedless plants, including bryophytes (Graham 1996). Apical growth in *Coleochaete* has been suggested as an adaptation useful in competing with other shallow-water periphytic species for substrate space (Graham 1982, 1993), and apical growth in benthic charaleans is essential to the formation of an axial thallus that maximizes light-capture surface area.

Putative Placental Transfer Cells

The placentae of bryophytes, characterized in most cases by transfer cells having extensive arrays of wall ingrowths (Ligrone and Gambardella 1988) are similar in location (at the junction between haploid and diploid generations), structure, and presence of resistant compounds (see later) in walls of mature parental cortical cells that surround zygotes of *Coleochaete* (Graham and Wilcox 1983; Delwiche et al. 1989). The *Coleochaete* "placenta" is viewed as morphological evidence for the occurrence of a maternal nutritional support system for the earliest developmental stages of the next generation (matrotro-

phy) (Graham 1996). The adaptive advantages are proposed to be the same as that postulated for the embryophyte placenta—amplification of spore production per fertilization event and, in the case of *Coleochaete*, facilitating more rapid early-season establishment than competitors (Graham 1984, 1993). However, homology of placental transfer-like cells of *Coleochaete* to the placenta of embryophytes (particularly that of liverworts) has yet to be tested at the biochemical or molecular level.

Acid Hydrolysis–Resistant, Autofluorescent Phenolic Cell Wall Polymers in Charophyceans

Autofluorescent, hydrolysis-resistant compounds that are distinguishable from sporopollenin by solubility and absorbance characteristics occur in vegetative cell walls of the desmid *Staurastrum* (and some other desmids) and are correlated with decay resistance (Gunnison and Alexander 1975a,b; Kroken et al. 1996). Although fossil desmids are extremely rare, there are occasional pre-Holocene occurrences such as the Middle Devonian *Paleoclosterium leptum* (Baschnagel 1966; but Edwards and Riding 1993:18 suggest that this material "may be more realistically assigned to the acritarchs"). The hydrolysis-resistant cell wall compounds of modern desmids, if also present in ancestral forms, might explain such fossils. Similar compounds, whose deposition is highly regulated, occur in transfer-like cells surrounding mature (though not developing) *Coleochaete* zygotes (see preceding). These are hypothesized to function in decay resistance, and possibly also to interrupt the flow of solutes into zygotes at a developmentally appropriate time. Similar function is proposed, though not demonstrated, for indistinguishable compounds in mature placental transfer cells of embryophytes (Kroken et al. 1996). Inheritance of such compounds and functions is hypothesized to have occurred from charophycean ancestors (Graham 1996).

Zygote Sporopollenin

Putative sporopollenin similar in solubility and absorbance characteristics to that of embryophytes has been demonstrated to occur in several charophyceans, where it is found in an inner layer of the zygote wall (see reviews: Graham 1990a, 1993). Such sporopollenin is speculated to have been ancestral to that of the spore exine of embryophytes, but this hypothesis requires testing at the molecular level, as discussed earlier. Sporopollenin is not known to occur in charophyceans that do not reproduce sexually (*Mesostigma*, *Chlorokybus*, and *Klebsormidium*), but it has been cited to occur in walls of *Chlamydomonas* zygotes (Van Winkle-Swift and Rickoll 1997), and the nonmotile resting cells (phycoma stages) of certain marine prasinophyte green flagellates and fossils believed related to them (see review of prasinophytes: Sym and Pienaar 1993), which are not included within the charophycean lineage. [A polymer once regarded as sporopollenin, which occurs in the walls of certain dinoflagellates and their cysts, is now known to be so distinct that it is termed dinosporin (Fensome et al. 1993), and a resistant polymer of chlorophycean and trebouxiophycean algae, including *Pediastrum*, *Scenedesmus*, and *Chlorella*, once thought to be sporopollenin, is now regarded as a distinctly different compound, algaenan (Blokker et al. 1997)]. Although embryophyte sporopollenin is probably quite archaic, its early evolutionary history is indistinct, hampered by difficulties in chemical analysis resulting from extreme insolubility.

There is no definitive evidence that sporopollenin (or other wall modifications) of algal resting cells prevents cellular water loss over extended periods of time, although it is possible that it functions as a short-term barrier to diffusion, as compact sporopollenin is regarded as essentially impermeable to water (Heslop-Harrison 1979). Rather, sporopollenin in cell walls of charophyceans (and other aquatic algae) may provide protection from varied environmental extremes, where resistance to degradation of cytoplasmic contents is essential, such as during obligate periods of zygote dormancy. A sporopollenin-impregnated wall in zygnematalean zygospores and charalean oospores might provide limited protection to protoplasts against atmospheric desiccation during putative aerial dispersal from one water body to another (Hoshaw et al. 1990; Grant 1990). The heavy oospores of charaleans are probably not transported by wind but more

likely by insects or waterbirds, and sporopollenin might provide protection against enzymatic degradation in the digestive tracts of birds (Proctor 1962, 1967; Grant 1990). It should be noted that some charaleans and *Coleochaete* have not been demonstrated to be transported aerially. Presence of a thick sporopollenin investment might impede rehydration during germination of charophycean zygotes. We conclude that desiccation protection in aerial distribution, if any, is probably a secondary function of sporopollenin in charophycean algae.

Desiccation Resistance

Although the desiccation resistance of algae is not as well studied as it should be, and drought stress genes have not as yet been investigated in charophyceans, there is considerable evidence that various charophyceans possess desiccation tolerance/anhydrobiosis adaptations similar to those found in bryophytes of seasonally dry habitats (Proctor 1982; Bewley and Krochko 1982) and postulated to have been essential for earliest embryophytes (Gray 1984, 1985). For example, resting structures of a *Klebsormidium* (reported as *Hormidium*) germinated after being air dried in soil for 36 months (Davis 1972), and vegetative stages of a number of desmids can survive extended periods in dried mud (Brook and Williamson 1988). Akinetes of *Zygnema* can germinate following total drying for at least 12 months (although unmodified vegetative cells of this genus were unable to survive even short periods of desiccation) (McLean and Pessoney 1971). These authors also found that *Spirogyra* zygotes and "parthenospores" were desiccation resistant, whereas desiccated cells from cultures that had not undergone conjugation (the sexual reproductive process) did not recover. Proctor (1967) reported germination of *Chara* oospores after four years of dry storage, and a second citation from this worker of *Chara* oospore germination after 12 years' desiccation can be found in Coleman (1975).

The most desiccation tolerant structures in charophytes and embryophytes share similar roles in conservation of the species in time and space by ensuring germination only when environmental circumstances are favorable to vegetative survival. They allow organisms to persist through unpredictable perturbations: the more unpredictable, the greater the potential for extended dormancy and the more useful the desiccation-tolerance adaptations. The presence of desiccation-resistance adaptations in a wide variety of organisms (Hinton 1968) (including charophyceans) suggests that at least some aspects of anhydrobiosis in seedless plants are likely to be ancient capacities (i.e., they are plesiomorphic). It is, however, possible that some aspects represent innovations that first appeared in earliest embryophytes. One example is isoprene emission, which first appears in terrestrial plants in mosses, occurs commonly in mosses but sporadically in tracheophytes, and is absent from liverworts and hornworts (Hanson et al. 1999). Isoprene is believed to protect the photosynthetic apparatus from heat stress damage, and it could be regarded as one aspect of protection from the effects of desiccation.

Why Charophyceans?

Why did the charophycean algae, rather than some other group of algae, give rise to land plants? Chlorophycean and other algae may be desiccation tolerant or possess desiccation-resistant stages, various chlorophyceans generate acid hydrolysis– and decay-resistant (though nonphenolic) cell wall compounds (algaenans), and at least some chlorophyceans produce sporopollenin (Van Winkle-Swift and Rickoll 1997). We suggest that only in charophyceans are such essential features combined with the morphological and reproductive features (including phragmoplasts, plantlike plasmodesmata, prehistogenetic meristematic cell divisions, and matrotrophy) that were necessary for the embryophytic radiation.

THE EPHEMERAL HABIT OF VARIABLE UNFAVORABLENESS: IMPETUS TO TERRESTRIALIZATION IN ANCESTRAL EMBRYOPHYTES?

Modern members of the Charales and Coleochaetales, regarded as the extant charophyceans closest to embryophyte ancestry, occupy varied nonmarine aquatic, particularly shallow-water, habitats (Graham 1990b) and some charaleans also occur in brackish waters (Grant 1990). Charales sometimes live in deep-water lakes close to the limits of the photic zone; this may be a post-Cretaceous refuge, since charaleans are not competitive with aquatic angiosperms and rarely share freshwater habitats with them. A few charaleans occur in saline waters with salt content exceeding that of seawater, as noted later.

A survey of charalean fossil history (detailed later) shows substantial decrease in species diversity over time. Charales thus appear to have come to an evolutionary dead end, occupying largely unchanged environments that they have tracked through time, while slowly losing ground to extinction. This suggests that habitats now occupied can be viewed as paths to specialization and extinction rather than routes to enhanced adaptive radiation in new environments.

Coleochaetales are primarily periphytic in shallow waters of perennial lakes and ponds of variable persistence times, although some species live within the walls of deep-water charaleans (Graham 1990b). There are relatively few species, suggesting extensive past pruning of the clade, with modern representatives, as with Charales, representing "living fossils." Graham (1993; 1996) has argued that shallow freshwaters selected for important pre-adaptive features, namely decay-resistant vegetative cell wall polymers, prehistogenetic meristems, and matrotrophy, that can be observed in modern *Coleochaete* and Charales. However, environments occupied by modern Charales and Coleochaetales would not appear to foster expression of long-term desiccation resistance in any portions of the thallus except zygotes (oospores).

We argue that ephemeral aquatic habitats more like astatic waters of playa lakes, or waters that undergo dramatic changes in level but do not disappear every year (Hartland-Rowe 1972; Belk and Cole 1975), such as prairie marshes of the U.S. Midwest, rather than shallow-water shorelines of perennial lakes, are the environments most likely to have led to terrestrialization among ancestral charophyceans. We hypothesize that unknown Early Paleozoic charophyceans may have occupied ephemeral aquatic environments of variable unfavorableness, particularly in terms of moisture content and temperature variation (with salinity variation as an additional possibility).

The effects of water stress are commonly linked to evolutionary change. For example, Yancey et al. (1982) argued that the strongest selective pressures exist for organisms that experience environmental water stress, whether desiccation, freezing, or high or fluctuating salinity. Wiggins et al. (1980) also recognized that single-site temporary pools experiencing consistent annual cycles with a very wide range of ecological conditions have strongly stimulated evolutionary change in many animal taxa. Such were termed environments of "predictable unfavorableness" by Greenslade (1983), who also proposed the concept of "adversity selection" for severe, but predictably astatic, environments that experienced periods of cyclically recurring physical and chemical unfavorableness regardless of the time span. Environments such as seasonal ponds in northern or high-altitude as well as warm, humid regions are stable in the sense that they are predictable; they have a nearly regular regime, with the wet phase occurring at about the same time each year (Belk and Cole 1975). Organisms that are specialized for predictable or recurring habitats, whether favorable or unfavorable, can be viewed as narrow or specialized stenotopes.

In contrast, irregularly appearing, astatic ponds, such as occur in some arid regions, may fill and dry once, or more than once, within a season or year, or may remain dry for one or more years (Belk and Cole 1975). These ponds lack periodicity in occurrence, and as such are unpredictable. Organisms specialized for survival in highly variable and unpredictable or nonrecurring habitats can be viewed as generalized eurytopes.

The collective adaptations that enable survival in environments of predictable unfavorableness are specializations to invariate environments rather than pre-adaptations leading to adaptive radiations into new environments. Ecologically predictable habitats may be sites of evolutionary change (as argued by Wiggins et al. 1980), but they lead to morphological and/or physiological specializations for predictable variables, and thus to stabilizing selection. There are many examples in the fossil record, such as charaleans. Organisms in invariantly harsh and unfavorable temporary environments are no less extinction prone in the face of environmental change than organisms living in any invariate environments. If organisms specialized for survival in temporary harsh invariate environments do not become extinct over time, it is because they are able to maintain contact with their required environment, which has existed continuously over time.

In contrast, ecologically unpredictable habitats (for which fossil examples are difficult to define) can be predicted to lead to directional evolutionary change beyond those related to the immediate habitat. In the case of the origin of terrestrial embryophytes, directional selection may have acted on favorable mutations in environments of variable and erratic moisture to effect change from wholly aquatic to wholly terrestrial. For aquatic organisms in unstable, ephemerally aquatic environments to adaptively radiate into terrestrial environments, they needed to have favorable mutations that were directionally selected for terrestrial existence, added to pre-adaptations such as a capacity for anhydrobiosis (such as occurs in charophycean zygotes). Thus, we argue that ephemeral freshwater aquatic habitats, sites of strong evolutionary pressures related to the effects of drying and low relative humidity and unpredictable swings in temperatures, most likely provided the evolutionary impetus to terrestrialization in ancestral charophyceans, rather than predictably moist habitats. More specifically, the type of environment in which the transition from charophyceans to early terrestrial embryophytes is most likely to have occurred is variably unpredictable ephemeral aquatic habitats found in arid and semiarid environments of unpredictable rainfall, rather than variably predictable ephemeral habitats of predictable unfavorableness of some monsoonal climates or regions of predictable seasonal rainfall.

Since Ordovician terrestrial habitats, lacking the water cycle influences of tree transpiration (Shukla et al. 1990), were likely quite limited in moisture, such ephemeral and environmentally variable bodies of water, including springs, may have been more common than perennial lakes, especially at low elevations. In such inconstant, inhospitable environments, pre-adaptation for dormancy (metabolic arrest) inherited from charophyceans with similar physiological adaptations in zygotes, would have ensured dispersal and survival. Vegetative phases of early embryophytes most likely avoided, rather than tolerated, desiccation by having brief life cycles tied to ephemerally moist microhabitats (Gray 1985). It should be noted that ephemeral waterbodies of unpredictable unfavorableness are unlikely to have generated sediments readily detectable as sources of fossil remains, or to have contributed substantially to nearshore marine sediments. Thus, the inhabitants of such habitats are unlikely to be found as fossils, except under unusual circumstances.

CHAROPHYCEAN ANCESTRY FOR EMBRYOPHYTES: AMBIGUITIES OF THE FOSSIL RECORD

A chronological gap of more than 200 million years exists between the first known fossil charophyceans [female reproductive structures attributed to the scycidialeans from Llandovery–Wenlock, Early to Late Silurian boundary rocks (Mamet et al. 1992)] and an estimated origin at 600 million years, near the Precambrian–Cambrian boundary, based on neontological evidence (Waters and Chapman 1996). Such a gap makes it necessary to hypothesize an extended cryptic evolution.

In the Paleozoic, there are a variety of extinct taxa based on calcified gyrogonites (female reproductive structures) with stratigraphically restricted records (Tappan 1980; Ishchenko and Ishchenko 1982; Shaikin 1987; Feist and Grambast-Fessard 1991). The relationship of these Paleozoic taxa to the Charales is obscure; no evidence links them with Mesozoic–Cenozoic groups, nor do they appear to be on the

same lineage with modern charophyceans (Shaikin 1987; Mamet, pers. comm., 1997). Moreover, there are reasons, discussed later, to regard these extinct Paleozoic taxa, as well as some of the earliest charaleans, as marine.

The earliest evidence for charaleans is based on extinct Paleozoic taxa, although the first undoubted modern Characeae (Charales) are from the Early (Shaikin 1987) or Middle Jurassic (Feist and Grambast-Fessard 1991). Paleozoic charaleans include the Eocharaceae (Middle Devonian: Tappan 1980; Shaikin 1987; Feist and Grambast-Fessard 1991; Mamet, pers. comm., 1987) and unassigned reproductive organs from France (earliest Devonian: Feist and Feist 1997). The lineage arising from the Eocharaceae is believed to include the Palaeocharaceae (Carboniferous) and the Porocharaceae (Pennsylvanian to Early Tertiary), which generated two extinct Mesozoic–Tertiary families and extant Characeae. The Characeae reached their peak of diversity in the Late Cretaceous and Paleogene, with reduction in diversity throughout the Neogene (Shaikin 1987). The modern Charales consists of a single family and six to seven genera; only *Chara* and *Nitella* are abundant and diverse. One possible explanation for this decline might be competitive exclusion during radiation of the angiosperms into shallow freshwater habitats.

The environmental habitats of the extinct Paleozoic charophyceans and early charaleans have been the subject of much discussion (see Feist and Grambast-Fessard 1985, for example, who conclude that Paleozoic charophyceans occupied more varied environments than do modern taxa). While calcified gyrogonites of modern freshwater charophytes, and presumably those of extinct taxa, are readily transported, the extensive worldwide occurrence of Devonian–Early Carboniferous gyrogonites often many kilometers from shorelines, the absence of evidence of abrasion expected in extended transport, and the absence of coeval nonmarine records suggest that Paleozoic charophyceans were wholly marine (Mamet, pers. comm., 1997). Mamet (pers. comm., 1997) notes that nearly all Devonian–Carboniferous charophytes come from "open-marine ramp to very shallow water littoral facies. Very few from lagoonal restricted facies. None from freshwater lacustrine facies." Moreover, the stratigraphic and environmental distribution indicates a progressive shoreward migration, with the first evidence for lagoonal–littoral facies charophyceans in the Late Devonian or Early Carboniferous.

With few exceptions, authentically nonmarine fossil charophyceans are unknown before the Mesozoic and all fresh- and brackish-water taxa are related to the Characeae. This brackish/freshwater group may stem from Carboniferous *Palaeochara* (Bell 1922), a probably nonmarine ancestral characean (Mamet, pers. comm., 1997). Potential nonmarine taxa such as Devonian *Palaeonitella*, *Eochara*, and unattributed specimens described by Feist and Feist (1997) are either taxonomically or environmentally ambiguous [the uncalcified *Palaeonitella*, for example, lacks the diagnostic gyrogonites of charophytes (Kidston and Lang 1921b; D. S. Edwards and Lyon 1983), and specimens described by Feist and Feist (1997) are found in marine shoreline deposits with associated conodonts].

In summary, there is no unequivocal evidence for Paleozoic nonmarine charaleans or charophyceans apart from Carboniferous *Palaeochara*, and brackish and freshwater charophyceans related to nonmarine Characeae do not occur before the Mesozoic. Consequently, ecological evidence derived from the fossil record conflicts with neontological evidence, discussed earlier, that the charophycean clade underwent a primary freshwater radiation.

Moreover, no matter how one interprets the ecology of extinct charophyceans, the known stratigraphic record begins in the mid Silurian much later than microfossils linked to earliest embryophytes. This stratigraphic discrepancy indicates that none of the known fossils were ancestral to embryophytes.

To deal with this anomaly, Graham et al. (1994) argued for the occurrence of a hypothetical group of freshwater, pre-mid-Ordovician, poorly calcified charophyceans that lacked fossilizable structures. Some precedence exists for unmineralized offshoots of mineralized marine invertebrates, and this could have been the case with charophyceans. The Early Devonian

Palaeonitella cranii, originally *Algites* (*Palaeonitella*) *crani* from the Rhynie Chert, Scotland (Kidston and Lang 1921b; D. S. Edwards and Lyon 1983), preserved under exceptionally favorable circumstances, might provide evidence for such a group of uncalcified nonmarine charophyceans, but its charophycean status is ambiguous.

The Graham et al. (1994) hypothesis for a major earliest-Paleozoic nonmarine group would require migration of the ancestors of Paleozoic scycidialeans, trochiliscaleans, and charaleans from freshwater into the sea, and, following marine radiation, a subsequent remigration of charaleans back into fresh and brackish waters. That *Chara evoluta* can live in waters with a salt content of up to 40 parts per thousand (Grant 1990), and species of *Chara* and *Lamprothamnium* live, but do not reproduce, in coastal and interior saline lakes in Australia with a maximum of 70 parts per thousand (Burne et al. 1980) indicates the capability for survival of some charaleans in saline water habitats. However, the evolutionary history for algae provides no support for the concept that any higher taxon evolved in freshwater, invaded and radiated in the sea, and then reentered and adaptively radiated in the nonmarine environment.

An alternative hypothesis is the existence of uncalcified or poorly calcified, pre-latest-Silurian, ancestral marine charophyceans, which may have generated unmineralized nonmarine Cambrian and Ordovician charophyceans. The possibility of a hypothetical nonmarine softbodied charophycean that preceded the first evidence of embryophytes documented by the spore record, is one that cannot be discounted, even if it cannot be demonstrated from the known charophycean fossil record.

ARE ORDOVICIAN–SILURIAN SPORE TETRADS, "CUTICLES," AND NEMATOCLASTS THE REMAINS OF BRYOPHYTE-LIKE PLANTS?

Gray and Boucot (1977, 1978) first suggested that Ordovician and Early Silurian spore tetrads represented evidence for embryophytes, possibly at the bryophyte grade. This concept was elaborated on by Gray (1984, 1985, 1991) based on varied kinds of evidence: (1) the spore tetrad as consistent with an origin by meiosis from a sporocyte; (2) persistent spore walls, in some cases with an enclosing envelope, as consistent with the presence of sporopollenin; (3) a juvenile tetrad stage in modern embryophyte spores (and microspores) as consistent with an ancestral tetrad in embryophytes; and (4) the presence of spore tetrads, both obligate and facultative, among liverworts (such as early-divergent Sphaerocarpales), as consistent with neoteny and retention of a juvenile character. Based on the more common occurrence of obligate tetrads among spores of liverworts than other bryophytes, Gray (1985) suggested a potential hepatic relationship, among other possibilities, for the spore tetrads. This interpretation was worded to suggest that the spore tetrads might be evidence for bryophyte-like plants, not necessarily evidence of the direct ancestors of modern bryophytes or modern liverworts.

Interpretation of Ordovician and Early Silurian spore tetrads as bryophyte-like remains has been controversial within the paleobotanical community. The terms *cryptospore, spore-like palynomorphs,* and *enigmatic sporopolleninous bodies* in the literature of early spore tetrads expresses this uncertainty as to their affinities. One reason for slow acceptance is that occurrence of spore tetrads as the dispersal unit within modern embryophytes is relatively rare, although tetrads occur during development of all embryophyte spores. Another reason is that only recently has a modern spore tetrad producer (the liverwort *Sphaerocarpos*) been conclusively identified as an early-divergent embryophyte (Lewis et al. 1997). An additional complication is the enigmatic "tetrad envelope" (Gray 1985) that invests many of the early tetrads. Although resistant envelopes do not occur around tetrads of modern spore tetrads, possible homologues (such as outer exine or perispore) occur around both single spores and anomalous dyads and tetrads in cases where the products of meiosis have failed to separate during meiosis (Gray 1991).

The major reasons for lack of acceptance of the Ordovician spore evidence include (1) the prevailing view that homoiohydric features are essential to the existence of earliest land plants (which, as we earlier argued, is untenable), (2) the late appearance of unequivocal megafossil evidence for bryophytes—the mid-Paleozoic *Pallaviciniites* is commonly regarded as the oldest fossil liverwort [although it has also been argued that the problematic Late Silurian *Sporogonites* may have hepatic affinities (Krassilov and Schuster 1984; Oostendorp 1987)], and (3) the predominant occurrence of the earliest spore records in shallow-water, nearshore marine deposits. We suggest that the nearshore deposits represent transported terrestrial remains, and we cite compelling evidence that embryophytic liverwort-like plants actually lived much earlier than the megafossil record indicates.

Additional evidence that the Ordovician spores represent early embryophytes includes (1) the fact that no extant algal group (including charophyceans) produces sporopollenin-walled meiospores (Gray and Boucot 1977; Gray 1991; Graham 1993), (2) ultrastructural similarities between walls of Early Silurian dyads and those of modern sphaerocarpalean liverworts (Taylor 1995a, 1996), (3) the existence of a Lower Devonian megafossil having features similar to modern liverworts that generated obligate spore tetrads similar to earliest known spore remains (Edwards et al. 1995b), and (4) the fact that sporangial epidermis of at least some modern liverworts and other early-divergent bryophytes is resistant to decay and hydrolysis, and morphologically resembles some of the earliest cellular scraps and banded tubes (phytodebris) that are sometimes associated with spores (Kroken et al. 1996).

Phytodebris consists of (1) enigmatic scraps with cell wall outlines, sometimes referred to as dispersed cuticles, but that do not closely resemble authentic cuticles; and (2) banded tubes, also known as nematoclasts, originally cylindrical, with circular or spiral wall thickenings (Gray and Boucot 1977, 1978; Gray et al. 1982; Gray 1985; Gensel et al. 1990; Edwards 1982, 1986; Edwards and Rose 1984; Edwards and Wellman 1996; Wellman 1995; Edwards et al. 1996a). These remains appear somewhat later than the earliest resistant-walled spore tetrads; the earliest known "cellular" sheets are from the Caradoc; and banded tubes, referred to by Gray and Boucot as tracheidlike, do not appear before the Silurian (Gray and Boucot 1978; Gray et al. 1982), results since confirmed by others. The cellular scraps are more common in Silurian through Devonian strata; the banded tubes also extend into the Devonian (Gensel et al. 1990). The absence of a coeval fossil record for "cellular" sheets and banded tubes indicates that they cannot all be attributed to a single origin.

In some cases, namely the Caradoc sheets described by Gray et al. (1982: figure 7), the scraps are apparently composed of a unistratose array of cells, in which all cell walls have survived. In other, later cases, the scraps appear to be a single layer of resistant material that is variously patterned with pores and/or raised flanges (Edwards 1982, 1986; Edwards and Rose 1984; Edwards, Abbott, and Raven 1996). Edwards (1982) thought that the cell-like patterns in her collections of "dispersed cuticles" might represent cell impressions rather than actual cells, and she therefore referred to them as units.

The provenance of these cellular scraps and banded tubes, like that of the spore tetrads, was initially controversial (Banks 1975a,b). Gray and Boucot (1977) considered and rejected a variety of possible sources suggested by Banks. They favored the possibility of plants at the bryophyte grade and nematophytes—a poorly understood group of terrestrial plantlike(?) organisms that have left no apparent modern representatives. Later, Wellman (1995) suggested nematophytes or "higher" land plants as possible sources for these scraps, but conceded (Edwards and Wellman 1996) that some of them might have been derived from bryophyte-like plants.

Potential Modern Analogues for "Cuticular" Scraps and Tubular Remains

Kroken et al. (1996) reported that sporangial epidermis of the liverworts *Lophocolea heterophylla* and *Conocephalum conicum* (representing each of the two major liverwort clades identified by Lewis et al. 1997), and of several species of

the peat moss *Sphagnum*, is resistant to high-temperature acid hydrolysis and application of the same strong acid procedures used to isolate microfossils from rocks (Gray 1965; Gensel et al. 1990) (figure 8.2A–D). Kroken et al. (1996) also observed that cell walls of acid-hydrolyzed bryophyte sporangial epidermis were highly autofluorescent in violet and UV excitation, suggesting the presence of phenolic compounds similar to those previously reported to occur in walls of the desmid *Staurastrum* (Gunnison and Alexander 1975a,b) and mature "placental" cells of two species of *Coleochaete* (Delwiche et al. 1989). These resistant, wall-bound compounds, which have not yet been chemically defined (due to low solubility), probably explain the hydrolysis and decay resistance of bryophyte sporangial epidermis.

At the light-microscopic (LM) level, acid hydrolysis–resistant scraps of the unistratose sporangial epidermis of the early-divergent (Mishler et al. 1994) moss *Sphagnum* (figure 8.2C) were similar, in terms of color and cell or "unit" arrangement, shape, and size, to Ordovician monostromatic cellular sheets described by Gray et al. (1982). Scanning electron microscopy revealed that most internal, external, and anticlinal walls of the sporangial epidermis of *Sphagnum* species resisted hydrolytic dissociation (the walls of pseudostomatal "guard cells" were sometimes exceptions—their dissolution leaving funnel-shaped pores through the epidermal layer) (figure 8.2D) (Kroken et al. 1996).

It is also reported here, for the first time, that unistratose tissue scraps with persistent cell walls also result from high-temperature acid hydrolysis (by the techniques of Good and Chapman 1978) of the sporangial epidermis of the early-divergent (Lewis et al. 1997) liverwort *Sphaerocarpos* (figure 8.2E). Adherent spore tetrads (figure 8.2F) were also retrieved, but other thallus tissues did not survive this harsh treatment. Significantly, acetolyzed sporangial epidermal remains of *Sphaerocarpos* (unlike those of *Lophocolea*, *Conocephalum*, or *Sphagnum*) were transparent and therefore almost completely invisible when viewed by bright-field LM. They were detected via fluorescence microscopy in violet or ultraviolet excitation, suggesting that extracted microfossils mounted in a nonquenching medium should be routinely examined by fluorescence microscopy. The spore tetrads of *Sphaerocarpos*, even those with an incompletely formed exine (figure 8.2F), remained adherent after acetolysis and appeared remarkably similar to previously published images of Ordovician spore tetrads (see Gray et al. 1982).

In contrast to sporangial epidermis of *Sphaerocarpos* and *Sphagnum*, wherein all cell walls are acid hydrolysis resistant, the cell walls of acetolyzed sporangial epidermis of the liverwort *Lophocolea* were incompletely preserved (figure 8.2B) (Kroken et al. 1996), and in this regard they were very similar to certain published photographs of Silurian dispersed "cuticles" (see Edwards 1982: figure 47). Internal walls and the innermost portions of anticlinal walls of *Lophocolea* sporangial epidermis were not preserved, but surficial walls and outermost portions of anticlinal walls were preserved, giving the appearance of "flanges" like those appearing in the Silurian scraps. In addition, the outermost wall of hydrolyzed *Lophocolea* sporangial epidermis was regularly perforated with a circular pore, one per cell. Similar conditions characterize a number of the fossil fragments described in the literature (e.g., Edwards 1982, 1986; Gensel et al. 1990). In the case of *Conocephalum conicum*, a member of the complex thalloid lineage of liverworts (Lewis et al. 1997), the hydrolyzed sporangial epidermis tended to fall apart into individual tubular cells (or small groups of such cells) ornamented with regularly spaced bands. These cells appeared remarkably similar to certain fossil nematoclasts (such as those illustrated in Gensel et al. 1990). In contrast to the mosses and liverworts tested, the sporangial epidermis of two species of hornworts was not acid hydrolysis resistant but rather appeared to be covered with a layer of cuticle-like material (Kroken et al. 1996).

Kroken et al. (1996) concluded that at least some of the most ancient cellular scraps and banded tubes could represent the remains of resistant sporangial epidermis of bryophyte-like plants, and the earliest known remains of the land-plant sporophyte generation. The results obtained by acid hydrolysis of *Sphaerocarpos* that have been presented here support this conclusion. The *Sphaerocarpos* data, together with earliest scrap fossil remains (Gray et al. 1982) and results from *Sphagnum* (earliest-divergent of extant mosses) also suggest that sporangial epidermis constructed of cells having simple, unornamented, polygonal cells, all of whose walls resist acid hydrolysis, may be the plesiomorphic (earliest-appearing) type for embryophytes. The more highly ornamented, fragile, and less-completely resistant sporangial epidermis of *Lophocolea* and *Conocephalum*, representative of derived liverwort lineages (Lewis et al. 1997), may represent derived conditions. If these speculations are correct, they may explain why putative sporangial epidermal fossils similar to those of *Sphaerocarpos* and *Sphagnum* occur earlier in the record than do fossils that resemble acetolyzed *Lophocolea* or *Conocephalum* sporangial epidermis. Resistance of such liverwort and moss sporangial epidermis may explain observed co-occurrences of Ordovician–Devonian spores and cellular scraps. It is common for at least some spores to remain associated with sporangial remains after most of the spores have been dispersed from modern bryophytes.

The limited potential for complete preservation of sporangial epidermal tissues and cellulose-dominant vegetative portions of bryophyte thalli provides a possible explanation for the appearance of resistant-walled spore tetrads in the fossil record prior to the earliest known cellular scraps and banded tubes, as well as bryophytic megafossils. Selective deposition of resistant compounds within thalli of most extant bryophytes—limited to sporangial epidermal and placental tissues, along with a few other cell types such as moss rhizoids and liverwort elaters (Kroken et al. 1996)—might explain the absence of megafossil remains of earliest embryophytes. If earliest embryophytes were like modern bryophytes in this respect, as seems reasonable, they would have tended to fragment and decay into nearly unrecognizable pieces during transport, sedimentation, and diagenesis. Ordovician remains of earliest plants would thus be expected

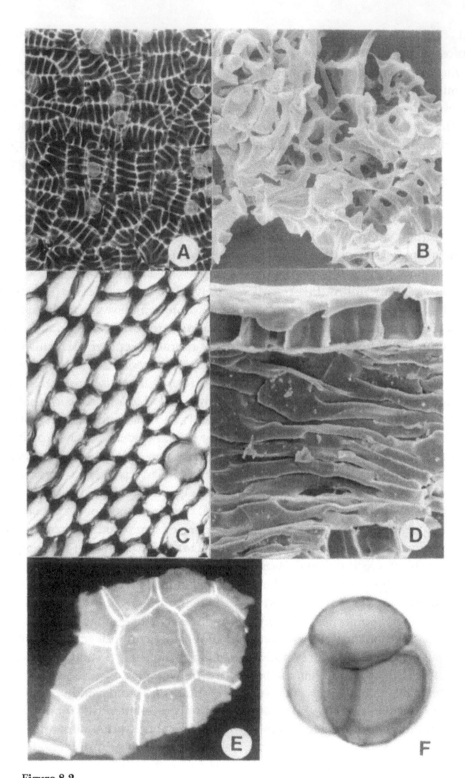

Figure 8.2.

Sporangial remains of bryophytes subjected to high-temperature acid treatment resemble some fossils attributed to early embryophytes. **A:** Fluorescence microscopic view of a fragment of sporangial epidermis of the leafy liverwort *Lophocolea heterophylla* revealing autofluorescence of hydrolysis-resistant, ornamented walls. × 360. **B:** Scanning electron microscopy (SEM) of sporangial epidermis fragment of *Lophocolea heterophylla* showing flangelike remains of anticlinal walls, and circular pores in other walls. × 1,000. **C:** Bright-field light micrograph of the sporangial epidermis of the early-divergent moss *Sphagnum*, wherein cell wall remains form a to occur only under exceptional preservational conditions, such as an autochthonous petrifaction (permineralization of plants in their growth environment). Gray and Boucot (1977) and Gray (1984, 1985) provide additional discussion of this and questions related to "time of evolution" versus "time of appearance in the fossil record."

Results cited here, obtained by comparison of modern organisms known to represent early-divergent groups to microfossil phytodebris, are consistent with the fossil spore record and its interpretation, and with noncontroversial molecular evidence that bryophytes are the earliest-divergent extant embryophytes. Additional studies of chemical and microbial decay-resistance of the sporangial epidermis and other tissues of additional bryophytes are advisable, as it is probable that other apparent matches with ancient microfossil debris will be made. Production of resistant carbon by modern liverworts and mosses and the Ordovician microfossils suggests that putative bryophyte-like, pretracheophyte early land plants could have contributed to carbon burial for at least 20 (and perhaps as much as 100) million years prior to the first appearance of lignified tracheophytes, which makes them relevant to models of Paleozoic carbon cycling (for example, Berner 1997).

SUMMARY

1. Ultrastructural, biochemical, and molecular evidence supports the concept that modern charophycean algae are closely related to bryophytes and tracheophytes, and that embryophytes are a monophyletic group that diverged from charophycean algae in the Early Paleozoic. A number of putatively pre-adaptive charophycean morphological and biochemical attributes are hypothesized to have been ancestral to characters of terrestrial embryophytes, and potentially pre-adaptive to land life. These include histogenetic-like, spirally arranged, asymmetric cell divisions; placental transfer-like cells; resistant compounds in walls of nonreproductive cells; sporopollenin in zygote walls; and desiccation-tolerant propagules.

2. We postulate that unpredictable or randomly drought-prone environments, termed environments of unpredictable unfavorableness, are most apt to have provided the selective pressures that led to terrestrialization in aquatic algae in the Early Paleozoic. We also suggest that the evolutionary paths in shallow persistent waterbodies (ecologically predictable habitats), presently occupied by most modern charophyceans, are apt to lead to morphological and physiological specializations and ultimately to extinction.

3. Although neontological evidence is consistent with a charophycean ancestry for embryophytes, a supporting fossil record for a continental aquatic group of charophyceans linked to the ancestry of embryophytes is not known early enough in time. The chronological schism between the evidence for the oldest known calcified charophyceans (at the Wenlock–Llandovery boundary), and varied lines of neontological evidence favoring charophycean ancestry for embryophytes in the Early Paleozoic is one involving potentially several hundred million years. Further, Paleozoic charophyceans, including early charaleans, appear to have been primarily marine, whereas neontological evidence indicates a primarily freshwater radiation for charophyceans. Thus, an extended cryptic record for charophyceans well below the first fossil evidence both in the marine and the nonmarine environment is postulated. The putatively fresh, ephemeral, continental waterbodies postulated to have been the cradle of embryophytes are unlikely to have left substan-

(*continued*) robust monostromatic sheet (all of several species examined were similar in this respect). × 330. **D:** SEM of *Sphagnum fimbriatum* sporangial epidermis viewed from a broken edge, showing that all cell walls resist acid hydrolysis. × 500. **E:** Fluorescence microscopy view of a fragment of sporangial epidermis of the early-divergent liverwort *Sphaerocarpos*, revealing a monostromatic layer of unornamented polygonal cells, wherein all walls appear to have survived acid hydrolysis. × 300. **F:** An immature spore tetrad of *Sphaerocarpos*; mature spores having a well-developed exine also remain associated in tetrads (i.e., they do not dissociate), even after high-temperature acid hydrolysis. × 400.

tial *in situ* sedimentary remains, or to have contributed to nearshore marine deposits, possibly explaining the lack of known charophycean fossils of the appropriate age, or remains attributable to transitional forms.

4. The earliest direct evidence that can be interpreted as the presence of terrestrial embryophytes consists of fossil spores in the mid Ordovician. These are argued to be related to, or to share vegetational and reproductive similarities with, modern hepatics (liverworts), even though megafossil remains interpreted as liverworts are not known earlier than the mid Devonian. A growing body of evidence derived from comparative morphological studies of Ordovician–Silurian microfossils believed to derive from embryophytes and modern, early-divergent embryophytes (bryophytes) supports the contention that a terrestrial bryophyte flora preceded the earliest tracheophytes by millions of years. Molecular data also argue for the "bryophytes first" hypothesis.

5. Microfossil evidences of these earliest embryophytes essentially disappeared in the Late Silurian–Devonian, as they were displaced by pretracheophytes and tracheophytes. This turnover is documented by the change in dominance from spore tetrads to single trilete spores in the late Early Silurian (latest Llandovery) and marks the beginning of Gray's Eotracheophytic Epoch or Evolutionary level. This change in dominance may have been accomplished in part by the extinction of some of the spore-tetrad-bearing lineages, as well as by phyletic replacement within some of these lineages (Gray 1985, 1991, 1993). Change in dominance from tetrads to single trilete spores does not preclude the appearance of spore tetrads throughout the Silurian and into the Devonian, or of trilete spores throughout the Silurian and even into the Ordovician.

ACKNOWLEDGMENTS

We thank Professor Pat Gensel for providing samples of *Sphaerocarpos* from the greenhouses at the University of North Carolina, Chapel Hill, and Dr. Lee Wilcox for producing figure 8.1. We also thank Bernard Mamet for his counsel concerning the fossil record of the Paleozoic charophytes and their habitat environments, and for references to the literature.

9

Biological Roles for Phenolic Compounds in the Evolution of Early Land Plants

Gillian A. Cooper-Driver

Edwards and Selden (1993) recognize four phases in the early colonization of the land. First there was colonization by Precambrian microbial mats comprising prokaryotes and later photosynthesizing protists. Second, in the Ordovician, the land surface was colonized by bryophyte-like plants, as evidenced by the presence of fossil spores and cuticles. Third, there appear in the fossil record small axial vascular plants, such as the rhyniophytoids in the Silurian (Gray 1985). These then diversified into taller vascular plants, the rhyniophytes, zosterophyllophytes, lycophytes, and trimerophytes, by the Late Silurian—Early Devonian.

The earliest fossil evidence for terrestrial arthropods and microorganisms is approximately synchronous with the earliest macrofossil evidence for nonvascular and vascular plants (Shear and Kukalova-Peck 1990; Edwards and Selden 1993; see chapter 3). Terrestrial arthropods, which were either detritivores or predators, predominate in these early paleoecosystems. Mutualistic, saprophytic, and parasitic fungi were also an important component of the Silurian—Devonian environment (Sherwood-Pike and Gray 1985; Stubblefield and Taylor 1988; Taylor 1990; Taylor, Remy, and Hass 1992a,b, 1994; Taylor, Hass, et al. 1995; Hass et al. 1994; Taylor and Osborn 1996). Thus, the first photosynthetic invaders of the land not only had to deal with the physical problems of water loss or desiccation, increased ultraviolet (UV) light, and the lack of readily available nutrients, but also had to compete and/or cooperate with a whole new set of coevolving organisms (Edwards, Selden, et al. 1995).

The ability for early land plants to successfully compete in colonizing the land required evolutionary innovations that were not only anatomical and morphological (Edwards 1993) but also biochemical (Swain and Cooper-Driver 1981; Niklas 1982; Cooper-Driver and Bhattacharya 1998). Striking changes occurred in the biochemistry of these early land plants, resulting in modifications to or extensions of existing biosynthetic pathways. Many of the newly synthesized compounds were important in the way they aided successful invasion of the land.

PHENOLICS—REACTIVITY

A class of chemical compounds that must have been of great importance to early land plants in their struggle for survival is the phenolics. These compounds are known to play important roles in extant plants as structural components, as light screens against intensive irradiation, as an influence on detritivores (and thus important in the regulation of nutrient cycling), and as a

159

defense against pathogens. Phenolics are "compounds which have an aromatic ring with one or more hydroxyl groups" (Waterman and Mole 1994). They are chemically very reactive, which is central to their many ecological roles.

The broad chemical reactivity of phenolic compounds arises from two major transformations, bond formation and oxidation (Appel 1993; Waterman and Mole 1994). Phenolics can form several different types of bonds with other molecules (figure 9.1).

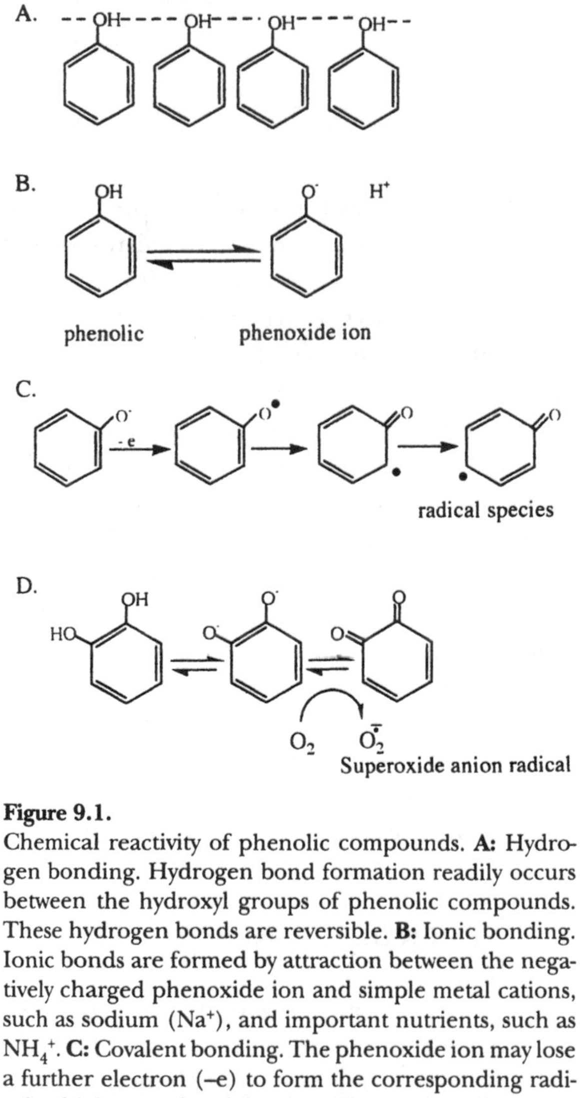

Figure 9.1.
Chemical reactivity of phenolic compounds. **A:** Hydrogen bonding. Hydrogen bond formation readily occurs between the hydroxyl groups of phenolic compounds. These hydrogen bonds are reversible. **B:** Ionic bonding. Ionic bonds are formed by attraction between the negatively charged phenoxide ion and simple metal cations, such as sodium (Na^+), and important nutrients, such as NH_4^+. **C:** Covalent bonding. The phenoxide ion may lose a further electron (–e) to form the corresponding radical, which can also delocalize. Two such radicals can undergo oxidative coupling to form carbon–carbon or carbon–oxygen bonds. **D:** Generation of superoxide anion radicals. Oxidation of phenolic compounds can result in the formation of superoxide anion radicals, which can generate additional radical species, such as hydrogen peroxide and hydroxyl radicals.

For example, the hydroxyl groups of phenolic compounds can readily form hydrogen bonds with other secondary metabolites (figure 9.1A) or with the amide carbonyl groups of proteins. Another important property of phenolic hydroxyl groups is their acidity, which results from the propensity of the bond between the oxygen and the hydrogen to break, forming the corresponding, negatively charged phenoxide ion (figure 9.1B). Ionic bonds are formed by attraction between the phenoxide and the cationic portions of other molecules such as sodium (Na^+) or organic cations like ammonium (NH_4^+). The phenoxide ion can delocalize (move the negative charge into the aromatic ring system), with the result that the charge resides on a carbon rather than on an oxygen. The phenoxide ion may lose a further electron and undergo oxidative coupling in which covalent carbon—carbon or carbon—oxygen bonds are formed with proteins, lipids, metals, or carbohydrates (figure 9.1C). When phenolic binding occurs with molecules that are nutrients, enzymes, or pathogens, binding can have a dramatic effect on digestion and disease resistance.

Phenolics are readily oxidized by enzymes and oxidants found in detritus, soil, and water and in the digestive tract of detritivores. Oxidation of phenolics can generate superoxide anion radicals (figure 9.1D), which are extremely dangerous to plant or animal cells, causing enzyme inactivation, membrane lipid peroxidation, and strand breaks in DNA. Superoxide anion radicals can generate additional radical species, including hydrogen peroxide or hydroxyl radicals.

Thus, the ecological activity of phenolics depends on the physicochemical conditions under which the phenolics occur and on whether they are protonated, ionized, or oxidized (Appel 1993).

PHENOLICS—BIOSYNTHESIS AND EVOLUTION

Phenolics are synthesized via three major biosynthetic pathways, all involving photosynthetic intermediates as precursors (Waterman and Mole 1994).

1. The Acetate or Polyketide Pathway

The acetate or polyketide pathway involves the combination of acetate groups to build polyketides (figure 9.2).

Of relevance to early colonizers of the land was the synthesis of lichen acids by lichenized fungi, via the polyketide pathway (Elix 1996). The lichen compounds (para-depsides, usnic acid, depsones, and depsidones) are formed by the bonding of two or three of the orcinol, or β-orcinol, type of mononuclear phenolic units through ester, ether, and carbon—carbon linkages. Anthraquinones, xanthones and chromones are all phenolic compounds probably formed by internal cyclization of a single folded polyketide chain.

2. Phenylpropanoids via the Shikimic Acid Pathway

Phenylpropanoids (C_6–C_3 compounds) are synthesized via the shikimic acid pathway (Douglas 1996). This pathway involves the deamination of phenylalanine to cinnamic acid and then hydroxylation to *p*-coumaric acid (figure 9.3). *p*-Coumaric acid is an important intermediate in the synthesis of lignans, coumarins, and lignin.

The shikimic acid pathway is particularly important to plants, as it not only leads to the synthesis of cinnamic acids (C_6–C_3), but also to cinnamyl alcohols, both precursors of lignin. Phenylpropanoids are important phenolic components of cutin, suberin, and sporopollenin, polymers of great structural and protective value for the existence of plants on land.

3. The Shikimate/Acetate Pathway

The shikimate/acetate pathway combines both the C_6–C_3 shikimate pathway and the C_3 acetate pathway, and it leads to the synthesis of flavonoids ($C_6C_3C_6$) and condensed tannins (polyphenolics) (figure 9.4).

The flavonoid precursors are *p*-dihydrocoumaric acid and malonyl-CoA (Shirley 1996). The bicyclic chalcone first formed converts into the flavanone naringenin. This flavanone then

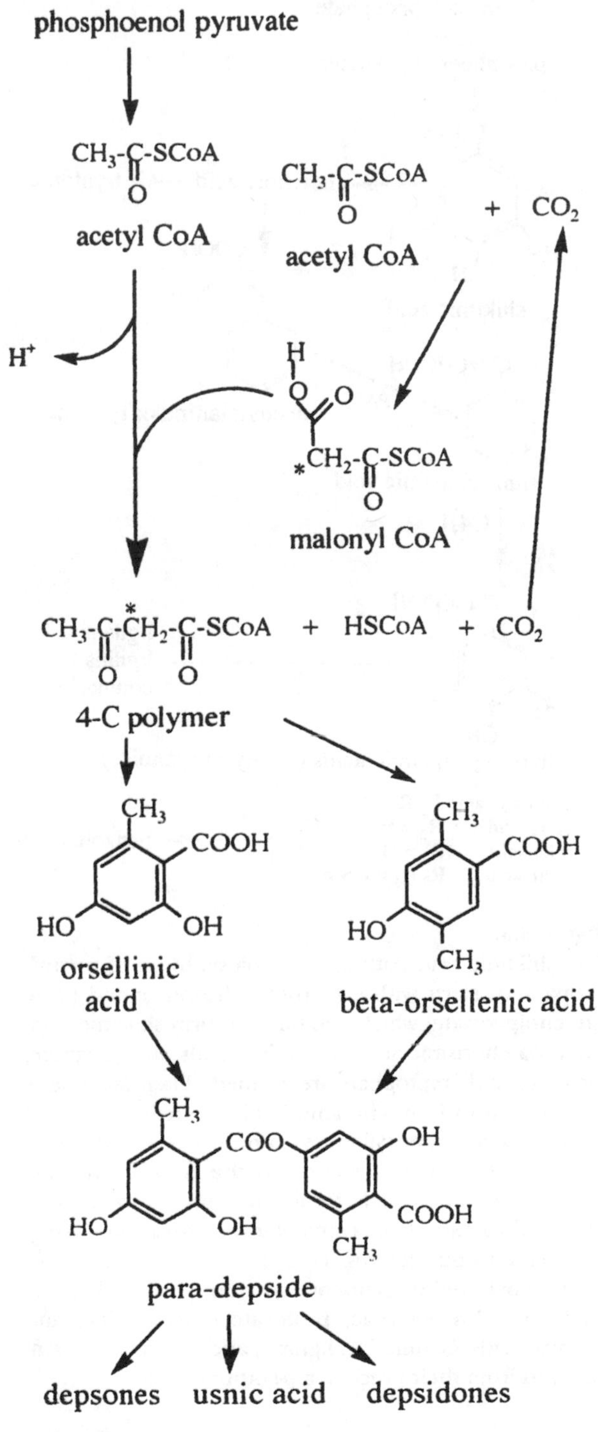

Figure 9.2.
Polyketide or acetate pathway. The condensation of two molecules of acetyl CoA is facilitated by first converting one molecule of acetyl CoA into malonyl CoA by the addition of carbon dioxide. During condensation between acetyl CoA and malonyl CoA, the carbon dioxide is lost and a 4C polymer is synthesized. This condensation is a repetitive process; the 4C polymer formed can condense with another acetyl CoA unit to build a 6C polymer, and so on. Lichen acids, such as depsones, depsidones, and usnic acid, can be synthesized via the polyketide pathway.

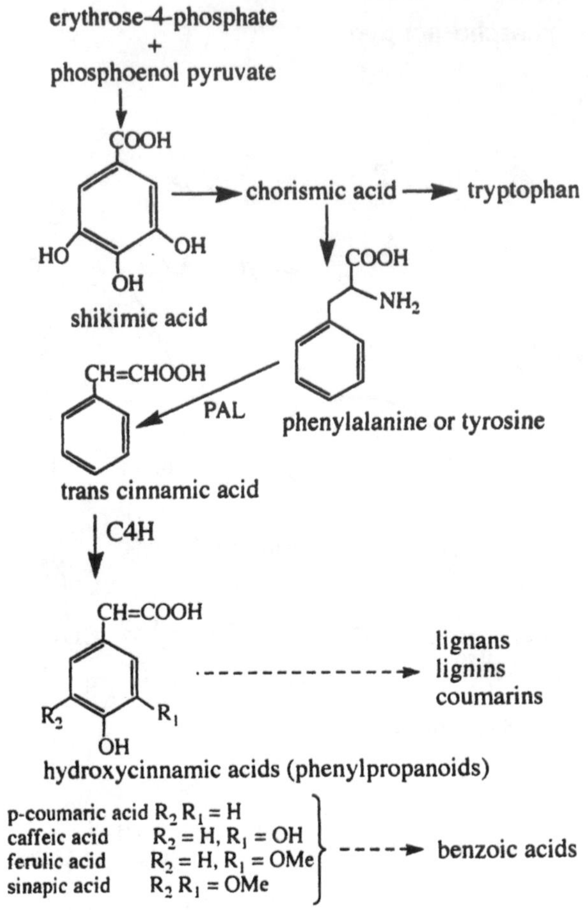

Figure 9.3.
The shikimic acid pathway. The biosynthesis of phenyl-propanoids starts with erythrose-4-phosphate and phosphoenolpyruvate, which combine to form shikimic acid; then, via chorismic acid, the amino acids phenylalanine, tyrosine, and tryptophan are formed. Phenylalanine is deaminated to form cinnamic acid, a reaction catalyzed by the enzyme phenylalanine-ammonia-lyase (PAL). Cinnamic acid is then converted to the hydroxycinnamic acids (*p*-coumaric, caffeic, ferulic, and sinapic) by the activity of hydroxylases [e.g., cinnamate-4-hydroxylase (C4H)] and O-methyltransferases. These four cinnamic acids can then be oxidized to their corresponding hydroxybenzoic acids—*p*-hydroxybenzoic, protocatechuic, vanillic, and syringic acids. Coumarins, lignans, and lignin can be synthesized from their precursor, *p*-coumaric acid.

undergoes modifications to the central pyran ring. Oxidation or reduction to the central pyran ring gives rise to the different classes of flavonoids, the flavones, isoflavones, 3-OH flavanones, and flavan-4-ols. The isoflavonoid skeleton arises through a shift of the aromatic ring from C-2 to C-3 of the pyran and biflavonoids (not shown) through oxidative coupling reactions. The introduction of the 3-hydroxyl group at the 3-position of naringenin to produce a 3-OH flavanone

leads to the synthesis of flavonols, flavan-3,4-diols, and anthocyanins. Flavan-3,4-diols polymerize to form proanthocyanidins (condensed tannins), one of the most important groups of deterrent compounds synthesized by vascular plants for protection against herbivores and pathogens (Bernays et al. 1989). Flavonoids have a diversity of roles in plants, and their roles as UV light absorbers and as microbial signaling compounds may have enhanced the competitive advantage of the invading plants.

Based on their known biosynthetic pathways, different phenolic groups are assumed to have appeared sequentially during plant evolution. Many structural genes in these biosynthetic pathways have now been sequenced and their gene products show homologies with enzymes from primary metabolism (Stafford 1991; Koes et al. 1994; Cooper-Driver and Bhattacharya 1998). While it is probable that the enzymes necessary for the synthesis of the first steps in these pathways were already present in many of the early terrestrial forms of plant life (Kubitski 1987), the structural diversity evidenced today is unlikely to have occurred at this geologic time. For example, the introduction of 3-hydroxylation to the flavanone C-ring (see figure 9.4), which led to the synthesis of flavonols and the corresponding flavan-3-ols (catechins) and flavan-3, 4-diols, does not occur in extant vascular plants before the evolution of the ferns. Also, the synthesis of colored pigments, such as the anthocyanins, rarely occurs except in angiosperms. This further elaboration of biosynthetic pathways came later with the changing environments.

PHENOLICS—STRUCTURAL AND MECHANICAL ROLE

Why were phenolic compounds so important to early land plants? Evidence for the evolution of the first vascular land plants comes from the findings of pieces of cuticle, lignified tubes or tracheids, and trilete spores with tough resistant walls of sporopollenin (Gensel and Andrews 1984; Niklas 1982). Some of these enigmatic mid Ordovician microfossils, cellular scraps,

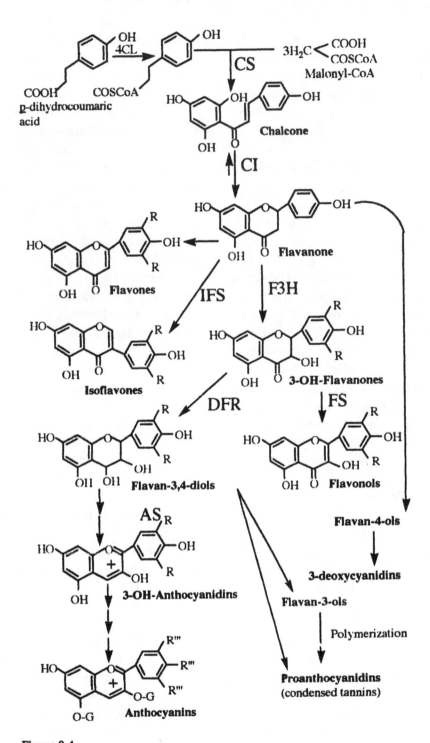

Figure 9.4.
Synthesis of the major classes of flavonoid compounds via the combination of the C_6-C_3 shikimate pathway and the C_3 acetate pathway. Enzymes: 4CL, 4-coumaroyl-CoA ligase; CS, chalcone synthase; CI, chalcone isomerase; F3H, flavanone 3-hydroxylase; IFS, isoflavone synthase; DFR, dihydroflavonol reductase; FS, flavonol synthase; AS, anthocyanin synthase. Substitutions: R = H or OH; R' = H, OH, or OCH_3; O-G = O-glycosidic groups.

and banded tubes may represent remains of the sporangial epidermis of organisms resembling modern liverworts and early divergent mosses (see chapter 8). Acid hydrolysis–resistant, phenolic cell wall polymers have recently been reported to occur in a variety of bryophyte tissues and also in extant charophytes, suggesting that the inheritance of such compounds may have occurred from the charophycean ancestors of land plants (Kroken et al. 1996).

There are also chemically unexplained resistant structures such as the peripheral thick-walled cortical tissues (the sterome) found in Devonian fossils of *Psilophyton* (Edwards, Ewbank, and Abbott 1997). Cutan/cutin in the cuticle, lignin in the tracheids, and sporopollenin in spore walls are structurally homologous polymers that contain phenolic compounds. Functionally, they are all polymers that restrict water movement, they are water repellent, and they reinforce specialized cell walls. Suberin, another biopolymer, associated with roots, has similar characteristics. They are also all highly resistant to breakdown by microorganisms.

1. Lignin

Lignification is an important component of the cell wall thickening process and probably played a key role in the successful colonization of the land by plants (Lewis and Davin 1994). Lignin is synthesized by an extension of the phenylpropanoid pathway (figure 9.5).

Lignin monomer subunits or monolignols, *p*-hydroxycoumaryl alcohol, coniferyl alcohol, and sinapyl alcohol are synthesized in the cytosol and then transported across the plasma membrane (probably as glycosides) to the apoplast of the cell wall (figure 9.6).

These aromatic alcohols may then become esterified to hydroxyl groups in the cell wall polysaccharides (this represents the first stage in the formation of lignin). Peroxidases may then generate hydrogen peroxide and monolignol radicals (Douglas 1996). Following the formation of phenoxy radicals, biomolecular radical coupling initially gives rise to dimeric structures; then, further oxidative coupling with monolignols provides the macromolecular lignins.

Unusual monomers can also be included in the classical lignin structure such as ferulic acid, cinnamaldehydes, dihydroconiferyl alcohol, and coumaric and sinapic acid (Boudet 1998).

While lignin per se may not have been present in the early bryophyte-like plants, lignans important to the evolution and biosynthesis of lignins (Lewis and Davin 1994) are present in both hornworts (Takeda et al. 1990) and liverworts (Cullmann et al. 1993). Both hydroxyphenolic acids (Rasmussen et al. 1996) and biflavonoids (Geiger 1990) are present in the cell wall of mosses.

Present in all vascular plants, lignin is a major structural component in the walls of sclereids and tracheary elements, such as tracheids, vessel elements, and xylem and phloem fibers. Its strength in the walls of tracheary elements facilitates the transport of water and nutrients and prevents thin-walled cells from collapsing. The role of lignin in the evolution of water transport must have taken place in stages—for example, *Aglaophyton major*, one of the first vascular plants, lacks the distinctive lignified thickenings characteristic of later tracheids (D. S. Edwards 1986; Remy and Hass 1996). While *Rhynia gwynne-vaughanii* has S-type thickenings, the G-type tracheids present in the zosterophylls and lycopsids are similar to protoxylem elements found in extant pteridophytes (Kenrick and Crane 1991).

Because of the irregularity of the structure, it must have been very difficult, if not impossible, for early microorganisms to digest lignin. It has been suggested that this inability of Paleozoic microorganisms to break down lignin may have been the cause of the large accumulation of organic carbon in the form of coal during the Carboniferous (Robinson 1990). Again because of difficulties of microbial breakdown, stress-induced lignin deposition may also have provided a mechanism for sealing off sites of pathogen infection and wounding. Recent data suggest that lignins play additional unsuspected functions. For example, the Casparian strip that interrupts passive diffusion across the root apoplast contains more lignin than suberin. Also, in some cases, lignin is deposited in pri-

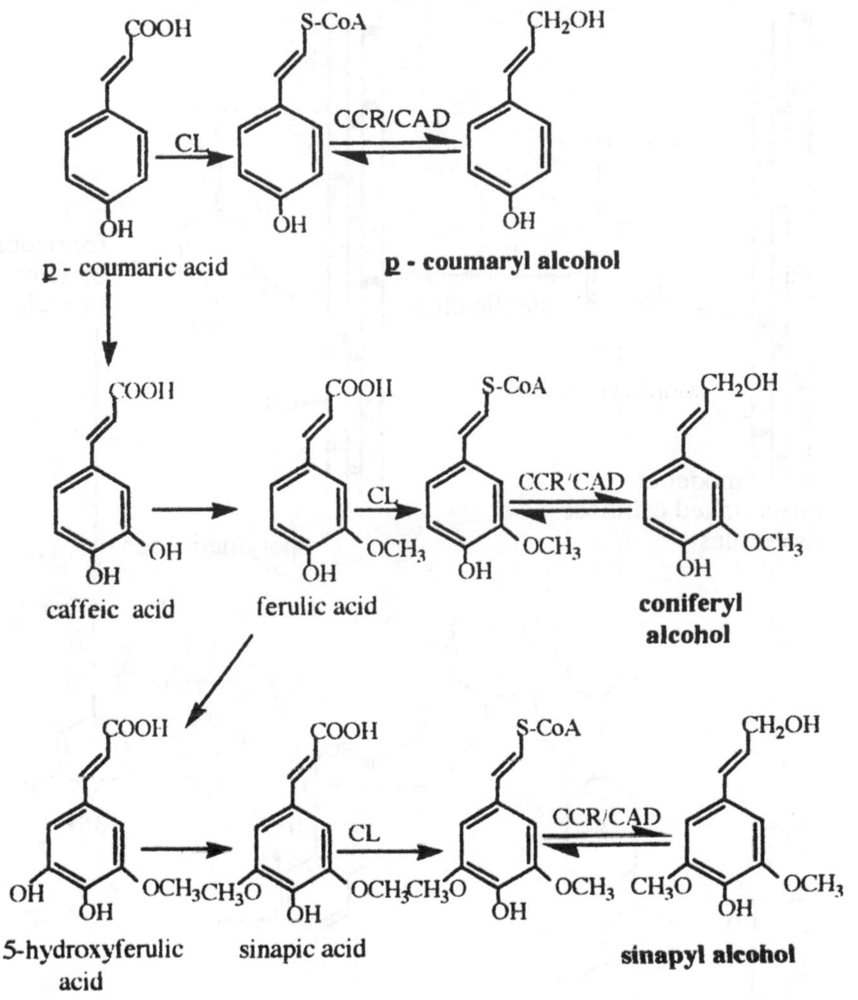

Figure 9.5.
Synthesis of monolignols. The conversion of cinnamic acid to the phenyl-propanoid acids (p-coumaric, caffeic, ferulic, and sinapic) via 5-hydroxyferulic, catalyzed by hydroxylases and O-methyltransferases, can take place as free acids or as CoA esters via hydroxycinnamate-CoA ligase (CL). Lignin monomer subunits, or monolignols (p-hydroxycoumaryl alcohol, coniferyl alcohol, and sinapyl alcohol), are derived from CoA esters of p-coumaric acid, ferulic acid, and sinapic acid, respectively, via a two-step reduction process catalyzed by cinnamoyl-CoA reductase (CCR) and cinnamyl alcohol dehydrogenase (CAD).

mary cell walls of young expanding cells, suggesting a role in the regulation of cell expansion (Boudet 1998).

2. Cutin

Unlike lignin, which is formed from monolignols, cutin is synthesized from both fatty acids and phenolics. The aerial parts of extant plants are covered by a cuticle that contains the polymer cutin, composed of dihydroxylated C16 and C18 fatty acids and simple cinnamic acids,

embedded in waxes (Kolattukudy 1984). Fatty acids are transferred through the plasma membrane by transfer proteins, where they are incorporated into the cuticle.

Analysis using Curie-point pyrolysis–gas chromatography techniques has shown the presence of a second type of biopolymer—cutan—which is an insoluble, nonhydrolysable polymethylenic biopolymer (Tegelaar et al. 1991; van Bergen et al. 1994, 1995). Cuticular matrices of extant vascular plants consist either of cutin, cutan, or a mixture of both polymers. Also, in fossil cuticles

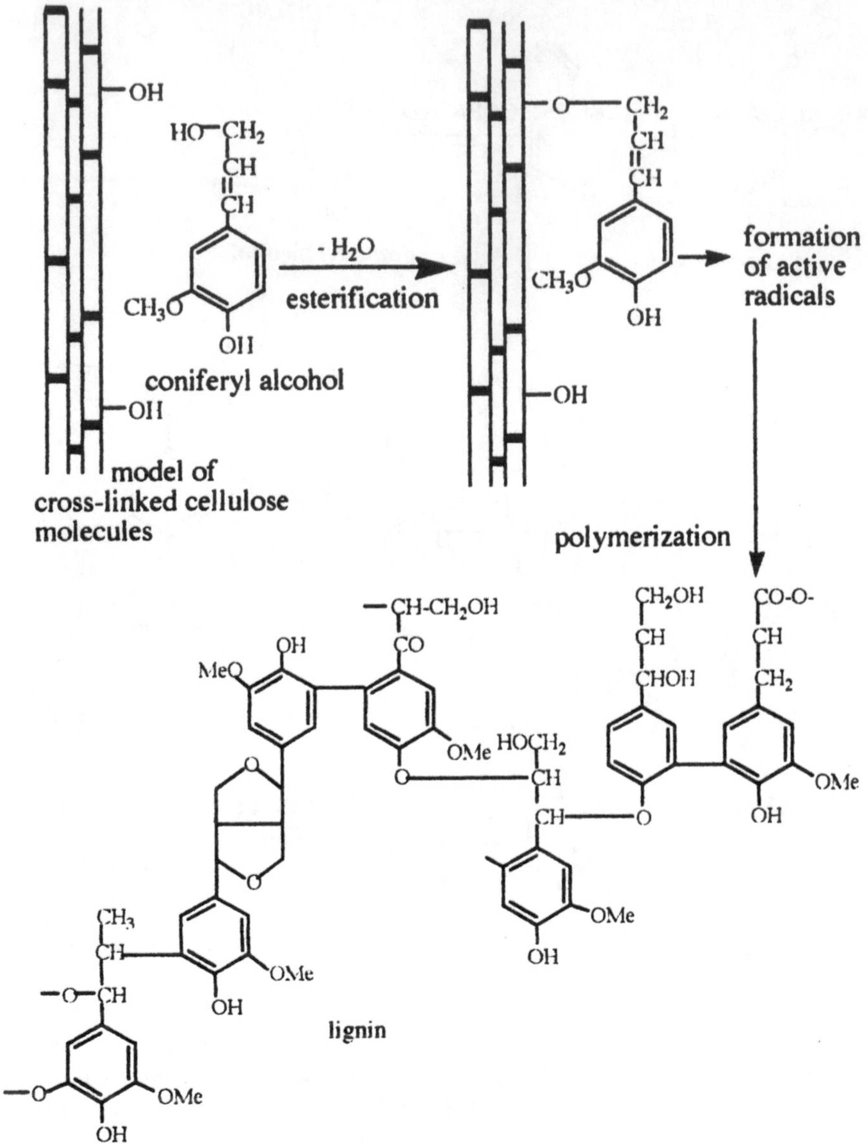

Figure 9.6.

Lignin formation. Lignin deposition in cell walls is a highly organized process. Monolignols (e.g., coniferyl alcohols) are transported across the plasma membrane to the apoplast of the cell wall and may become esterified to hydroxyl groups in the polysaccharides of the cell wall. Monolignols may subsequently be oxidized and polymerized to form lignin. The relative amounts of the three monolignols vary between plant species. It is still not possible to isolate lignin in its unaltered form.

there is an additional cuticular matrix type consisting of cutan- and cutin-derived material. The paleobotanical record is biased toward taxa originally having a significant amount of cutan in their cuticular matrix, and in all fossil taxa investigated, cutan is the major component of the cuticular matrix. Both cutan and cutin are water impermeable and play a role in preventing dehydration and penetration of the cuticle.

3. Suberin

Suberin is formed within the cell wall and outside the plasma membrane. It is probable that fatty acids are first esterified with aromatic cinnamic acids and the conjugates are subsequently incorporated into the polymers by peroxidase-catalyzed polymerization. In support of this sequence of synthesis is the finding of ferulyl esters of long-chain fatty acids (Kolattukudy 1984). As mentioned earlier, suberin is present in the Casparian strip of the root endodermis, and it also occurs in the outer cell walls of all underground organs and in the cork cells of periderm, where it forms an impermeable barrier. In extant plants, the synthesis of suberin also occurs as a stress response either to wounding or to nutrient stress.

4. Sporopollenin

Sporopollenin is the major component of the cell walls of spores and pollen. Sporopollenin is another biomolecule with physical strength, chemical inertness, and resistance to biological attack. The Ordovician and Silurian record of the first land plants is largely known only from spores and cryptospores (chapter 8). [13]C nuclear magnetic resonance (NMR) has shown that sporopollenin is not a unique substance but a series of related biopolymers derived from largely saturated precursors such as long-chain fatty acids and various levels of oxygenated aromatic rings (Guilford et al. 1988; Scott 1994). The phenolic monomers are coupled by ester bonds, as occurs in lignin and suberin. The chemical composition of sporopollenin varies among different plant groups (Hemsley, Scott, et al. 1996), and these chemical differences occur quite independently of diagenesis (Hemsley, Barrie, and Scott 1995).

Lignin, cutin, suberin, and sporopollenin form a family of biopolymers that are clearly biosynthetically related (figure 9.7). Fossils provide evidence that these biopolymers were all present in early land plants. Electron microscopy has shown that sporopollenin, cutin, and suberin all have a lamellar structure with similar physical characteristics, which suggests that these three polymers, with their high lipid content, may naturally undergo self-assembly into trilayered configurations. In all four of these biopolymers, the oxygenated aromatic rings (phenylpropanoids) serve as cross-linking agents that strengthen the plant cell.

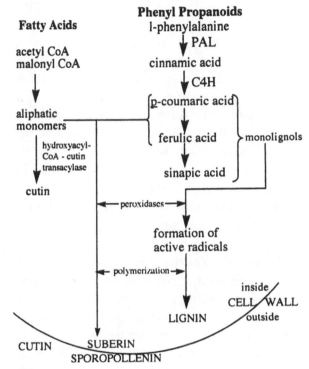

Figure 9.7.
A family of biopolymers present in early land plants. In the formation of lignin, the oxidation of phenolic alcohols yields mesomeric phenolic radicals, which rapidly polymerize and form covalent bonds with cell wall polysaccharides. Similarly, during the synthesis of suberin and sporopollenin, superoxide radicals are generated at the plasma membrane and form covalent bonds with acyl moieties. In the synthesis of cutin, the activated cutin monomers are transferred to the growing polymer with the help of the enzyme hydroxyacyl-CoA-cutin transacylase.

PHENOLICS AND PHOTOOXIDATION

In addition to the structural and protective roles played by these biopolymers, there were other roles for phenolics in promoting the competitive advantage of the first photosynthetic organisms on land. The absorption of sunlight in excess of photosynthetic usage poses a serious threat to terrestrial plants (Rozema et al. 1997). Excess energy can result in photooxidative damage to the photosynthetic apparatus as well as to a range of other essential cell components. Environmental stresses that lower a plant's photosynthetic rate, such as shade conditions, water stress, nutrient stress, or temperature stress, increase the degree to which absorbed light can be excessive, increasing the need for energy dissipation.

Lichens in the Rhynie chert played an important role in the potential establishment of the Lower Devonian ecosystem (Taylor and Osborn 1996). In this ancient association, the photobiont is represented as a cyanobacterium, while the affinity of the fungus may be a zygomycete. Recent small-subunit rRNA data suggest that lichens have arisen several times during the course of evolution, from both fungal parasitic groups and saprobes (Gargas et al. 1995). The success of this symbiosis is strictly dependent on protection of the autotroph's photosynthetic membranes (Galloway 1995). So how did early organisms, such as the lichens, cope with excess sunlight probably exaggerated by stressful environmental conditions? Present in the upper cortex of extant lichens are phenolic compounds, lichen acids, which are synthesized mainly via the polyketide pathway (see figure 9.2) by the mycobiont. These lichen acids absorb ultraviolet light and can act as light screens, regulating the solar irradiation reaching the algal zone in the upper cortex. Lichen acids therefore provide protection from photodestruction of the photosynthetic membrane of the autotrophic component (Galloway 1995). Quilhot et al. (1992) from Chile have shown that a number of lichen metabolites not only provide protection from UV light but also stimulate photosynthesis. Acids, such as atranorin, usnic acid, and norstictic acid, absorb short-wavelength UV radiation (UV-B) and reemit it as fluorescence at longer wavelengths that can then be transferred to photosynthetic pigments, thus increasing photosynthesis (Galloway 1993). Prior to the evolution of the xanthophyll cycle, lichen acids may have been extremely important to these early colonizers of the land by providing protection against light stress (Demming-Adams and Adams 1996).

PHENOLICS—SHIELDING DNA FROM UV-B DAMAGE

Other phenolic compounds, for example cinnamic acid esters and flavonoids synthesized by the acetate–shikimate cycle, may also have played an important role in shielding DNA from UV-B damage (Shirley 1996; Rozema et al. 1997) and must have acted as effective light screens in Paleozoic plants. UV-B has a number of detrimental effects on plants, causing damage to DNA, photosystem II, and plant hormone levels (Rozema et al. 1997).

In recent years, there has been renewed interest in UV-B effects on plants because of the deterioration of the earth's ozone layer. Markham et al. (1990) have shown that absolute levels of photoprotective flavonoids in mosses from the Ross Sea region of Antarctica correlated directly with measured levels of ozone between 1965 and 1989, and decreased ozone resulted in a higher synthesis of flavones in the upper epidermal layer. While the absence of flavonoids from perhaps 50 to 60 percent of all bryophytes (flavonoids have never been found in species of the Anthocerotales) sets this group of plants apart from all other land-based groups of green plants, where simple phenolics and phenylpropanoids and flavonoids are ubiquitous, mosses of the order Bryales synthesize flavone C- and O-glycosides, biflavones, aurones, isoflavones, and 3-deoanthocyanins but lack flavonols, proanthocyanidins, and 3-hydroxyanthocyanins (Markham 1990). This may indicate that the pathway initiated by 3-hydroxylation of a flavonone and leading to dihydroflavonols, flavonols, and anthocyanins is essentially inoperative in mosses (Markham 1990). On the other hand, flavone C- and O-glycosides, flavonols, dihydroflavones, dihydrochalcones, and aurones have all been isolated from one or more liverworts, whereas isoflavones, chalcones, biflavones, anthocyanins, and proanthocyanidins thus far have not been found.

Diversification of the zosterophylls and lycopsids took place during the Devonian. The lycopsids were mainly herbaceous in the Givetian–Frasnian, followed by arborescent species in the Upper Devonian (DiMichele and Skog 1992). Evolution in the early sphenopsids and pteropsids led to the common occurrence of 3-hydroxylated flavanones, flavonols, flavan 3-diols, flavan-3, 4-diols, and the condensed tannins. In every ecosystem studied, there is a positive relationship between the level of phenolic compounds and increased light intensity (Dustin and Cooper-Driver 1993; Waterman and Mole 1994), suggesting that even with the evolution of more sophisticated mechanisms of light protection, phenolics still play a role as light screens. Cinnamic acids and flavonoids must have been particularly important during the period of early land-plant evolution, when the atmospheric O_3 layer was lower and the solar UV-B radiation higher than at present (Robinson 1990).

PHENOLICS AND NUTRIENT CYCLING

Bryophytes, lichens, bacteria, fungi, and cyanobacteria formed the biological covering of the mid-Paleozoic land surface in coastal mudflats and marsh environments (Edwards and Selden 1993). While these pioneering plants, like extant bryophytes and lichens, may have had little effect on the physical environment (see chapters 2, 5, 12, and 13) because of the shallow depths of their roots and rooting structures, nevertheless they must have increased insulation on the soil surface, enhanced nutrient cycling, and provided food and stability for soil invertebrates, such as myriapods and large collembolans (Brown and Brown 1991; Longton 1992). In these early paleoecosystems, animals and plants were largely "decoupled" trophically, with most primary productivity flowing through detritivores and fungivores, which is characteristic of extant soil and litter communities (see chapter 3).

Phenolics have an important influence on the rates of nutrient cycling in both extant terrestrial and aquatic systems (Palm and Sanchez 1991; Schlesinger 1991). The phenolic compounds that enter the soil through leaching or plant remains slow down rates of litter decomposition. Phenolics may inhibit the activity of microbial enzymes that are important in the breakdown of detritus, such as the cellulases,

hemicellulases, or pectinases, by binding to the enzyme proteins. As microbial decomposers secrete phenol oxidases and peroxidases, the phenolics in the detritus are oxidized to active radicals that inhibit microbial activity. The formation of cellulosic and lignolytic residues may also reduce digestibility of plant remains and make them less attractive to herbivores.

But at the same time as they are promoting detritus buildup and humus formation, phenolic compounds also play a role in mineral retention in the soil. The capacity of phenolic compounds to ionize is important for the role they play in nutrient binding in the soil. Quinones oxidized by microbial enzymes or mineral catalysts covalently bind with amino acids, sugars, and minerals to form a matrix recalcitrant to microbial digestion (Appel 1993). Soil lignins are only slowly degraded by fungi and bacteria, and partly degraded lignins and lignin phenols may also contribute to the formation of humus (Sainz-Jimenez and de Leeuw 1986). Phenolic acids not incorporated into humus regulate the cation exchange capacity of soil and water by weathering and complexing minerals (McColl and Pohlman 1986). Phenolics increase the availability of soluble phosphorous to terrestrial plants by competing for anion adsorption sites on humus and clay and by binding with soluble aluminum, iron, and manganese that would otherwise bind to phosphate.

Phenolic compounds, leaked either from rootlike structures or directly from the living or dead plant body, must have been just as important in ancient ecosystems as they are today for the role they play in slowing litter decomposition while at the same time promoting nutrient retention.

PHENOLICS AND SIGNALING BETWEEN MICROBES

As nutrients were in such short supply, it would clearly make sense for the first land plants to develop cooperative relationships with the microorganisms already present in the paleoecosystem. It is known from recent studies that phenolics are involved in "conversations" between microbes and plant roots in the soil (Peters and Verma 1990). This interchange of chemical signals between microbes and plants is a common feature in symbiotic relationships—for example, between plants and plant roots in mycorrhizal associations, between nitrogen-fixing symbionts and their plant hosts, and also between plants and microbial pathogens (Lynn and Chang 1990).

Mycorrhizal associations are implicated in the establishment of terrestrial plants (Stubblefield and Taylor 1988), and their evolution was a crucial step in the evolution of terrestrial flora (Pirozynski 1981; Taylor 1990). Remy et al. (1994) and Taylor, Remy, et al. (1995) report unequivocal evidence for arbuscules being present in an endomycorrhizal symbiosis from the 400-million-year-old Rhynie chert. The primary benefit for the plant is the increase in phosphorous and other nutrient uptake from the soil, although protection from plant pathogens, increased production of plant hormones, and increased solubility of soil minerals also occur. Modification of gene expression during arbuscular mycorrhizal synthesis has been shown to involve changes in flavonoid chemistry (Harrison 1997). Nitrogen is often the most limiting nutrient in the soil. A much-studied symbiosis is the relationship between *Rhizobium* and the roots of leguminous plants. In the early stages of the establishment of this symbiosis, the root hairs of the legume curl to facilitate the entry of the bacterium into the plant root. The curling is controlled by *nod* genes. The induction of *nod* gene expression is in response to flavones, flavonones, and the isoflavone diadzin synthesized by the plant root (Fisher and Long 1992).

While such bacterial/fungal and plant interactions will probably always be impossible to document from the fossil record, these important phenolic interactions must have evolved during the Paleozoic. Many of the phenolic compounds involved in signaling are incorporated into the plant cell walls, and it has been suggested that this outermost protective boundary for the plant cell, like the plasma membrane of animal cells, provides a source of signal molecules (Lynn and Chang 1990).

PHENOLICS AND DISEASE RESISTANCE

Phenolic compounds have a role in the active expression of resistance of plants to plant pathogens (Nicholson and Hammerschmidt 1992). Extant plants generally respond to attack from pathogens in several different stages. One of the most rapid responses is an oxidative burst of reactive oxygen species including H_2O_2, hydroxyl radical (.OH), and the superoxide radical O_2^- (Mehdy 1994). This may be followed by the accumulation of toxic phenols such as benzoic acids and hydroxycinnamic acids (see figure 9.3) at the infection site. When a plant tissue is damaged, such hydroxycinnamic acids and their derivatives are thought to contribute to the discoloration and autofluorescence of the host tissues at the site of infection—causing the hypersensitive response. The formation of lignin or suberin generally blocks further penetration. Infection may also be followed by the erection of physical barriers such as appositions or papillae.

There is evidence from Rhynie chert fossils of damage to tissues (Stubblefield and Taylor 1988) and a host response. The earliest examples of a host response involving fungal parasites come from fungal interactions with the green alga *Palaeonitella* (Taylor, Remy, and Hass 1992b). The presence of large hypertrophied algal cells represents a direct host response. Another parasitic interaction detected in the Early Devonian paleoecosystem involves a host response caused by mycoparasites (Hass et al. 1994). In this interaction, distinct thickenings, papillae, or lignotubers are formed on the inner surface of thick-walled chlamydospores. Banks (1981), Trant and Gensel (1985), and Banks and Colthart (1993) have described a wound-response tissue in *Psilophyton* that anatomically resembles necrophylatic periderm.

In present-day ecosystems, phenolic compounds may also act as feeding deterrents to herbivores (Bernays et al. 1989). Although there is sparse evidence for herbivores until the Permian, insects with stylets or sucking mouth parts are evident in the Paleozoic (Kevan et al. 1975; Scott et al. 1992; see chapter 3). These sucking insects were probably avoiding the phenolic compounds already present in the cell wall tissues. Animals could have been spore eaters or detritivores similar to modern millipedes (Edwards, Selden, et al. 1995). It is possible that widespread herbivory did not arise until the Late Paleozoic because it took time to evolve mechanisms and enzymes to overcome the deterrent effects of phenolic compounds (Edwards, Selden, et al. 1995).

THE FUTURE

This chapter has shown how, as a result of their chemical properties, phenolic compounds have played a central role in the evolution of the first photosynthetic plants on land. While presently most of our understanding of the evolution and the role of phenolics is based on our knowledge of extant flora, this is beginning to change. Phenolics are evident in the structural components of ancient fossils from the Silurian and Early Devonian dating from 425 million years ago. They were present as part of lignin, cuticle, and suberin and as spore coverings of sporopollenin. New developments in chemical techniques have enabled a much greater understanding of the chemical composition of both living and fossil cutan/cutin and lignin (Ewbank et al. 1996; Edwards, Abbott, and Raven 1996). The use of NMR has enabled some exciting work to be carried out on fossil sporopollenin and possible self-assembly of these biomolecules (Hemsley, pers. comm. 1996).

Molecular genetic studies are providing new insights into the mechanism of phenolic evolution and new information on the evolution of phenolic enzymes and their relationship to enzymes involved in primary metabolism. The lignification process was undoubtedly a critical innovation for the structural development of large terrestrial plants, and some of the genes responsible for lignification are similar to flavonoid pathway genes (cinnamoyl-CoA reductase), suggesting that the lignification genes have evolved from the flavonoid genes (Lacombe 1997) or that divergence from a common ancestral gene associated with primary metabolism has occurred.

Recent work described in this chapter has also resulted in a much greater understanding of the development of early terrestrial ecosystems and the range of plant–animal and plant–microbe interactions that occurred during the establishment of plants on land. We have seen how fossils show evidence of plant damage and also evidence of chemical repair.

Phenolics are presently involved in many different environmental roles. They afford protection of plant cells against intense atmospheric radiation and UV light. They affect rates of litter decomposition and mineral retention in the soil. They are involved in communication between plants and microbes and in the establishment of nutritionally beneficial relationships. They act as important defenses against pathogens as part of an active defense response.

Over the course of evolution, phenolics have gradually taken over new and more diverse functions. For example, lignins and intermediates in their synthesis appear to be involved not only in providing mechanical strength but perhaps also in the Casparian strip and the movement of ions, and also in cell elongation (Boudet 1998). If the phenolic compounds that were present in these early pioneering plants lack the sophistication and structural variation of present-day terrestrial plants, it is because of their more limited environmental role. Angiosperms require highly colored petals to attract insects and other pollinators to the flower; similarly, the dispersal of fruits requires the more visible coloration of the methoxylated flavonols, flavones, chalcones, and aurones. Early land plants were also not likely to have had the sophistication of present-day angiosperms for preventing herbivory and disease through the synthesis of phytoalexins and induced systemic resistance, both direct responses to infection. They also probably lacked advanced mechanisms for signaling to microbes, as seen in the *Rhizobium*–legume association. Nevertheless, even if they lacked the sophistication found in the evolutionarily advanced modern-day taxa, phenolic compounds clearly played a vital role in terrestrialization of the land.

It appears that the synthesis of phenolic compounds is an extremely malleable biosynthetic system that is capable of adjusting quickly and easily to new environmental pressures. Through increased knowledge of the early Paleozoic paleoecosystem, we will be better able to appreciate the role that these compounds have played in the diversification of plants.

10

The Effect of the Rise of Land Plants on Atmospheric CO_2 During the Paleozoic

Robert A. Berner

On a multimillion-year time scale, the major process affecting atmospheric CO_2 is exchange between the atmosphere and carbon stored in rocks. This long-term, or geochemical, carbon cycle is distinguished from the more familiar short-term cycle that involves the transfer of carbon between the oceans, atmosphere, biosphere, and soils. In the long-term cycle, loss of CO_2 from the atmosphere is accomplished by the burial of organic matter in sediments and the conversion of atmospheric CO_2 to dissolved HCO_3^- in soil and groundwater via the weathering of Ca—Mg silicate minerals. Silicate weathering is followed by the transport of Ca^{++}, Mg^{++}, and HCO_3^- via rivers to the ocean, where the ions are precipitated as Ca and Mg carbonates, and carbon is thereby removed from the surficial system. Release of CO_2 to the atmosphere in the long-term carbon cycle takes place via the oxidative weathering of old organic matter and by the thermal breakdown of buried carbonates and organic matter (via diagenesis, metamorphism, and volcanism), resulting in degassing to the earth surface.

The preceding description can be represented by succinct overall chemical reactions. The reactions (Urey 1952; Holland 1978; Berner 1991) are as follows:

$$CO_2 + CaSiO_3 \Leftrightarrow CaCO_3 + SiO_2 \qquad \text{(i)}$$
$$CO_2 + MgSiO_3 \Leftrightarrow MgCO_3 + SiO_2 \qquad \text{(ii)}$$

$$CH_2O + O_2 \Leftrightarrow CO_2 + H_2O \qquad \text{(iii)}$$

The double arrows in reactions (i) and (ii) refer to silicate weathering plus sedimentation of marine carbonates when reading from left to right, and thermal degassing when reading from right to left. The double arrow in reaction (iii) refers to weathering (or thermal decomposition plus atmospheric oxidation of reduced gases) when reading from left to right, and burial of organic matter (the net of global photosynthesis over respiration) when reading from right to left.

CARBON CYCLE MODELING AND THE ROLE OF PLANTS

A model (Berner 1991, 1994) of the long-term carbon cycle, named GEOCARB, has been constructed; it attempts to quantify the various processes that affect weathering and degassing over Phanerozoic time. Factors affecting weathering include (1) the evolution of the sun as it affects global warming, (2) the uplift of mountains as they affect relief, climate, and silicate weatherability, (3) the rise of large vascular land plants and their spread to upland areas as they affect both silicate weathering and organic carbon burial, (4) changes in continental size and position as they affect temperature and river runoff from the continents, and (5) variations of atmospheric CO_2 as they affect plant growth and global temperature plus river runoff (via the atmospheric greenhouse effect) and thereby serve as a negative feedback mechanism for stabilizing both CO_2 and climate. Factors affecting degassing include (1) changes in seafloor spreading rate as an indicator of global degassing from ridges, subduction zones, and the continents, and (2) changes in the distribution of carbonates between shallow platforms and the deep sea as they affect the amount of carbonate accompanying the subduction of ocean floor.

Of particular interest in this book is the treatment, via the GEOCARB model, of land plants as they affect atmospheric CO_2. This includes the burial of plant-derived organic matter in sediments and the acceleration of Ca–Mg silicate weathering by the activities of growing plants, subjects to be discussed later. It does *not* include the uptake of CO_2 in plants, or its release from plants, which is only a part of the short-term carbon cycle. On a million-year time scale, the amount of carbon that is present in living plants (or in soils) is tiny compared to the amount of organic carbon that is stored in sedimentary rocks or that is transferred to and from the rocks over millions of years. An example is illustrative. The amount of carbon present in all life is about 0.5×10^{18} grams. By comparison, the amount of organic carbon stored in rocks is roughly $15,000 \times 10^{18}$ g, or about 30,000 times that present in the biosphere. Further, in just 5 million years, the amount of organic carbon buried in sediments is over 50 times that present in life (e.g., see Berner 1991).

The global rate of burial of organic matter as a function of geologic time is calculated in GEOCARB modeling through the use of data on the carbon isotopic composition of seawater as recorded in limestones and carbonate fossils. A first-order plot of the $^{13}C/^{12}C$ ratio of paleo-seawater (expressed as $\delta^{13}C$), as used in the modeling, is shown in figure 10.1. From these data and the theoretical approach of Garrels and Lerman (1984), which is included in the GEOCARB model, one can readily calculate organic carbon

burial rates over time, and the results closely mirror the curve for carbon isotopes. Qualitatively, high rates of organic burial are marked by rises in oceanic $\delta^{13}C$ due to the extra removal of isotopically light carbon from the oceans as organic matter. Thus, just by observing figure 10.1, one can state that a very large rate of organic matter burial must have taken place beginning in the Devonian and hitting a peak during the Carboniferous and Early Permian. This burial need not have occurred just in the oceans. Burial of much ^{12}C-enriched organic matter on land, and in swamps and lakes, should cause the remaining residual carbon, present as atmospheric CO_2 and dissolved HCO_3^-, to be heavy and depleted in ^{12}C. Delivery of this residual heavy carbon to the oceans via rivers and exchange with the atmosphere would then result in heavier carbon in seawater and the formation of heavier marine carbonates.

What could have caused the high organic burial rate of the Permo-Carboniferous? Most likely it was the rise of large vascular plants. The plants provided a new source of organic matter to be buried both in terrestrial sediments and in marine sediments along with marine organic matter. This

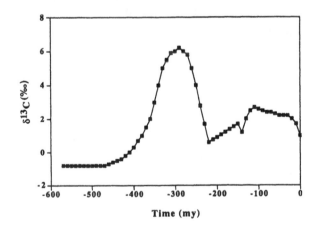

Figure 10.1.
Generalized plot of $\delta^{13}C$ of seawater over Phanerozoic time. The plot is a very low frequency fit to actual data as recorded in limestones and calcareous fossils (see Berner 1991). Because of depth and areal variations within the oceans, nonequilibrium isotope fractionation by some organisms, diagenesis, and short-term variations with time, the plot is only a crude first-order approximation to the actual record. The parameter $\delta^{13}C = (^{13}C/^{12}C \text{ sample } / ^{13}C/^{12}C \text{ standard } - 1)\ 1,000$.

is especially true of lignin, which is found only in terrestrial plants (Killops and Killops 1993). Lignin is resistant to microbial decomposition and is a major source of peat and coal. Most of the increase in carbon burial, mirrored by the ^{13}C maximum in figure 10.1, was probably the result of the burial of ligniferous terrestrial plant material, both in coal basin sediments and in marine sediments after transport of the terrestrial organic matter to the oceans by rivers. Calculations of total organic carbon burial rates in Permo-Carboniferous sediments (Berner and Canfield 1989), including both coal and dispersed carbon in coal basin sediments and kerogen in marine sediments, agree rather well with independent calculations of total global burial rates based on carbon isotopic modeling (Berner 1994).

Enhanced burial of organic matter should result, all other factors being held constant, in a drop in atmospheric CO_2, as shown by reaction (iii) reading from right to left. The rela-

tive importance of organic matter burial as it affects atmospheric CO_2 is shown in figure 10.2. Here, the sensitivity of CO_2 to the changes in oceanic carbon isotopic composition (and thus organic burial rate) during the Devonian to Permian is illustrated, based on calculations of the GEOCARB II model (Berner 1994). One curve is shown for the standard formulation, where $\delta^{13}C$ is given by the plot of figure 10.1, and the other for the situation where $\delta^{13}C$ is held constant, 1.0 percent for all time. The deepness of the CO_2 "valley" for Permo-Carboniferous time is a function of enhanced organic matter burial at that time; however, most of the drop in CO_2 during the Devonian and Early Carboniferous cannot be accounted for in terms of organic matter burial. What is the principal cause of this drop?

The major cause of the Devonian Carboniferous decrease in atmospheric CO_2, according to the GEOCARB modeling, is the acceleration by

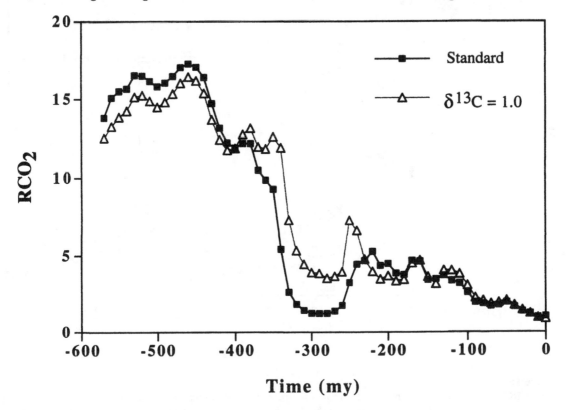

Figure 10.2.
Plot of RCO₂ versus time for the standard, best estimate formulation of the GEOCARB II model (Berner 1994) compared to the result for holding the carbon isotopic composition of the oceans constant at $\delta^{13}C$ = 1.0‰. The parameter RCO₂ represents the ratio of the mass of atmospheric carbon dioxide at any past time to that at present (assuming a preindustrial value of 300 ppm).

vascular plants of the rate of Ca–Mg silicate weathering. This is shown in figure 10.3. Here, results for various values of the effect of plants on the rate of weathering are shown. The weathering rate effect refers to the ratio of weathering rate for vascular plants to that in their absence (plant/bare in figure 10.3). It is assumed that plants invaded upland areas around 380 million years ago (Algeo et al. 1995) where deep and extensive rooting systems could accelerate the decomposition of silicate minerals. It is clear that plants have a very large effect. The best estimate or standard value, chosen for the plant-weathering effect (plant/bare) is a value of about 7. This is

based on present-day field studies as will be discussed.

It should be kept in mind that the mid-Paleozoic drop in CO_2 was not due simply to the removal of CO_2 from the atmosphere by plant-enhanced weathering. Although often misunderstood, the actual rate of CO_2 removal from the atmosphere by weathering, on a multimillion-year time scale, is not an independent parameter and must be essentially equal to the supply of CO_2 from degassing (e.g., see Berner and Caldeira 1997). Thus, an acceleration of weathering, in this instance by the rise of vascular land plants, must have been matched by some decel-

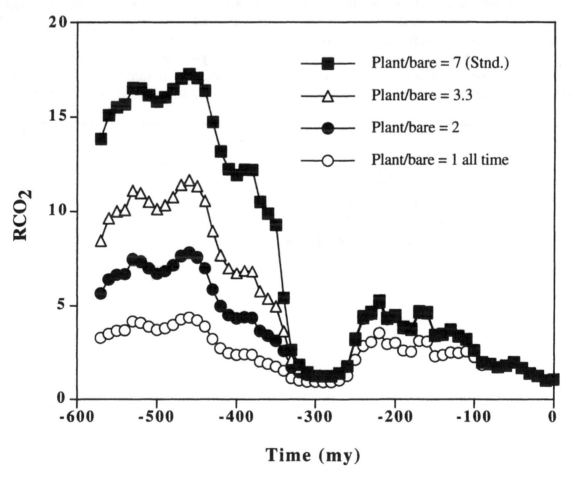

Figure 10.3.
Plot of RCO_2 versus time for the standard formulation compared to results for variations in the effects of vascular plants on weathering. The ratio plant/bare (for values of 7, 3.3, and 2) represents the effect of Paleozoic vascular plants on the rates of weathering of Ca and Mg silicates as compared to the previous early Paleozoic situation of essentially "bare" continents, possibly populated by primitive organisms such as algae. The curve labeled "plant/bare = 1 for all time" is the result obtained if vascular plants had never existed.

erating process, since there is no evidence of any equivalently large changes in degassing rate at this time. The decelerating process was the drop in atmospheric CO$_2$. Lower CO$_2$ via the atmospheric greenhouse effect, brought about lower global temperatures and less river runoff, which had a decelerating effect on global weathering rate and helped to balance the accelerating effect of the plants.

STUDIES OF THE QUANTITATIVE EFFECT OF PLANTS ON WEATHERING RATE

How could vascular plants have accelerated weathering rate during their rise in upland areas in the Devonian and Carboniferous? Studies of modern ecosystems suggest that rock weathering should be accelerated by plants for the following reasons:

1. Rootlets (plus symbiotic microflora) with high surface area secrete organic acids/chelates, which attack minerals to gain nutrients (e.g., Griffiths et al. 1994; Baham and Caldwell 1994).
2. Organic litter decomposes to H$_2$CO$_3$ and organic acids provide additional acid for weathering.
3. On a regional scale, plants recirculate water via transpiration followed by rainfall and thereby increase water–mineral contact time. There is greater rainfall in forested regions than there would be in the absence of the trees (e.g., Shukla and Mintz 1982).
4. Plants anchor clay-rich soil against erosion, allowing retention of water and continued weathering of primary minerals between rainfall events (e.g., Drever 1994).

To evaluate quantitatively how plants affect the rate of rock weathering, it is necessary to hold all other factors affecting weathering, such as bedrock lithology, climate, slope, aspect, and soil permeability, constant. This means that it is very difficult to perform an ideal experiment. However, some progress has been made by studying small geographic areas where these other factors are reasonably constant. This includes studies of the chemistry of waters draining forested and nonforested portions of the southern Swiss Alps (Drever and Zobrist 1992), a small high-elevation area of the Colorado Rocky Mountains (Arthur and Fahey 1993), and the Skorradalur area of western Iceland (Moulton and Berner 1998). By comparing the fluxes of dissolved Ca^{++} and Mg^{++} in solutions exiting from forested and minimally vegetated ("bare") portions of each area (including storage in young growing trees in the case of Iceland), results indicate an enhancement of weathering by trees, by factors ranging from about 3 to 10. This is in general agreement with results found for a controlled experiment, where plants were grown in adjacent plots of identical size, bedrock lithology, aspect, and microclimate, at the Hubbard Brook Experimental Forest Station (Bormann et al. 1998). Here, the fluxes of Ca^{++} and Mg^{++}, lost in drainage waters, accumulating in growing pine trees, and precipitated in soil was found to be about 10 times higher in a pine-covered plot than the flux losses in solution for an adjacent "barren" plot populated sporadically by, for example, bryophytes and lichens. A compilation of the results of the various studies is shown in table 10.1.

In table 10.1, it can be seen that the available data indicate an accelerating effect of trees on the rate of calcium and magnesium silicate weathering of a factor of roughly 2 to 10 times, with an average value of about 6. Thus, the choice of 7 as the plant-acceleration factor, as the standard or best estimate in figure 10.3, is close to results based on modern field experiments. That this is a good choice is further shown by the agreement between the GEOCARB model standard CO$_2$ plot and independent estimates of atmospheric CO$_2$ based on the study of paleosols (Yapp and Poths 1996; Mora et al. 1996). This is illustrated for the Paleozoic in figure 10.4.

Table 10.1.   **Ratio of weathering fluxes for forested and minimally vegetated portions of different areas**

Area	Ions	Ratio of Ions (Trees/Bare)
Southern Swiss Alps* Drever and Zobrist 1992	Ca^{++}, Mg^{++}	8
Western Iceland** Moulton and Berner 1998	Ca^{++}	3
" " "	Mg^{++}	4
Colorado Rocky Mountains Arthur and Fahey 1993	Ca^{++}, Mg^{++}, Na+, K+	4
Hubbard Brook, New Hampshire*** Bormann et al. 1998	Ca^{++}	10

The ratio of weathering fluxes is based on fluxes of dissolved ions in drainage waters (corrected for input from precipitation) plus storage in growing trees and soil where available. The term *trees* refers to forested land, and *bare* to adjacent land minimally vegetated with primitive organisms (e.g., bryophytes, lichens). All values are for the ratio of fluxes except for the Swiss Alps, where only the ratio of concentrations in stream waters is listed.

* Data corrected for temperature difference between high and low elevations.

** Data include storage in trees.

*** Data include storage in trees and in soil weathering products.

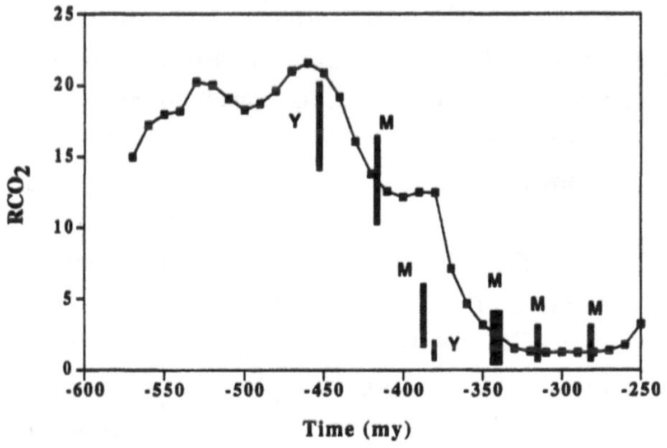

Figure 10.4.
Plot of RCO$_2$ versus time for an updated version (slightly altered from figures 10.2 and 10.3) of the standard GEO-CARB II formulation compared to recent independent estimates of CO$_2$ based on the study of paleosols. Y = Yapp and Poths (1996); M = Mora, Driese, and Colarusso (1996).

CONCLUSION

The large drop in CO$_2$ that occurred during the Devonian and Early Carboniferous was due predominantly to the rise of vascular land plants and their spread to upland areas. The plants provided new microbially resistant organic mat-ter for enhanced burial in sediments but, more importantly, caused an acceleration of the rate of weathering of calcium and magnesium silicate minerals. Together, these effects brought about a very large decrease in atmospheric CO$_2$, which, because of the greenhouse effect, led to mid-Paleozoic global cooling and ultimately to the Permo-Carboniferous glaciation, the longest and most areally extensive glaciation of the entire Phanerozoic.

ACKNOWLEDGMENTS

The author's research on plants and weathering has been supported by NSF Grant EAR-9417325, DOE Grant DE-FG02-95ER14522, and ACS-PRF Grant 29132-ACS. Acknowledgement is made to the donors of the Petroleum Research Fund, administered by ACS, for partial support of this research. The author acknowledges the musical inspiration of the opera Merry Mount, by Howard Hanson, and its rare performance by Gerard Schwarz in October 1966.

11

Early Terrestrial Plant Environments: An Example from the Emsian of Gaspé, Canada

C. L. Hotton, F. M. Hueber, D. H. Griffing, and J. S. Bridge

The origin of embryophytes in the Ordovician and their subsequent radiation in the Silurian and Devonian were profoundly important events in the history of life. Land plants ameliorated an initially hostile terrestrial environment, paving the way for colonization by other organisms. Rhizoid and root activity reduced substrate instability, promoted chemical weathering, and increased nutrient availability; the establishment of a plant canopy dampened fluctuations in humidity and temperature, greatly augmented primary productivity, and opened up niches for exploitation by other organisms (Beerbower 1985). Increased incorporation of CO_2 into organic carbon by land plants over the course of the Siluro-Devonian was apparently a primary factor in precipitating a sharp drop in atmospheric CO_2 (Berner 1993, 1994; Mora et al. 1996; see chapter 10), which in turn was likely responsible at least in part for the Permo-Carboniferous glaciations. Even the Frasnian—Famennian marine mass extinctions have been attributed to coastal marine eutrophication fueled by tracheophyte radiation and evolution of deeply penetrating root systems over the course of the Devonian (Algeo et al. 1995; Algeo and Scheckler 1998).

Despite the significance of early land plant paleoecology, detailed studies of pre-Carboniferous plant assemblages are rare. A few paleoenvironmental studies of specific plant assemblages have been conducted (Andrews et al. 1977; Gensel and Andrews 1984; Edwards and Fanning 1985; Scheckler 1986b; Kasper et al. 1988), of which the most detailed is that by Schweitzer (1983). However, these studies are limited by absence of sedimentological detail. Because the degree to which a fossil assemblage resembles the parent plant communities depends in large part on sedimentological and taphonomic processes, understanding the sedimentological context of fossil assemblages and associated facies is crucial to the reconstruction of their living environment. Hence, we have undertaken a detailed study of the sedimentology of fossil land plant assemblages in the Emsian age Cap-aux-Os Member of the Battery Point Formation, exposed along Gaspé Bay, Quebec, Canada. The diverse Gaspé flora has been studied for over a century, beginning with the pioneering work of J. W. Dawson (1859, 1871). Many subsequent studies have established it as one of the best-described floras of Early Devonian age (Hopping 1956; Daber 1960; Kräusel and Weyland 1961; Banks and Davis 1969; Banks et al. 1975; Hueber 1967, 1971, 1983b, 1992; Gensel 1976, 1979, 1984; Granoff et al. 1976; Remy et al. 1993; Stein et al. 1993). However, no paleoecological study of this flora has been attempted.

Although the fossil record suggests that by

the Emsian, embryophytes had occupied the land for roughly 70 million years (Gray 1993; Strother et al. 1996), terrestrial ecosystems at that time are still considered primitive. Absence of tetrapods and the plesiomorphic nature of arthropod faunas at this time suggest that plant–animal interactions would have been limited (Selden and Edwards 1990; DiMichele et al. 1992). Plants are believed to have occupied sites in rare sexual reproductive events and then formed monospecific stands through extensive vegetative growth (the "turfing-in" model of Knoll et al. 1979; Niklas et al. 1980). Interspecific plant interactions consequently would have been limited as well, and early land plants are considered to have exhibited relatively little niche specialization (DiMichele et al. 1992). They are further believed to have been restricted by their free-sporing life cycle and relatively inefficient water conducting systems to moist, mesic sites, unable to occupy high-stress (e.g., hydric or xeric) habitats (Scheckler 1986b; Bateman 1991; Algeo and Scheckler 1998). Some authors, nevertheless, place them in salt marshes, a physiologically stressful habitat (Schweitzer 1983; Retallack 1990; Edwards and Selden 1993). In our opinion, these propositions are best treated as null hypotheses about early land plant ecology, to be tested by independent data. It should be kept in mind that this generalized model is more applicable to early stages of land plant evolution. By Emsian time, land plants had diversified into at least three major tracheophyte clades (lycophytes, trimerophytes, and rhyniopsids *sensu* Kenrick and Crane 1997a) as well as, most likely, one or more bryophyte clades, suggesting concurrent ecological differentiation.

Our goal here is to test hypotheses about early land plant ecology against data derived from detailed biostratinomic analysis of the Cap-aux-Os sediments. We first present data on floristics and taphonomy, then a summary of sedimentological evidence for overall depositional environment, as well as more detailed description and interpretation of depositional environments of plant megafossil assemblages. We have chosen to address the following issues in this

paper: (1) to characterize types of environments occupied by plants; (2) to search for consistent associations between plant taxa and environments; (3) to characterize the degree to which plant assemblages were in fact monodominant; (4) to investigate the extent to which early land plants were capable of occupying high-stress environments, focusing specifically on brackish water habitats. Finally, we document the taxonomic and diversity disjunction between megafloral and palynofloral records within the Cap-aux-Os Member, and we discuss its implication for estimating actual plant diversity and ecosystem complexity.

MATERIAL AND METHODS

More than 186 meters of stratigraphic section were logged in centimeter-scale detail along sea cliffs exposed near d'Aiguillon, Québec (figure 11.1). The strike and dip of the strata are such that lateral exposure of most stratigraphic units is limited to a few hundred meters. However, erosional irregularities in the coastline provide three adjacent exposures of one plant-rich mudstone interval near the base of the measured section. Three sections of this interval were measured (our Seal Rock Landing, Seal Rock West, and First Cove West localities) to construct a more accurate three-dimensional depiction of the strata (figure 11.1). Each rock unit was examined for grain size and color variation, primary sedimentary structures, pedogenic features, paleocurrent indicators, vertebrate and invertebrate faunal assemblages, and ichnofabrics. Samples of critical rock types and plant-bearing strata were prepared for petrographic examination and taphonomic and ichnologic analysis at Binghamton University. Photomosaics of key stratigraphic intervals were constructed to facilitate description of lateral variation in the strata. Polaroid images annotated in the field were later combined with the photomosaics to generate lateral bedding diagrams and facies distribution information. Paleocurrent orientations were corrected for tectonic tilt of strata using the stereonet method outlined in

Lindholm (1987). The sedimentary facies recognized were compared with those in modern sedimentary environments and comparable ancient deposits, to interpret depositional environmental conditions directly associated with plant habitats and burial settings. Only a brief synopsis of this work will be provided here. Full description and interpretation of the Cap-aux-Os Member facies will be published separately (Griffing et al. 2000).

Plant megafossils were collected from every productive horizon within the measured section. A total of 58 megafossil sites were collected. Where possible, large blocks were quarried, but in some cases only relatively small hand samples could be collected. Identification of plant taxa from each site was made first on whole-rock specimens in the laboratory, after degaging if necessary. A representative subsample from each horizon was selected and bulk macerated in hydrofluoric acid for spore samples and to identify additional small, fragmentary, and minor components of the horizon. Some identifiable material from bulk macerations was subjected to additional oxidative treatment, either with Schulze's solution (saturated potassium chlorate in concentrated nitric acid) or warm dilute potassium hydroxide. In many cases, naturally retted material was sufficient for observing cellular details. Selected specimens were dried from 100 percent ethanol and mounted on stubs for observation and photography on Hitachi S-570 and Leica Stereoscan 440 scanning electron microscopes (SEM).

The digested material from every bulk maceration was prepared for dispersed spores (heavy liquid separation, residues bleached for 6 to 8 minutes, and mounted in glycerine jelly). In addition, every unoxidized horizon within the measured section was collected and prepared for dispersed spores. Approximately 200 to 600 grains were counted and identified per sample. Information on dispersed spore assemblages in this paper is based on 60 samples from the lower half of the measured section. A paleoecological analysis of the dispersed spore assemblage of the Cap-aux-Os Member will be presented elsewhere. For comparison of *in situ*

spores with dispersed spores, sporangia were mounted with no additional treatment on stubs for SEM. For light microscopy (LM), some sporangia were subjected to Schulze's solution followed by ammonium hydroxide, washed, and mounted in glycerine jelly. Other sporangia were mechanically crushed, bleached for about 7 minutes, and mounted, to standardize comparison with dispersed spores.

Additional information on the sedimentology and paleontology is drawn from observation of laterally equivalent sediments exposed along the beach west of Gros Cap-aux-Os in Forillon National Park and in correlative units exposed on the south shore of Gaspé Bay (figure 11.2). Of particular interest is an extensive sandstone unit on the south shore with abundant plant remains permineralized by $CaCO_3$. This unit correlates with the Cap-aux-Os Member on the north shore (McGregor 1973, 1977), and in this chapter it will be referred to as the permineralized sandstone plant assemblage.

REGIONAL GEOLOGICAL SETTING

The Battery Point Formation is part of the Gaspé Sandstone Group, a coarsening-up clastic wedge developed on the southern edge of Laurussia, 10° to 20° S latitude in Emsian times (Scotese and McKerrow 1990). As a consequence, climate was presumably tropical (Witzke 1990). In the absence of reliable biological climate markers, sedimentology alone can contribute little to understanding of climatic conditions during the deposition of the Battery Point Formation, except that rainfall was moderate, judging from the absence of calcretes and sedimentological evidence for perennial rivers. The Battery Point Formation is exposed on both north and south shores of Gaspé Bay (figure 11.2); however, detailed correlation between strata on the two shores is currently not possible because of facies differences. On the north shore, three members have been recognized (Brisebois 1981). Our measured section is from the middle, mudstone-dominated Cap-aux-Os Member, as it is richest in plant fossils (figure 11.2). The age of the

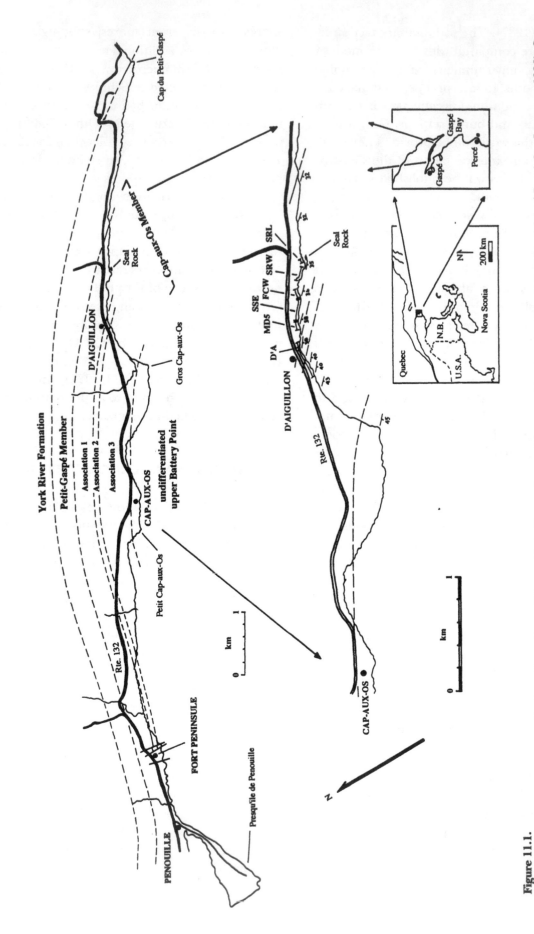

Figure 11.1.
Map of study area, depicting distribution of members of Battery Point Formation, and measured sections on the north side of Gaspé Bay (after Lawrence 1986). Sections referred to in text: *D'A*, d'Aiguillon; *MD5*, Mudstone 5; *SSE*, Sandstone E; *FCW*, First Cove West; *SRW*, Seal Rock West; *SRL*, Seal Rock Landing.

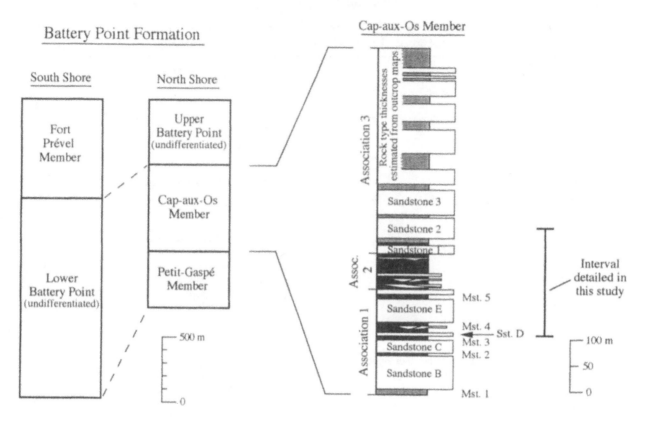

Figure 11.2.
Inferred correlation between exposures of Battery Point Formation on the north and south shore of Gaspé Bay, and generalized measured section on north shore of Gaspé Bay, based on the internal stratigraphy of Lawrence (1986) and the study described in this chapter. *Stipple*, red mudstone intervals; *black*, gray mudstone intervals.

Battery Point Formation is early Emsian to early Eifelian, based on data from brachiopods (Boucot et al. 1967) and spores (McGregor 1973, 1977; Richardson and McGregor 1986). The measured section falls within the early to late Emsian *annulatus–sextantii* spore assemblage zone of Richardson and McGregor (1986).

FLORISTICS AND TAPHONOMY

Approximately 18 species of embryophyte megafossils have been recognized within the Cap-aux-Os Member, including two possible bryophytes, two probable gametophytes, six zosterophyll species, one lycopsid, two species of rhyniopsid, and at least five species of trimerophytes (table 11.1). Details of the taxa may be found in the appendix to this chapter and the associated figures. Common taxa include *Psilophyton*, *Pertica/Trimerophyton*, *Huvenia*, *Crenati-*

caulis, *Sawdonia*, and *Sciadophyton*. Zosterophylls, even as fragments, can be identified to species level on the basis of their distinctive cuticle. Trimerophytes are readily recognized as a group from fragmentary vegetative material, although identification to species level requires relatively complete specimens. Two other organisms, *Prototaxites* and *Spongiophyton*, are also important constituents of the flora. We interpret both as eumycetous fungi. *Prototaxites* has recently been reinterpreted as a gigantic perennial fungal body analogous to a modern bracket fungus (Hueber 1996). Anatomically preserved specimens of *Spongiophyton* from the south shore permineralized sandstone assemblage, consisting of septate hyphal tubes surrounded by a thick resinous "cuticle" of apparently agglutinated hyphae, have been interpreted as lichens (Stein et al. 1993).

Consideration of taphonomy of fossil assemblages is essential to any reconstruction of their growth environment. Assemblages reflecting

**Table 11.1. Megaflora of Cap-aux-Os Member
and associated dispersed spore taxon (where known)**

Plant megafossil	Dispersed spore taxon
Embryophyta *incertae sedis*	
New genus A (?hepatic)	?
Sciadophyton spp.	N/A
New genus B	N/A
New genus C	?
Rhyniopsida	
Huvenia sp. nov.	*Retusotriletes* or *Calamospora*
"*Taeniocrada dubia*" (new genus D)	?
Lycophytina	
Drepanophycus spinaeformis Goeppert	*Retusotriletes* or *Calamospora*[a]
Crenaticaulis verruculosus Banks et Davis	*Retusotriletes* spp.
Renalia hueberi Gensel	*Retusotriletes* spp.
Sawdonia ornata (Dawson) Hueber	*Retusotriletes* spp.
Zosterophyllum sp.	*Retusotriletes* spp.
New genus E	?*Retusotriletes*[b]
"*Bathurstia*" sp. (new genus F)	?
Trimerophytes	
Psilophyton dawsonii Banks, Leclercq et Hueber	*Apiculiretusispora arenorugosa* morphon
P. forbesii Andrews, Kasper et Mencher	*A. arenorugosa* morphon
P. princeps (Dawson) Hueber[c]	*A. arenorugosa* morphon
Pertica varia Granoff, Gensel et Andrews	*A. arenorugosa* morphon
Pertica sp.	*A. arenorugosa* morphon
Trimerophyton robustius (Dawson) Hopping	*A. arenorugosa* morphon

?, unknown, or questionable; N/A, not applicable.

[a] Spores have been extracted from fertile specimens of *Drepanophycus spinaeformis* from New Brunswick (Li et al. 2000).

[b] Based on spore masses from a monodominant assemblage of new genus E.

[c] Reported by other authors but not found in this study.

minimal transport [autochthonous or parautochthonous—i.e., transported but still within life environment (Bateman 1991)] are essential to interpret growth habitat, habit, and associations. We judge that roughly 35 of the 58 megafossil assemblages in the Cap-aux-Os Member are either autochthonous or parautochthonous. Evidence of minimal transport includes evidence of anchorage, axes oblique or perpendicular to bedding plane, and preservation of complete axes with fine detail. Other authors have noted that early land plant assemblages are often preserved in parallel alignment along the bedding plane, as though plants were rooted at one end in the process of burial by the enclosing sediment (Andrews et al. 1977; Edwards 1979a). Observation of axes undulating in parallel suggests that they were anchored (presumably rooted) at one end, leaving the other ends free to become aligned to water currents. Regular spacing of aerial axes, especially characteristic of large trimerophytes, again suggests attachment to a horizontal stem or rhizome that is no longer preserved. That axes branch through the bedding plane is another strong indicator of *in situ* burial, since this position is not energetically stable under most conditions of transport. Examples of this include aerial axes of *Sawdonia ornata* at the type locality, and stalked gametophores of *Sciadophyton* and new genus B that extend through the bedding plane into superjacent sediments. The preservation of delicate structures such as distal branch tips, buds, and attached sporangia suggests limited transport, although the degree of resistance to fragmentation and degradation that these plants may have displayed is unknown because of the lack of taphonomic analyses of appropriate analogues.

Most important, among the major classes of tracheophytes identified from the measured section (lycophytes, trimerophytes, rhyniopsids), none appear to have exhibited significant differences in preservation potential.

SEDIMENTOLOGY

Facies Associations

Lawrence (1986) recognized three facies associations within the Cap-aux-Os Member (figure 11.2). The lowermost of these associations (Association 1) contains 4- to 60-m-thick, multistory sandstone bodies separated by thinner, red- and gray-mudstone-dominant intervals. Association 2 is dominated by gray mudstones at d'Aiguillon but also contains decimeter- to meter-thick sandstone sheets, and lenses and meters-thick, single-story sandstone bodies. Relatively coarser-grained, multistory sandstone bodies with uncommon, thinner red mudstone intervals characterize the uppermost association (Association 3) of the Cap-aux-Os Member. Paleocurrent indicators within the sandstones in the measured section at d'Aiguillon suggest unidirectional flow to the north and west, but a few strata indicate opposing paleoflow directions (Lawrence 1986; see following paragraph and figure 11.3). Most of the *in situ* and *in loco* plant megafossil horizons lie within the mudstone-rich portions of upper Association 1 and throughout Association 2 (figure 11.3), whereas highly fragmented assemblages occur in the multistory sandstone bodies.

Mudstone-dominant intervals contain a variety of distinctive features, including (1) meters-thick sandstone sheets and lenses that fine or coarsen upward, and that indicate paleoflow directions to the northeast to southeast, as well as to the north and west; (2) centimeter-scale, lenticular, wavy, and flaser bedding with wave-ripple and current-ripple marks (figure 11.4A); (3) centimeter- to decimeter-thick sandstone sheets containing wave-ripple marks, *Diplocraterion-*, *Skolithos-*, and *Phycodes*-like traces, and disarticulated cephalaspid fish skeletons (figure 11.4B,C); (4) desiccation-cracked mudstones with articulated lingulid brachiopods (figure 11.4D); and (5) dark gray shales and siltstones bearing acritarchs, small bivalves, or small, thin-shelled lingulid, orthid, and rhynchonellid brachiopods.

Overall Environmental Interpretation

The Cap-aux-Os Member is interpreted as fluvial and delta-plain deposits (Lawrence 1986; Lawrence and Williams 1987; Bridge et al. 1998; Griffing et al. 2000). Paleocurrents suggest that the rivers transported sediment seaward to the northwest. Whereas Lawrence (1986) interpreted most of the multistory sandstone bodies in Association 1 as deposits of high-sinuosity river channels, marine or marginal-marine trace fossils (i.e., *Diplocraterion*) in the upper portions of Lawrence's Sandstones B, C, and D (Association 1, figure 11.2) suggest close proximity to the sea. Furthermore, the single-story sandstone bodies in Association 2 strongly suggest deposition in the form of channel bars and fills very close to the tidal limit of tidally influenced channels (Bridge et al. 1998; Griffing et al. 2000). Features of the mudstone-dominant intervals of Associations 1 and 2 just listed also suggest a coastal plain setting, with levees and crevasse splays, freshwater marshes and lakes, lacustrine deltas, brackish marshes and interdistributary bays, and sandy and muddy tidal flats (Griffing et al. 2000). The easterly directed paleocurrents are interpreted as flood tidal currents. These types of marginal marine deposits have also been described from the Catskill clastic wedge in New York and Pennsylvania (Bridge and Droser 1985; Halperin and Bridge 1988; Bridge and Willis 1994).

Plant-Bearing Facies

Most *in situ* and *in loco* plants in the Cap-aux-Os Member lie within two types of facies sequence: (1) at or near the base of 1- to 2-meter-thick, coarsening-upward sandstone sheets or lenses (Facies A), and (2) within the uppermost portion of 4- to 5-meter-thick, fining-upward single-story channel sandstone bodies (Facies B). Examples of both types of sequence occur

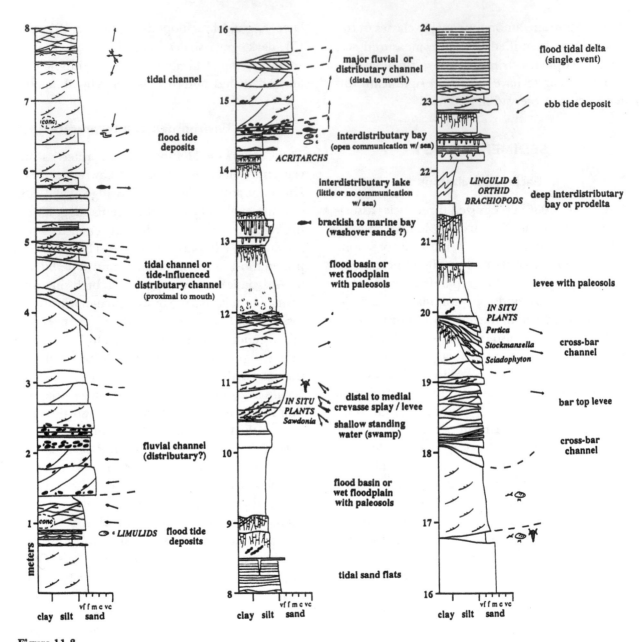

Figure 11.3.
Detailed log of base of Association 2 (top of Association 1 is at 3 m mark). See text for further details. See p. 187 for key to symbols used.

within the lowermost 19 meters of Lawrence's Association 2 section at d'Aiguillon: the *Sawdonia ornata* type (Hueber 1971) and the *Pertica varia* type localities (Granoff et al. 1976), respectively (figure 11.5). In addition, a distinctive transported assemblage of *Spongiophyton* and *Prototaxites* occurs within the fine-grained portions of the multistory channel sandstone bodies in Associations 1 and 3 (Facies C).

Description of Facies A

The type *Sawdonia ornata* horizon is situated at the base of a coarsening-upward sequence of dark green-gray, silty claystones, to light green-gray, fine-grained sandstones (figure 11.3A, meters 10 to 12), like several other *in situ* zosterophyll occurrences in Association 1 and 2. This sequence is 1.7 m thick at the base of the exposure and tapers laterally upslope to about 0.5

KEY

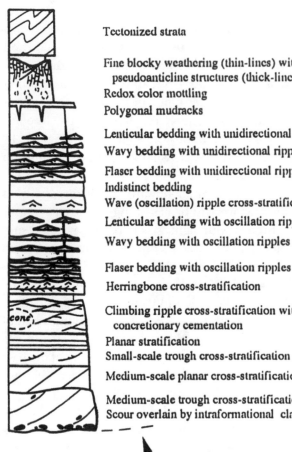

Tectonized strata

Fine blocky weathering (thin-lines) with
　pseudoanticline structures (thick-lines)
Redox color mottling
Polygonal mudcracks

Lenticular bedding with unidirectional ripples
Wavy bedding with unidirectional ripples

Flaser bedding with unidirectional ripples
Indistinct bedding
Wave (oscillation) ripple cross-stratified
Lenticular bedding with oscillation ripples

Wavy bedding with oscillation ripples

Flaser bedding with oscillation ripples
Herringbone cross-stratification

Climbing ripple cross-stratification with
　concretionary cementation
Planar stratification
Small-scale trough cross-stratification

Medium-scale planar cross-stratification

Medium-scale trough cross-stratification
Scour overlain by intraformational clasts

Large-scale cross-stratification
(lateral accretion) surface

Unidirectional paleocurrent orientation

Bidirectional paleocurrent orientation

Surface trails and tracks

Skolithos -like vertical burrows ("piperock")

U-shaped burrows

Phycodes- like stellate burrows

Burrow homogenized beds

Fish skeletal fragments

Pterygotus eurypterid fragments

Small bivalve - *Cyricardella*

Small bivalve - *Modiolopsis*

Irregular horizontal to sub-vertical organic
　bands (fossilized plant rhizomes?)
Large plant fragments

In situ plant megafossils

"Gametophyte establishment surfaces"

---

m thick (figure 11.5). Dusky red-gray siltstones and dark green-gray mudstones underlie the *Sawdonia* horizon. The dusky red-gray siltstones have blocky fabric with slickensides and weakly developed pseudoanticline structures (i.e., vertic paleosol features). These siltstones contain carbonized, boudin-like structures that lie parallel to one another on the bedding plane but produce short lateral branches that plunge several millimeters into the bedding plane (figure 11.6A). We tentatively interpret these as plant rhizomes. Overlying the rhizome-bearing beds are dark green-gray mudstones that contain weakly developed blocky fabric (also interpreted as paleosol features; see also Elick et al. 1998b). A discontinuous, extensively weathered claystone interval

occurs at the base of the plant accumulation; it resembles the underclays of Carboniferous cyclothems in North America and Europe (Dulong and Cecil 1989). Waxy, dark green-gray mudstones above the underclay contain dense accumulations of toppled *Sawdonia* axes. These mat-like accumulations of horizontal axes extend along the entire outcrop and create localized paper coals (figure 11.6B).

A sheet-like to lenticular, coarsening-upward sandstone body directly overlies the *Sawdonia* mat. The lowermost 30 cm of this body grade upward from mudstone to small-scale cross-stratified, fine-grained sandstone. Numerous anchored plant axes that protrude upward through the siltstones and sandstones indicate unidirectional flow to

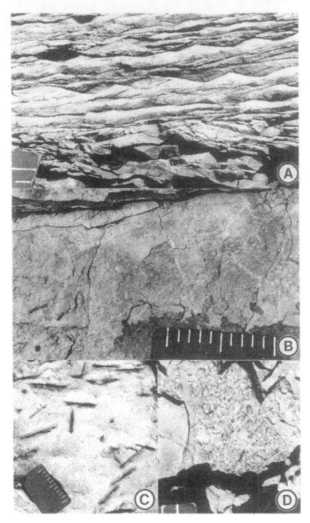

Figure 11.4.

A: Wave-rippled flaser-bedded sandstones, showing net migration direction (leftward) opposite to that of underlying medium-scale tabular cross-strata, Association 2 at d'Aiguillon. (1 cm scale, *lower left.*) **B:** Bedding-plane view of a muddy siltstone containing *Skolithos*-like vertical tube traces (white spots) and desiccation-crack polygons, Lawrence's Mudstone 4, Association 1 at exposures on the west side of Seal Rock Point (SRW) (1 dm scale). **C:** Upper surface of a white to buff medium sandstone bed displaying protrusive U-shaped burrow identified as *Diplocraterion*, Lawrence's Mudstone 4, Association 1 at First Cove West (FCW) west of Seal Rock Point (1 dm scale, *lower left*). **D:** Articulated but transported lingulid brachiopods forming one of several thin concentrations within gray mudstones with large polygonal mudcracks, lower portion of Lawrence's Mudstone 5 interval, Association 1 east of d'Aiguillon. (1 cm scale, *lower left.*)

the east and southeast (figure 11.6C). Some of the longer axes display several abrupt bends to a shallower angle at the base of sandstone beds,

before curving upward again. *In situ Sawdonia* axes terminate approximately 35 cm above the clay seam at the base. This coincides with the coarsest sandstone within the unit. Axes of a large trimerophyte with laterals stripped, and fragments of *Sawdonia, Crenaticaulis, Zosterophyllum* sp., *Drepanophycus,* and other taxa are common within the upper portion of the coarsening-upward sequence. These very fine to fine-grained sandstones are dominated by climbing-ripple cross strata (angle of climb up to 13°).

The coarsening-upward sandstone body is directly overlain by (1) fining-up, dusky red-gray siltstones with light green redox mottling and blocky fabric; (2) dark green-gray mudstones with weakly developed blocky fabric; (3) decimeter-thick beds of dusky red and green-gray siltstone featuring *Skolithos* and U-shaped traces, and disarticulated fish bone; and (4) a 32-cm-thick bed of dark gray shale with wave-rippled lenses of siltstone, containing numerous acritarchs at the base and millimeter- to centimeter-sized bivalves (?*Cypricardella*) throughout (figure 11.3).

We have identified at least 10 additional *in situ* plant horizons associated with this type of coarsening-upward sequence. Like the type *Sawdonia* sequences, most of these coarsening-upward sheets and lenses vary from decimeters to meters in thickness. One such sequence in Mudstone 4, Association 1 (see figure 11.2) has a triangular cross section. The upper portions of some of these sequences also include features not displayed in the *Sawdonia* example, such as meter-scale channel forms, wave ripple cross-stratification, and extensive disruption by burrows.

Interpretation of Facies A

Environmental interpretations of the lower d'Aiguillon section are provided with the stratigraphic log (figure 11.3). Basal mudstones display features typical of floodbasin muds with weakly developed paleosol horizons (e.g., vertisols, inceptisols, and entisols: see also Elick et al. 1998b). The underclay and waxy claystone underneath the *Sawdonia* mat and other plant assemblages most likely represent swampy gley

Figure 11.5.
Facies A and B at d'Aiguillon. Outcrop of Facies A siltstone/sandstone beds associated with *Sawdonia ornata* type locality (*center*), overlain by Facies B sandstones associated with the *Pertica varia* type locality (*upper left*). [2 m ranging pole
(*far left*) for scale.] This and other photographs of outcrops are rotated so that beds appear to lie horizontal; actual
dip about 30°.

soils with the O horizon preserved (Burrman
1975; Retallack 1990). The overlying wedge-
shaped to lenticular, coarsening-upward sequence possibly represents the episodic progra-
dation of a crevasse splay/levee into an adjacent
floodbasin, which would have buried plants in
place. An alternative origin, in view of the easterly directed paleocurrents, may be prograda-
tion of a coastal washover. At the *Sawdonia* site,
the bends and successive upward curvature of
axes through the strata point to persistent
growth through several episodes of burial. Wave-
rippled and/or burrow-modified beds in other
coarsening-upward sequences of Facies A imply
standing water for longer periods of time after a
flood event. The overlying mudstones represent
deposition of floodbasin muds punctuated by
brief incursions of brackish or marine standing
water. The dark gray shales at the top of the mudstones overlying the *Sawdonia* assemblage (figure
11.3, 11 to 14 m) are interpreted as interdistributary bay deposits with limited transport of
coarser sediments during storms or river floods.
The brief appearance of acritarchs in these dark
gray shales indicates a short-lived marine incur-
sion into the muddy bay.

Facies B: Type *Pertica varia* Site

In situ accumulations of *Sciadophyton*, *Huvenia*,
and the type occurrence of *Pertica varia* (Granoff
et al. 1976) are found within the uppermost
beds of the 5.5-meter-thick, fining-up channel
sandstone body directly above the type *Sawdonia*
sequence (figure 11.3, 14 to 20 m. The basal
medium-grained sandstones contain abundant
intraformational mudclasts and cephalaspid
and acanthodian fish bone fragments. Most of
the sandstone body comprises a single set of
large-scale inclined strata (i.e., a single story),
where each decimeter- to meter-thick, large-
scale stratum fines upward and laterally as it
climbs to higher levels in the sandstone body.
Associated with these fining trends within the
large-scale inclined strata, medium-scale trough
and planar cross-stratifications grade into small-
scale trough cross-stratified fine-grained sandstones (figure 11.5). Paleocurrents are directed
to the north in the lower half of the sandstone
body.

The upper portion of the sandstone body consists of lenticular beds of small-scale trough cross-
stratified, fine to very fine grained sandstones
with a dominant paleocurrent orientation to the

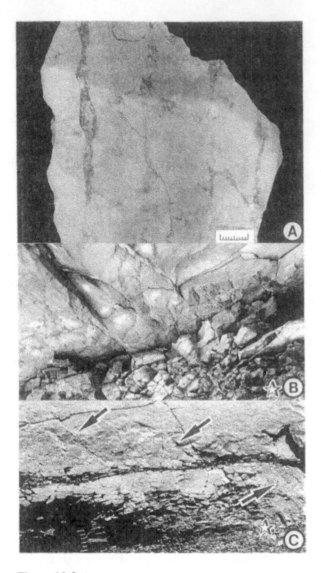

Figure 11.6.

A: Subparallel boudin structures covered with a thin carbonaceous film, preserved parallel to the bedding plane and bearing short lateral branches that plunge through the bedding plane, interpreted as rhizomes. Stratigraphic location of specimen is noted in figure 11.5 ("rhizome" bed). (1 cm scale, *lower right.*) **B:** Oblique underside view of basal sandstone bearing *Sawdonia ornata* at type locality. Note the subparallel axes of *Sawdonia* protruding upward through sandstone (toward SE, *upper left*). Base of plant occurrence highlighted by star. (Frame of photograph approximately 1 m across.) **C:** The same interval as shown in (B) from a cross-sectional view. Base of plant occurrence in claystone is highlighted by star. Notice traces of axes (*arrows*) bend to the left (SE) above a thin mud drape. (5 cm scale, *lower left.*)

east. These fine-grained sandstones also contain abundant fragmentary zosterophyll and trimerophyte axes, as well as disarticulated modiolopsid and mytiloid bivalves and fragmented

eurypterid cuticle. Small channel forms (40 to 80 cm deep, 1 to 3 m wide) transect the small-scale cross-stratified sandstones in the uppermost part of the sandstone body. Channel fills typically contain very fine to fine-grained sandstone beds or dune-shaped lenses separated by medium green-gray shaly partings that are often occupied by *Sciadophyton*, and more rarely, by *Huvenia* (figure 11.7A). Plant axes bend in a preferred easterly direction. Large axes of *Pertica varia* also appear to be anchored in mudstone and are toppled in the same preferred easterly orientation.

The sandstone body is directly overlain by dusky red-gray siltstones containing redox mottling and desiccation cracks, and by dark gray to dusky red-gray mudstones with blocky fabric and pseudoanticlines. A strongly tectonized black to dark gray shale containing acritarchs, lingulids, and poorly preserved articulate brachiopods caps the sequence (figure 11.3, 21 to 22 m).

Facies B: Pyritic Sandstone at Seal Rock

A similar fining-upward sandstone sequence crops out on the east (SRL) and west (SRW) sides of Seal Rock Point and along the west side of the first cove to the west (FCW) (see figure 11.1). Like the type *Pertica varia* occurrence, *in situ* and *in loco* plant occurrences in the pyritic sandstone are located in the upper finc to very fine grained sandstone beds capping a channel-form sandstone body. *Drepanophycus, Pertica* and *Trimerophyton, Huvenia, Crenaticaulis,* and *Sciadophyton* occur on shaly drapes separating small-scale cross-stratified sandstone beds within small channels. *Sciadophyton* is undoubtedly *in situ;* its radially directed axes often display evidence of current reorientation in a preferred direction (figure 11.7B). Other plants within this sequence are most likely *in loco;* growing along channel margins, they were dissociated from the substrate and transported a short distance as attached clumps, often with axes displaying current reorientation (figure 11.7C).

The pyritic sandstone sequence is overlain by heterolithic mudstone and sandstone sheets that vary considerably in thickness among the

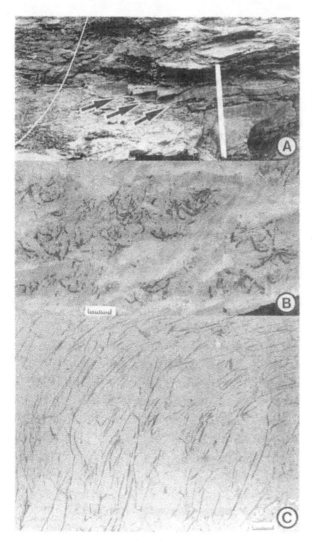

Figure 11.7.
A: Uppermost portion of single-story channel sandstone at the *Pertica varia* type locality. *Sciadophyton* occurs on clay drapes within shallow medium-scale cross-strata (*arrows*) (1 dm color bands on the ranging pole for scale). **B:** Plan view of *Sciadophyton* concentrated on lee face of small dune form, upper portion of pyritic sandstone, First Cove West. Note that axes, normally arranged in a radial pattern, show preferred orientation in the direction of the current activity that buried them (1 cm scale, *lower left*). **C:** Axes swirled in response to current action (?*Renalia* or ?*Eogaspesiea*) from Facies B, upper pyrite sandstone, Lawrence's Mudstone 4, Association 1, Seal Rock Point (1 cm scale, *lower right*).

three sections. These strata display wave-rippled flaser to lenticular bedding and desiccation cracks (see figure 11.4B). Centimeter- to decimeter-thick sandstone sheets within this interval display small-scale cross-stratification with wave ripples and abundant U-shaped traces like *Diplocraterion* (figure 11.4C) and densely

packed *Skolithos* traces. Lingulid-rich siltstones are also common.

Facies B: Type *Renalia hueberi* Site (Fort Péninsule)

A variant of Facies B is exposed within a red mudstone interval of Association 3 near Fort Péninsule, west of d'Aiguillon (figure 11.1). *In situ* axes of *Renalia hueberi* (Gensel 1976) occur in a 3-m-thick channel fill in the uppermost portion of a 6-m-thick, fining-upward sandstone body. Unlike the *Pertica varia* and pyritic sandstone sequences, this fining-up sequence exhibits an abrupt change from medium- and fine-grained sandstone with medium-scale cross-stratification to 1- to 3-cm-thick alternations of purple-gray, small-scale cross-stratified coarse siltstone and mudstone in the channel fill.

Anchored *Sciadophyton* occurs on the top surface of almost every coarse siltstone bed in the channel fill. Elongate axes of *Renalia hueberi* protrude up through the intercalated mudstones, many displaying a weak preferred orientation to the north or northwest, but commonly reoriented to the east or northeast at the top of each mudstone interval, directly below the next siltstone bed. The brick red–brown mudstones above the sandstone body have extensive disruption and fine blocky weathering, as well as redox color mottling, slickensides, and pseudo-anticlines. No beds in this sequence contain invertebrate concentrations or traces commonly associated with marine environments.

Interpretation of Facies B

Facies B strata represent laterally migrating channel-bar and channel-fill deposits, and each set of large-scale inclined strata represents deposition from a discrete flood event (Griffing et al. 2000). At the *Pertica varia* type locality, the upper fine-grained sandstone strata and small channel fills containing *in situ* plants represent bar-top and cross-bar channel environments. Burial of plants by flood flow deposits was followed by establishment of plant communities in low, wet areas. Dune surfaces with mud drapes within cross-bar channel fills display little evidence of long-term exposure (e.g., desiccation cracks or extensive burrowing), and in places, they display small wave ripples, indicating very shallow standing water. Paleocurrent and lithofacies variations within this type of sandstone body indicate the influence of tidal currents (see also Lawrence and Rust 1988). The lower parts of the channels were dominated by river currents and by ebb-tidal currents, whereas the upper parts were dominated by flood-tidal currents. This type of current segregation is typical of the strongly asymmetrical tidal currents expected in the channels near the tidal limit of estuaries and tidally influenced deltaic distributaries. The distribution of grain size and sedimentary structures is also typical of tidal channel bars near the fluvial–tidal transition: medium-scale cross strata limited to lower (subtidal) parts of bars; dominance of small-scale cross strata and relatively high proportions of mud in the upper, intertidal parts. The plant-bearing strata contain virtually no acritarchs or marine trace fossils, even though immediately adjacent strata do (see later discussions). Therefore, although the plants were probably buried *in situ* by deposits of tidal flood currents, there is no evidence of fully marine conditions in this facies.

In situ plant horizons in both the pyritic sandstone and *Pertica varia* type examples of Facies B are bounded by obvious tidal facies, which is once again suggestive of near-coastal, tidally influenced channels. The type *Renalia hueberi* site represents a muddy channel fill that formed after channel abandonment. The siltstone/mudstone alternations probably represent periodic flood deposition, when relatively weak unidirectional flows carried a small amount of fine-grained bedload through the abandoned channel. However, the *Renalia hueberi* channel fill is overlain by a paleosol, indicating a subsequent floodplain setting.

Description of Facies C:
Spongiophyton/Prototaxites Sandstones

Fragments of *Spongiophyton* and *Prototaxites*, along with fragmented plant "hash," commonly occur within the upper, finer-grained portions of 10- to 50-m-thick, very coarse to fine-grained sandstone bodies in Lawrence's Associations 1 and 3. These bodies are multistoried and the stories are bounded by relatively major erosional surfaces with extraformational conglomerates and intraformational breccias. The *Spongiophyton* assemblage occurs in two stratigraphic settings: (1) as paper-thin drapes along medium-scale cross strata, and (2) in centimeter-thick intervals of planar stratified and small-scale cross-stratified, very fine to fine sandstone capping some sandstone stories. Plant fragment concentrations commonly develop a limonitic stain with weathering and were presumably partially pyritized. This assemblage is notably absent from bar-top facies in mudstone-dominant intervals (such as Association 2), yet it is common in upper Battery Point strata exposed on the south shore of Gaspé Bay (e.g., the permineralized sandstone assemblage). Its frequency of occurrence increases upsection in the Cap-aux-Os Member, along with an increase in average grain size and abundance of extraformational pebbles above basal erosion surfaces.

Interpretation of Facies C

The multistory sandstone bodies of Associations 1 and 3 represent channel-bar and channel-fill deposits of the main river channels that migrated across the coastal plain. The fragmentary *Spongiophyton* assemblage was transported from fully terrestrial settings inland and was deposited in the channels from decelerating floodwaters.

PALEOBOTANY AND PALYNOLOGY: RESULTS

Facies–Plant Associations

Facies A, B, and C are each characterized by a distinctive plant association (table 11.2; note that not all plant occurrences could be categorized within one of these facies). Zosterophylls and lycopsids are especially characteristic autochthonous components of Facies A. Of 11

Table 11.2.   Facies/plant associations

	Taxon			*Facies*
	A	*B*	*C*	*unclassified*
Sawdonia ornata	4	1	—	—
New genus E (zosterophyll)	3	—	—	—
New genus B (gametophyte)	3	—	—	1
Drepanophycus spinaeformis	2	2	—	—
Renalia hueberi	—	3	—	—
Crenaticaulis verruculosus	—	4	—	—
Trimerophytes	—	7	—	1
Huvenia sp. nov.	—	4	—	—
Sciadophyton spp.	2	5	—	—
New genus A (?hepatic)	—	—	—	1
Spongiophyton sp.	—	—	15	—
Prototaxites sp.	—	—	5	—

examples of Facies A, eight are occupied by *Sawdonia*, new zosterophyll genus E (see appendix, figure 11.16C,D), *Drepanophycus*, or *Renalia*, and three others are occupied by new genus B, a putative gametophyte (see appendix, figure 11.12A, C–E). Trimerophytes occur in the coarser, transported fraction of Facies A, in the crevasse-splay and levee sandstones that bury autochthonous zosterophylls. We have noted four examples of this kind of occurrence, including the sandstone assemblage of the *Sawdonia ornata* type locality. Facies B is characteristically associated with trimerophytes, especially *Pertica* and *Trimerophyton*, as well as with *Huvenia* and *Sciadophyton*. Of 20 plant occurrences in Facies B, seven are of trimerophytes, four of *Huvenia*, four of *Crenaticaulis*, three of *Sciadophyton*, and one each of *Sawdonia* and *Renalia*. Facies C is characterized by *Spongiophyton* and occasional fragments or (rarely) large logs of *Prototaxites*. Dozens of examples of this association have been observed. Recognizable plant material is usually absent from this association, with the exception of the south shore permineralized sandstone assemblage, which contains abundant large-axis fragments of *Psilophyton* and occasional *Gothanophyton* (inferred anatomy of *Pertica/ Trimerophyton*), along with abundant permineralized *Spongiophyton* and occasional large fragments of *Prototaxites*. There is a correlation between size of *Spongiophyton* and *Prototaxites* fragments and inferred proximity to source sediments: the most complete pieces of *Spongiophyton* and large *Prototaxites* logs occur in the most

proximal sediments. Conversely, these elements are rare and very fragmentary in the mudstone-dominated facies.

Surfaces occupied by autochthonous gametophytes represent potential sites for establishment of stands of sporophytes, thus affording direct observation of microhabitat during this crucial stage of the life cycle. We term such gametophyte-occupied horizons establishment surfaces. *Sciadophyton* characteristically occurs on fine-grained mud drapes of migrating dunes, or in mud-draped troughs between ripple crests (figure 11.7B). The surfaces were likely wet but subaerial, and apparently represent an influx of sediment affording a fresh, unoccupied surface on which to germinate. Sites episodically shifted laterally, so recolonization of portions of the site after partial burial probably occurred frequently. Vegetative dispersal may have been effected by means of small, thin, round to oval structures resembling moss gemmae, which have been found associated with *Sciadophyton* at several localities (see appendix, figure 11.14E). Cover on *Sciadophyton*-bearing surfaces is relatively limited, ranging from 10 percent to 35 percent. New genus B tends to occupy the finest-grained portions of Facies A, on top of weakly developed gley (water-saturated) soils. None of the establishment surfaces display evidence of significant subaerial exposure; mudcracks and burrows, for example, are virtually absent. These gametophyte assemblages are generally monodominant: five out of six well-exposed *Sciadophyton* horizons, and three of four assemblages with new genus B, are monotypic. Two exceptions to this rule are noted later.

Stand Composition and Patch Size

Each species (or group of species) was tabulated as abundant (greater than approximately 10 percent) or rare (less than approximately 10 percent) in a given assemblage for all plant-bearing horizons (figure 11.8). Trimerophytes were not further subdivided because of the difficulty of distinguishing species. The megaflora of the Cap-aux-Os Member exhibits low equability (i.e., only a few taxa are abundant). Undoubtedly this is due in part to the lumping of trimerophytes, but even were they recognizable and included as individual species, the species distribution would still be skewed toward the left, reflecting the fact that most taxa in this flora are rare. Trimerophytes as a group are abundant at the largest number of sites; however, zosterophylls tabulated as a group (excluding *Renalia*) are common in nearly as many sites (figure 11.8). Among the zosterophylls, only *Sawdonia ornata* is abundant at a number of sites, whereas the rest tend to be uncommon, or at low abundance, in a given site (figure 11.8). The fact that species distribution at low abundance in a given site roughly mirrors that at high abun-

dance suggests that there is indeed no significant difference in preservation potential among the known fossil megaflora.

Most of the assemblages judged to be autochthonous or parautochthonous on sedimentological grounds are dominated by one taxon, commonly with minor fragmentary components of one or more additional taxa, the latter probably transported elements. However, a few *in situ* assemblages with more than one taxon have been noted. At one site, *Sciadophyton* co-occurs with new genus B, and at another site it intermingles with new genus A, a thalloid plant (see appendix, figure 11.12B). Assemblages that on sedimentary grounds appear to be allochthonous consist of axis fragments and may contain two to eight recognizable taxa in varying abundance. No correlation between taxonomic composition of transported assemblages and sedimentological or inferred paleoenvironmental differences was detected.

The lateral extent of autochthonous assemblages is hard to assess within the measured section because of the steep dip of beds, but several stands of zosterophylls appear to be quite large. For example, the autochthonous *Sawdonia ornata* bed at its type locality extends at least 25 m laterally up to the top of the cliff. The new zosterophyll genus E also forms a distinctive bed that extends at least 30 m up the cliff. Furthermore, this bed is exposed at two sites along the beach that are about 800 meters apart, suggesting a very extensive stand if the bed is in fact continuous. Similarly extensive trimerophyte stands have not been observed.

In Situ Spores and the Dispersed Spore Record

Although the parent plants of very few species of Devonian *sporae dispersae* are known, the converse (that the spores of very few megafossils are known) does not also hold true. Thirteen of 16 species described from the Cap-aux-Os Member have been at least tentatively matched with spores, either from published literature or in the course of this study (see table 11.1). Only one sporangium with spores has so far been

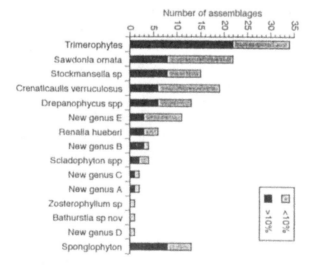

Figure 11.8.
Distribution of megafossil taxa by assemblage (*black*, abundant in a given assemblage; *gray*, less than approximately 10 percent in a given assemblage).

recovered from *Huvenia* sp. nov. These spores, possibly immature or abortive, are smooth and apparently thin-walled, with hints of *curvaturae perfectae* (see appendix, figure 11.14C,D). Spores of *Zosterophyllum* sp. (see appendix, figure 11.15D) and *Crenaticaulis verruculosus* (see appendix, figures 11.15E, 16A,B) are both smooth, with scattered, irregular gemmae, *curvaturae perfectae*, and a darkened patch of exine surrounding the central trilete aperture. Both can be compared to *Retusotriletes rotundus* and similar species. No fertile axes of new genus E have been found, but associated spore masses recovered from a monodominant assemblage of that taxon are also ascribable to species of *Retusotriletes*, such as *R. simplex* (see appendix, figure 11.16E,F). A few spores have been recovered from specimens closely resembling *Drepanophycus spinaeformis* from New Brunswick; these spores are smooth, trilete, and covered with irregular gemmae (Li et al. 2000) All species of trimerophyte collected from the Cap-aux-Os Member, including specimens from the inferred type locality of *Trimerophyton* (*Psilophyton*) *robustius*, have yielded spores referable to the *Apiculiretusispora arenorugosa* morphon (McGregor and Playford 1992). This species complex displays relatively uniform ultrastructure at the SEM level, with a variably attached exoexine comprising an irregular reticulum topped by biform or multiform elements (see appendix, figure 11.17C,D).

The dispersed spore record of the Cap-aux-Os Member displays much greater diversity in both species and morphotype compared to the megafossil record. McGregor (1977) reports 53 species from within the Cap-aux-Os Member. However, direct comparisons of dispersed spore and megafloral species are misleading, because dispersed spore taxonomy is notoriously inflated by reliance on variable, often taphonomically induced characters. A more reliable measure of diversity is number of distinct morphotypes, defined as distinct morphological forms that display little or no intergradation. By this measure, at least 25 distinct morphotypes, in addition to the *Retusotriletes/Calamospora* and *Apiculiretusispora* forms, are recognizable in the Cap-aux-Os Member (Hotton, unpublished data). With few exceptions, these forms exhibit a pattern of sporadic abundance or are very rare throughout. In contrast, the *Apiculiretuspora* complex comprises from 50 to 99 percent of a given sample, and species of *Retusotriletes* make up from 5 to 20 percent of a given sample, both forms displaying a high degree of constancy from sample to sample (Hotton, unpublished data).

Acritarchs comprise a small but environmentally important component of the Cap-aux-Os Member palynoflora. Approximately 10 species have been identified, including species of *Veryhachium, Helosphaeridium, Micrhystridium, Multiplicisphaeridium,* and *Gorgonosphaeridium.* Most acritarchs are thought to represent cysts of marine phytoplankton (Strother 1997); recent geochemical analyses suggest that many may represent dinoflagellates (Moldowan and Talyzina 1998). They are absent from horizons with autochthonous or parautochthonous plant fossils, but they are moderately abundant in horizons with marine indicators such as brachiopods (e.g., figure 11.3, 21.5 m). Acritarch species diversity is characteristically low in a given horizon within the Cap-aux-Os Member.

DISCUSSION

Niche-Partitioning and Life History Strategies

Sedimentological evidence suggests that plants within the Cap-aux-Os Member displayed incipient niche-partitioning along clade divisions (phyletic niche specialization of DiMichele and Phillips 1996). Many zosterophylls, including *Renalia,* preferentially occupy fine-grained, dysaerobic facies (also noted by Gensel 1986), characteristic of low-energy, water-saturated habitats such as marshes and backswamps (figure 11.9). In some cases, these environments display some organic accumulations. Trimerophytes, *Huvenia,* and *Sciadophyton* preferentially occupy coarser-grained facies suggestive of higher energy, more ephemeral habitats, such as upper bar deposits and cross-bar channels.

Figure 11.9.
Landscape reconstruction of Facies A. Swampy flood basins are covered with monotaxic stands of zosterophylls (e.g., *Sawdonia ornata*). Plant density and branching are illustrated as sparser than the likely actual habit, to demonstrate patterns of vegetative propagation and burial. A flood event (*in background*) initially buried plants with silts and fine sands (distal crevasse-splay deposits), followed by coarser sands containing large fragments of channel-bar and nearbank plants (proximal crevasse-splay deposits). Partially buried plants recovered and continued growing between flood episodes. See text for further details. Drawing by D. H. Griffing.

The occurrence of trimerophytes in the coarser fractions of Facies A suggests that these plants were growing near channel margins and were transported into the flood basin during flood events (figure 11.10).

The Seal Rock area represents a cross section of fluvial environments, with outcrops east of Seal Rock (SRL) representing backswamps, Seal Rock itself representing a channel, and exposures west of Seal Rock (FCW) representing near-channel overbank environments. Plants are distributed systematically among these environments. Large trimerophytes are most abundant in channel and channel margin sediments at Seal Rock. Here also occurs the highest diversity of plants in the Cap-aux-Os Member, where a variety of transported, autochthonous, and parautochthonous plants occur in localized lenses within the channel complex. Trimerophytes, *Huvenia*, and *Sciadophyton*, along with *Crenaticaulis*, are prevalent west of Seal Rock channel in strata that represent habitats slightly farther from the channel margin. Those sites representing backswamps at Seal Rock Landing are dominated by *Crenaticaulis*, *Psilophyton forbesii*, and *Sawdonia*.

Figure 11.10.
Landscape reconstruction of Facies B. Habitat illustrated consists of point bar along a delta distributary channel influenced only rarely by storm tidal transport. *Sciadophyton* occupies mud drapes in the swales of small dunes within a crossbar channel (*center foreground*) between flooding events. Episodic floods reorient free axes in downcurrent direction and bury plants in dune sands. Higher portions of the point bar and banks are inhabited by large trimerophytes (*right foreground*), which are toppled and buried by rippled and sheet sands in episodic floods. See text for further details. Drawing by D. H. Griffing.

Niche partitioning correlates with clade-level differences in inferred life history strategies and in morphology. Zosterophylls, although common overall, rarely occur in a fertile state within the Cap-aux-Os Member. Overall, only 5 out of 60 sites contain *any* fertile zosterophylls, excluding *Renalia* (figure 11.11). If numbers of fertile axes are counted, this disparity is increased by orders of magnitude. For example, in the laterally extensive stand of *Sawdonia ornata* at the type locality, only a handful of specimens have been found with sporangia. Given their low sporangial production, most of these zosterophylls were probably in a vegetative state for most of their life cycle. Furthermore, many zosterophylls retained the capacity to continue vegetative growth after sporangial production. The characteristic lateral and terminal circinate buds of zosterophylls suggest a stored potential for extensive vegetative growth within the constraints of a determinate growth system, which in turn suggests both dependence on vegetative growth to establish large stands and a capacity to remain in a site over an extended interval. That zosterophylls sometimes formed extensive stands is further evidence of their capacity for extended vegetative growth. The types of environments that zosterophylls preferentially occupied appear to have been sufficiently stable to allow some accumulations of organic matter. In addition, the xeromorphic aspect of their cuticle (tough, thick cuticle, sunken stomata) likely reflects oligotrophic

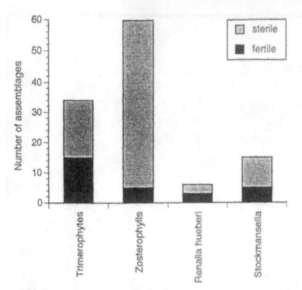

Figure 11.11.
Presence of fertile and sterile specimens in a given megafloral assemblage, by major taxonomic group (*black*, fertile; *gray*, sterile). A single fertile specimen in a given assemblage is sufficient to be included in the fertile category.

conditions in dysaerobic, water-saturated habitats, rather than xeric conditions.

In contrast, the morphology of trimerophytes suggests a more opportunistic and ephemeral life history strategy. Trimerophytes commonly occur in a fertile state, with about 45 percent of the assemblages containing at least some fertile axes (figure 11.11), suggesting that a significant part of their life cycle was spent in a fertile state. This in turn suggests that they grew rapidly and achieved reproductive maturity quickly. Their thin cuticle, which displays poor resistance to chemical oxidation in contrast to zosterophylls, is also consistent with rapid growth (although many other factors also influence cuticle thickness). The terminal sporangia of trimerophytes, in which apical meristems were apparently converted to fertile structures, suggests that aerial shoots at least were monocarpic. For example, in complete specimens of axes from the presumed *Trimerophyton robustius* type locality, both terminal apex and lateral branches terminate in a corymb-like arrangement of sporangia, all apparently at an equal stage of maturity (see appendix, figure 11.17B). It appears likely that the aerial shoots of trimerophytes completed their life cycle with a burst of reproduction, although they may have been produced season-

ally from a perennial rhizome. Rapid growth and abundant sporangial production would have been a selective advantage in the kinds of ephemeral environments that trimerophytes occupied. Sedimentological evidence also supports the inference of rapid development: Surfaces on which these plants apparently were anchored display evidence of reactivation during floods, suggesting that these plants may have had to complete their life cycle within a short time, perhaps as little as a few months. Although it may appear paradoxical that the tallest early land plants were fast growing and short-lived, the main selective force for height may have been to broadcast spores into the air stream, rather than for light interception (Tiffney 1981).

Not all plants of the Cap-aux-Os Member follow this simple model. *Renalia* occurs in the same kind of low-energy environments as zosterophylls, but it is comparable to trimerophytes in the frequency with which it is found fertile (figure 11.9). However, in other respects, it resembles zosterophylls in bearing sporangia laterally (on small lateral branches), and in possessing elongated axes with numerous "branch buds" suggestive of a clambering habit. *Huvenia* sp. occupies higher-energy environments and displays moderate levels of fertility (figure 11.11), but it bears sporangia on small lateral branch spurs; thus at least the potential for continued growth after spore production remains. *Crenaticaulis verruculosus*, a zosterophyll, characteristically occurs in Facies B; however, it tends to occupy the quiet water portions of upper bar environments.

Zosterophylls and trimerophytes occupy two of the three vertices in the life history model of Grime (1977). Zosterophylls fit the model of stress, or S-selected, species, occupying physiologically stressful habitats (for example, dysaerobic or nutrient-poor sites), whereas trimerophytes are ruderal, or R-selected, species, occupying sites subject to high levels of disturbance and displaying high reproductive output and an inferred short life span. Ruderal selection, suggested by the small size and terminal sporangia of the earliest polysporangiophytes, may well represent the ancestral embryophyte life history strategy (Niklas et al. 1980; Gray 1984). In contrast, stress selection may be a somewhat later development, involving acquisition of physiological and morphological traits for extended site occupation and resource acquisition (Beerbower 1985). Certain zosterophyll characters, such as resistant cuticle, continued growth of the axis after sporangial formation, and copious production of branch buds, may be interpreted as adaptations toward stress tolerance. These traits may, in fact, represent adaptations to hydric habitats, which sedimentological evidence suggests may have evolved very early within the lycophyte clade.

The dispersed spore record might be expected to mirror the megafloral pattern, but preliminary cluster and principal components analyses of the palynoflora have yet to reveal a clear pattern. This is probably because trimerophyte spores swamp practically every other form in almost all samples. Many of the dispersed *Retusotriletes* type spores probably also represent the inner bodies of *Apiculiretusispora*, since only a small fraction of dispersed *Retusotriletes* species closely resemble *in situ* zosterophyll forms. Variation probably occurred on a very small scale (meters to tens of meters), due to the low stature of the plants, and any consistent pattern is likely erased through transport. Likewise, transported megafloral assemblages display purely random variation, probably for the same reason.

Monodominant Assemblages

Megafloral data from this study reaffirm observations by many others that early land plants tended to form monodominant stands (Bateman 1991; DiMichele et al. 1992). Furthermore, some evidence from *in situ* gametophytes (primarily *Sciadophyton*) suggests that monodominance begins at the gametophyte stage; however, the sample is too limited to rule out the possibility of inter gametophyte competition resulting in sporophyte dominance (Edwards and Davies 1990). In any case, the tendency toward monodominance in early land plants is not a necessary attribute of "primitive" embryophytes, as witness the mixed species assemblages of modern moss microcommunities. Several factors could be controlling establishment in early land plants. For example, rapid clonal growth from a single or limited number of spore germination events may characterize the type *Sawdonia* locality, whereas rapid reoccupation of partially covered stands appears to occur at many of the Seal Rock localities, where successive horizons are filled with the same plants. In both cases, rapid mat-forming growth that excludes other species is the mechanism that results in monotypic patches. However, extreme patchiness of available habitats, rendering occupation by more than one species at a time unlikely, may also have been a factor, although probably less by the late Emsian. Monotypic assemblages, especially clones where the plants are genetic copies of one another, would presumably have experienced little intra- and interspecific competition (Tiffney and Niklas 1985; DiMichele et al. 1992). Despite this, land plants had by Emsian times apparently begun to divide the landscape, whether driven by chance or by competition at the margins of patches.

Coastal versus Fluvial Habitats

At first glance, the absence of embryophyte fossils and the predominance of *Spongiophyton* and *Prototaxites* in large fluvial channels appear to support the view that early land plants were restricted to coastal environments, with *Spongiophyton* and *Prototaxites* forming a transported assemblage from near-channel sites upstream. However, limited evidence from a few channel palynoflorules, as well as the south shore permineralized sandstone assemblage, suggests the presence of at least some tracheophytes in fully fluvial environments upstream, notably *Psilophyton* and other trimerophytes. Several reasons for the apparent rarity of embryophytes from such environments may be adduced. It is possible that they were in fact rare and patchily distributed on the landscape, therefore rarely preserved. However, we think it more likely that fluvial environments were much less conducive to plant preservation, as a result of such factors as high oxidation levels, and especially to the susceptibility of herbaceous plants to fragmentation under the high-energy regime in which these channels were deposited.

The sedimentological context of *Spongiophyton* and *Prototaxites* alone, regardless of their taxonomic affinities, unequivocally establishes them as terrestrial organisms with a marked preference for fully fluvial, rather than coastal, environments. Thick lenses of *Spongiophyton* on channel dunes suggest that it was growing in abundance along the channel margin. The fact that the larger fragments of *Prototaxites* occur in more proximal channels also strongly suggests that it was growing in far-inland sites and was occasionally rafted downstream into more coastal environments. The *Spongiophyton/Prototaxites* assemblage is puzzling not for its occurrence in fluvial channels but for its absence from plant-rich coastal environments. If these organisms were fungi, one might expect that they would find such nutrient-rich habitats congenial. Hypotheses for their absence may be proposed: For example, they may have been unable to tolerate salinity fluctuations or have been unable to occupy the muddy, shifting surfaces of tidal flats and backswamps. *Prototaxites* was capable of attaining immense size, suggesting a life cycle measured in years rather than months; perhaps it occupied more stable environments such as the edges of large, inland lakes. However, such possibilities remain speculation until the life habitats of this organism are characterized.

Occupation of High-Stress Environments

Were early land plants restricted to moist, mesic habitats? Sedimentological evidence points to occupation by both sporophytes and gametophytes of fine-grained, water-saturated substrates. The need for water by gametophytes is a given, and indeed is often cited as the reason for restriction of early land plants to such environments. There is some reason to believe that this may not have been insuperable. For example, the occurrence of cuticle, stomata, and a rudimentary conducting system in *Sciadophyton* suggests that it was much more tolerant of water stress than free-living gametophytes of modern

plants; the same may have been true of other early land-plant gametophytes as well. Nevertheless, fine-grained substrates and at least ephemeral water availability was almost certainly a requirement for establishment of gametophytes, if not for long-term occupation by sporophytes. Many zosterophylls apparently preferentially occupied moist, fine-grained substrates, and sites occupied by trimerophytes seem to have been only marginally less wet and more energetic. However, the limitations of sedimentary evidence must be kept in mind: Wet, dysaerobic environments are precisely the sites where fossils are likely to be preserved, whereas xeric sites are generally poor prospects for preservation. Evidence of occupation of more xeric sites comes from indirect evidence such as root traces. Millimeter-wide, obliquely directed traces that dichotomize downward occur in some of the more oxidized paleosols of the Cap-aux-Os Member (see also Elick et al. 1998a; see chapter 13), suggesting that at least some plants could occupy more xeric sites. However, strong taphonomic bias toward preservation of water-saturated environments remains a serious obstacle to fully characterizing early land plant habitats. More data, especially from paleosols where trace evidence of root and rhizome activity is preserved, are needed before we can draw conclusions about the capacity of early land plants to occupy xeric habitats.

Good records exist for one type of stressful environment in the Cap-aux-Os Member, namely brackish and saline habitats. Salt tolerance is a derived trait among embryophytes, requiring physiological and morphological modifications, and it is especially rare among bryophytes and pteridophytes (Schuster 1966; Brown 1983). Thus, we might expect early land plants to occupy primarily or exclusively freshwater sites. Within the limits of resolution, this appears to be the case for plants of the Cap-aux-Os Member. Autochthonous and parautochthonous assemblages of both sporophytes and gametophytes occur in sites lacking clear marine fossil and sedimentological indicators. Conversely, only plant fragments occur in clearly tidal or marine facies. Acritarchs, the most sensitive indicator of salinity within our study area, are absent or very rare from autochthonous and parautochthonous plant horizons, although they are moderately abundant in brachiopod horizons (figure 11.3). Even in clearly marine horizons with-in the Cap-aux-Os Member, acritarch diversity is low, which is especially characteristic of nearshore marine habitats (Staplin 1961; Wicander and Wood 1997 and references therein). These data do not preclude the possibility that plants were at times exposed to brackish conditions, especially given that they occupied tidally influenced environments that would have been subjected to periodic inundation from the sea. Equivalent modern environments represent a mosaic of brackish and freshwater habitats that may be separated at a meter-level scale. It is difficult to determine from the sedimentological record whether plants tolerated occasional inundation of brackish or saline water, or were killed by them. However, there is no evidence that any species preferentially or consistently occupied identifiably brackish habitats, and indirect evidence suggests that they did not grow in such environments. This suggests that members of the Cap-aux-Os flora, at least, had not acquired the types of physiological and morphological adaptations necessary for occupation of saline habitats (see also Edwards 1980).

Apparent versus Actual Floral Diversity

The megafloral evidence implies a low-diversity flora of simple structure, dominated by a few common and many rare species that formed monodominant stands of limited extent and probably limited interaction. The megaflora and palynoflora both suggest that the Cap-aux-Os flora was of low equability (i.e., it was dominated by a few common taxa with a high percentage of taxa characterized by few or single occurrences). Although this pattern may approximate the true species distribution, it must be kept in mind that the megafloral record is biased in a number of ways; thus, intrinsic properties of the dominant clades may tend to overestimate their actual importance on the landscape. For example, trimerophytes appear

overwhelmingly dominant; however, their high spore productivity and their occupation of near-channel habitats may overestimate their actual abundance on the landscape. Likewise, many zosterophylls tend to occupy wetland sites likely to be preserved, and thus they may be overrepresented in the fossil record relative to their true abundance.

That the megafossil record is biased certainly comes as no surprise, given that numerous modern taphonomic studies have established the propensity toward preservation of plants growing near sites of deposition (Spicer 1989). However, the dispersed spore record suggests that the early land plant megafloral record is particularly deficient in capturing actual floral diversity. Comparison of *in situ* spores and *sporae dispersae* of the Cap-Aux-Os Member shows that the two dominant megafossil clades, trimerophytes and lycophytes, account for only a small fraction of the morphological diversity of the *sporae dispersae*. In this respect, the Cap-aux-Os flora mirrors the disjunction observed from many other localities of similar age (Gray 1984, 1993; Fanning, Richardson, and Edwards 1991; Edwards 1980, 1996), and it pinpoints a central problem in Devonian plant paleoecology. The diversity of the dispersed spore record strongly suggests that many forms, perhaps whole classes of plants, have thus far escaped sampling in the megafossil record. Some of the rarest unknown morphotypes may be derived from extrabasinal plants (Allen 1980; Gray 1984). However, many unknown spore types are sporadically abundant at individual horizons, suggesting that they occupied the same coastal fluvial environments as the megafossils (although probably drawn from a wider range of environments that includes habitats unsuitable for megafossil preservation). The parent plants of many of these unknowns may be of bryophytic grade (Richardson 1985; Edwards 1996). Their apparent absence in the mega-fossil record could be attributable to low preservation potential; however, another cause could be their very small size, which renders them easily overlooked. The discovery of small forms in the Cap-aux-Os flora through bulk maceration, such as new genus C, lends support to the latter explanation. The rarity of the dispersed spore morphotypes could be due to any of a number of factors. The parent plants might sexually reproduce rarely or produce fewer spores, or they may be short statured, so that their spores do not disperse widely, or they may be ephemeral, or rare and patchily distributed on the landscape. In any case, the apparently simple Early Devonian landscape was likely more diverse, with more complex kinds of interactions, than may be inferred from the megafossil record alone.

CONCLUSIONS

The Cap-aux-Os Member of the Battery Point Formation was deposited within a low-lying fluvial–deltaic coastal plain. Plant megafossils are restricted to a relatively limited number of wet environments. Indirect evidence suggests that plants probably occupied primarily or exclusively freshwater habitats. Autochthonous and parautochthonous assemblages of both sporophytes and gametophytes were for the most part dominated by one species, suggesting limited interspecific competition among early land plants. Nonetheless, sedimentological evidence supports an inference of clade-related niche-partitioning; many zosterophylls appear to have preferentially occupied dysaerobic, wetland sites within interdistributary basins, whereas trimerophytes and rhyniopsids apparently occupied more ephemeral, near-channel environments. Both sedimentological and morphological evidence suggest a stress (S-) selected life history strategy for zosterophylls, in contrast to ruderal (R-) or disturbance selection in trimerophytes. These observations are based on a small sample of plants from a limited geographic area and time interval. They provide sufficient data to frame hypotheses; whether these hypotheses hold true requires comparably detailed sedimentological analyses of other early land plant localities.

At least certain early land plants, notably trimerophytes, as well as the probably eumyce-

tous *Spongiophyton* and *Prototaxites*, occupied fully fluvial (riparian) environments, but the plant fossil record remains heavily biased toward low-lying coastal communities. The extent of plant cover at this time remains a difficult question to address. Nevertheless, we think it unlikely that early land plants were restricted to moist mesic sites by their free-sporing life cycle (Algeo and Scheckler 1998). Certainly many extant bryophytes and pteridophytes occupy extremely hostile habitats such as bare rock substrates, and hot and boreal deserts; bryophytes in particular are common in pioneer habitats. The need for moisture for germination and fertilization in free-sporing plants is short-lived, and sexual reproduction need take place only rarely as long as vegetative means of propagation are available. Extant free-sporing plants adapt to harsh conditions in a variety of ways, through rapid completion of the life cycle, physiological resistance to desiccation, dormancy, or reduced resource allocation (Gray 1985). Early land plants faced an environment that was at least as hostile as any modern landscape, without soil or plant cover other than algal and bacterial mats. It would be surprising if they had not developed extensive physiological and morphological adaptations to harsh conditions that would enable them to cover a large portion of Earth's surface by Emsian times.

The dispersed spore record reminds us that embryophyte diversity, and consequently ecosystem complexity, was likely much higher than is reflected in the megafossil record, a point made forcefully by Gray (1985). Furthermore, the rarity of many of these morphotypes does not necessarily signify that their parent plants were rare; even if their reproductive output was low, they may yet have been important players on the landscape. We emphasize the apparent bias of the megafossil record not as a counsel of despair but rather as a call to action. Continued search, especially through bulk maceration of fine-grained sediments (Edwards 1996), and ultrastructural study of dispersed fossil spores are just two potentially fruitful approaches to close the gap between the mega- and palynofloral record.

ACKNOWLEDGMENTS

This research was supported by grants from Scholarly Studies Program grant #1233S40G to FMH and CLH, from the Charles E. Walcott and Roland A. Brown funds, from the Smithsonian's Evolution of Terrestrial Ecosystems program, and the Center for Evolution and the Paleoenvironment (Binghamton University). DHG and JSB thank Anne Hull for assistance with drafting and Richard Jacyna for help with preparation of rock specimens. CLH and FMH thank Susann Braden and Walt Brown for assistance with SEM, Jean Dougherty of the National Type Collection, Geological Survey of Canada, Ingrid Birker of the Redpath Museum in Montreal, and Bill Stein of Binghamton University for access to specimens. CLH thanks Bill Stein, Dick Beerbower, and Conrad Labandeira for helpful discussions, Bob Harberts for technical assistance, and Bobbie Friedman and Dick Beerbower for their hospitality over the course of this study. We thank Pat Gensel and Dianne Edwards for the invitation to participate in this symposium and for their Job-like patience in awaiting the long gestation of this manuscript.

Appendix

Descriptions of Plant Megafossils

Brief descriptions of new taxa as yet unpublished, as well as our perhaps idiosyncratic treatment of certain existing taxa, are presented here. This list is not to be considered a formal taxonomic treatment. Taxa that appear to us to be well founded and well described in the literature are not discussed. Classification of major clades follows Kenrick and Crane (1997a). All figured and type specimens are to be deposited in the National Type Collection, Ottawa, Canada.

EMBRYOPHTYA *INCERTAE SEDIS*

New Genus A

The material consists of elliptical, dichotomously lobed, thalloid structures roughly 2–3 cm × 3–5 cm (figure 11.12B), with little cellular detail visible. In gross morphology it resembles a thalloid liverwort such as *Marchantia*.

New Genus B

Specimens are cespitose, consisting of isotomously branched axes about 1 mm wide, terminated by paired reniform "cups" (figure 11.12A,C). Each cup bears scattered small, round, or oval structures on short stalks (figure 11.12D,E), which may be antheridia. This taxon is under study by F. M. Hueber and C. L. Hotton; similar plants, but lacking the paired cups, have been illustrated by Remy et al. (1993) and are under study by P. Gensel and associates.

New Genus C

Only distal portions of the plant, bearing sporangial trusses, have been recovered. Sporangia are borne in a cymose fashion—that is, each sporangium terminates a branch, which undergoes two close dichotomies subterminally to give rise to another sporangium, and so on (figure 11.13A). Axes display no evidence of conducting tissue. Sporangia range from 0.2–0.5 mm wide and 0.5–1.5 mm long; they narrow to a thickened pore at the apex and display no evidence of a dehiscence structure (figure 11.13B). This plant may represent a new major clade of polysporangiophyte, or alternatively it could represent a primitive moss or hepatic (Hotton and Newton, unpublished data).

RHYNIOPSIDA

New genus D
("*Taeniocrada dubia*" Hueber 1983)

This plant consists of broad strap-shaped axes 0.9–2.0 cm wide, with thin but resistant cuticle, often unbranched, or anisotomously or rarely isotomously branched (figure 11.13C). The surface is covered with conspicuous oval stomata. Larger oval scars consisting of a raised ring surrounding a depression and a central papilla, approximately 1–1.4 mm wide by 1.5–2.0 mm long, occur on some axes ["areoles" of Dawson (1871)] (figure 11.13F). We interpret these larger scars as sporangial attachment scars

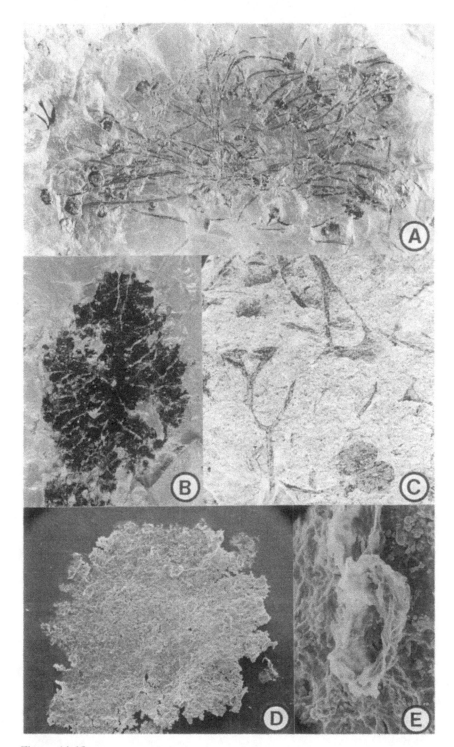

Figure 11.12.
A: New genus B. Probable gametophyte. Cap-aux-Os Member, Association 2. Compression of whole plant. Note cespitose habit and terminal "cups". DAA95-P7. ×
0.6. **B:** New genus A. Compression of possible hepatic. Cap-aux-Os Member, Association 2, DAA95-P7. × 1.2. **C:** New genus B. Compression/impressions of terminal cups. Note plan view of paired cups and lateral view of another specimen perpendicular to bedding plane. FMH 1967. × 0.6. **D:** New genus B. Upper surface of cup (SEM). Note scattered bowl-shaped structures on short stalks on surface, interpreted as possible antheridia. FMH 1967. × 15. **E:** New genus B. Detail of stalked structure (= ?antheridium) from another cup (SEM). FMH 1967. × 180.

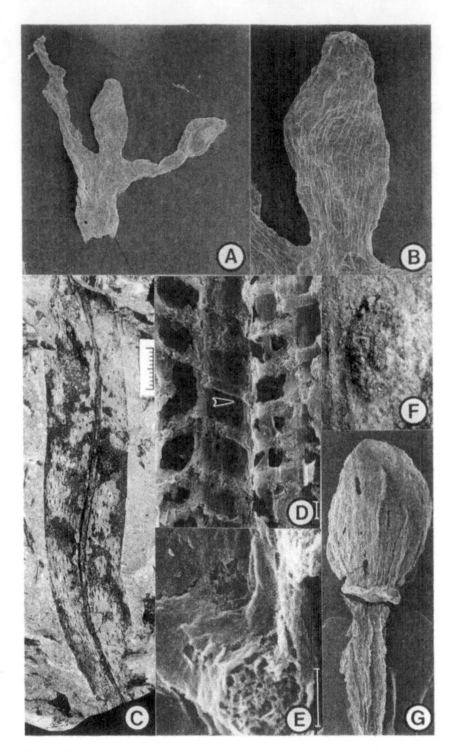

Figure 11.13.
A: New genus C (Embryophyta *incertae sedis*). Cap-aux-Os Member, Association 2, DAA95-P15. Distal sporangial truss, displaying "cymose" habit (SEM). × 60.
B: New genus C. Detail of central sporangium. Note apical knob with apparent pore (SEM). × 60 (B). **C:** New genus D, Rhyniopsida (= "*Taeniocrada dubia*" Hueber 1983). Battery Point Formation, Gaspé Bay. Axis compression with permineralized vascular strand. Note oval pits (stomata). South shore of Gaspé Bay, approximately 1 km NW of axis of Cap Blanc anticline. GSC 6322. Scale, 1 cm. **D:** New genus D. S-type tracheids from specimen in (C). Note change in gyre of thickening on tracheid at left (SEM). × 400. **E:** New genus D. Detail of tracheid

because of their resemblance to such structures in *Huvenia* sp. nov. (compare figures 11.13F and 11.14B). Permineralized axes from the south shore of Gaspé Bay, of the same size and morphology but lacking sporangial attachment scars, have a terete central strand with S-type tracheids (figure 11.13D,E). The anatomy and attachment scars are evidence that this plant is a member of the rhyniopsid clade (Kenrick and Crane 1997a). This species is currently under study by F. M. Hueber and C. L. Hotton.

Huvenia sp. nov.

This plant consists of axes about 0.5 cm wide and up to 20 cm long, with distinct longitudinal plications (figure 11.13G). Sporangia are round to ovoid, borne on shortened branchlets, with thickened, radially elongate cells at the base, forming a distinctive scar, or "collar" (figure 11.14A,B). Spore masses clinging to one sporangium (figure 11.14C) are roughly 60 μm in diameter, smooth, more or less spherical, apparently thin-walled, and trilete, with hints of *curvaturae imperfectae* (figure 11.14D). This species is currently under investigation by P. Gensel and students.

Small, round to oval discs (figure 11.14E), sometimes paired, have been found closely associated with, and possibly attached to, axes of *Sciadophyton* sp. We hypothesize that these functioned as vegetative disseminules of these plants.

ZOSTEROPHYLLOPSIDA

"*Bathurstia*" sp. (New Genus F)

Two specimens were recovered from the Seal Rock area that are identical to plants under

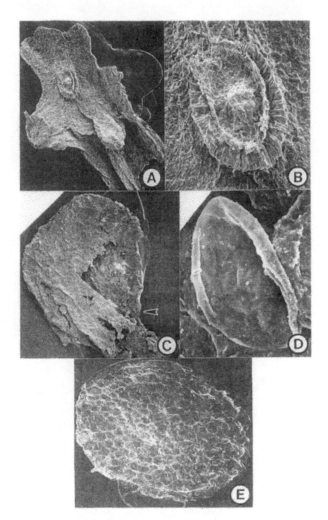

Figure 11.14.
A: *Huvenia* sp. nov. Cap-aux-Os Member, Association 1, FCW94-P15. Fragment of axis with sporangial bases borne on very short side branches (SEM). × 15. **B:** *Huvenia* sp. nov. Detail of sporangial base in figure 11.14A; compare with figure 11.13FA (SEM). × 47. **C:** *Huvenia* sp. nov. Sporangium with clumps of spores (SEM). × 25. **D:** *Huvenia* sp. nov. Proximal view of single spore from cluster near base of sporangium in (C) (*arrow*). Note indistinct trilete aperture and hint of *curvaturae* in upper part of photo (SEM). × 694. **E:** Small, flat disc (gemma-like structure), composed of probably no more than two cell layers, associated with *Sciadophyton* sp. Cap-aux-Os Member, Fort Péninsule, GSC 5411. × 70.

Figure 11.13 (*continued*)
from (D) (at *arrow*), showing alveolar internal layer and microporate covering (SEM). × 3,000. **F:** New genus D. Impression of oval scar on another axis: probable site of sporangial attachment. Cap-aux-Os Member, Association 1, Seal Rock area, pyrite sandstone. FMH 1967. × 18. **G:** *Huvenia* sp. nov. Note thickened "collar" of cells at base of sporangium and plicate axis. Cap-aux-Os Member, Association 1, FCW94-P15. × 11.

study from Campbellton, New Brunswick. These consist of isotomously branched axes, 3–6 mm wide, with two rows of short, acicular spines along the margins of otherwise smooth axes. Regions of tightly clustered sporangia are interspersed with vegetative regions (figure 11.15A). No spores have been recovered from any of these specimens. Although these specimens

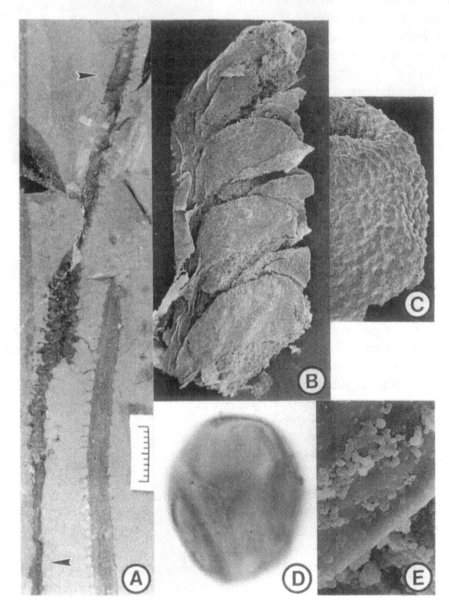

Figure 11.15.

A: "*Bathurstia*" sp. (new genus F) (impression). Note sporangia clustered in central portion of axis with vegetative portions above and below (*arrows*). A second vegetative axis displaying short narrow spines is to the right. Campbellton, New Brunswick. GSC 6267. × 1.4. **B:** *Zosterophyllum* sp. Cap-aux-Os Member, Association 2, 91-3z. Whole spike (SEM). Note alternating, bilateral position of sporangia. × 13.5. **C:** *Zosterophyllum* sp. Detail of central sporangium in (B). Note papillate thickenings on cell surface. × 90. **D:** *Zosterophyllum* sp. Proximal view of one of the spores freed from a sporangium by mechanical means. Note scattered gemmae, curvatural ridge in focus to left, and darkened central patch around trilete aperture. × 1,000. **E:** *Crenaticaulis verruculosus* spore, detail of spore ornamentation. Note irregular gemmae on both spore and sporangial surface (SEM). Compare to (D) and figure 11.16F. × 3,500.

were initially interpreted as a new species of *Bathurstia*, recent work by Kotyk (1997) has shown substantial differences in vegetative and reproductive morphology between the type species of *Bathurstia* and these new forms, suggesting that they should be placed in a new genus.

Zosterophyllum sp.

Our material consists of spikes of bilateral, alternating rows of sporangia approximately 2–3 mm wide (figure 11.15B). Tufts of narrow, isotomously branched axes approximately 1.5–2.5 mm wide are closely associated on the same bedding surface but not attached. Sporangia lack clear dehiscence lines, and the cells are distinctly papillate (figure 11.15C). Spores are irregularly gemmate, range from 50 to 100 μm in diameter, and conform to the *Retusotriletes* type, with distinct, narrow *curvaturae* paralleling the equator, and a darkened area surrounding the trilete aperture (figure 11.15D).

Crenaticaulis verruculosus Banks and Davis

A single fertile axis bearing three sporangia with abundant *Retusotriletes*-type spores clinging to the inner sporangial surface (figures 11.15E, 11.16A,B) adds new detail to this well-described species.

New Genus E

Axes are isotomously to anisotomously branched, 0.5–1.2 cm wide (figure 11.16C), some branches ending in circinate buds, with axillary "tubercles" just below each dichotomy (figure 11.16D). The cuticle is extremely tough, with regularly arrayed papillae forming a chevron pattern (figure 11.16D). Although no sporangia have been found attached, the aforementioned vegetative characters clearly place it among the zosterophylls. Reniform spore masses macerated from a monotypic assemblage of the plant are small (30–45 μm in diameter), with *curvaturae perfectae* and irregularly scattered gemmae (figure 11.16E,F). This species is currently under study by P. Gensel and students.

Eogaspesiea gracilis Daber

Extensive collections and maceration of material from the Cap-aux-Os Member have failed to recover any fossil resembling the description of *Eogaspesiea* [i.e., sparsely isotomous, narrow axes bearing single terminal fusiform sporangia (Daber 1960)]. Furthermore, new collections of material from the type locality of *Eogaspesiea gracilis* at Seal Rock have revealed sporangia and vegetative axes closely resembling *Renalia hueberi*. We offer the hypothesis that the original description of *Eogaspesiea* is in error, and that the two genera are similar or perhaps identical.

TRIMEROPHYTES

Plants attributable to this group are instantly recognizable even as fragments on the basis of (1) their paired, fusiform sporangia, often containing spores with an irregularly sloughing, apiculate outer exine, and (2) their rigid, often anisotomously branched axes with elongate epidermal cells, and (3) their thin, chemically unresistant cuticle. Distinguishing among species within this group is much more difficult. Three species of *Psilophyton* have been described from the Battery Point Formation: *P. dawsonii*, with smooth axes (Banks et al. 1975); *P. forbesii*, with ridged axes (Gensel 1979); and *P. princeps*, with stout, possibly glandular emergences capped by a cup-shaped structure (Hueber 1967). Only the first two of these taxa have been recognized in the course of this study.

Two genera of large trimerophytes with erect sporangia are usually recognized: *Trimerophyton*, with "trichotomous" branching (i.e., two close dichotomies) (Hopping 1956); and *Pertica*, with variably opposite-decussate, quadriseriate, or spiral branching (Kasper and Andrews 1972; Granoff et al. 1976; Doran et al. 1978). In our opinion, branching in this group is both far more complex than implied by these terms and highly variable across and within species. For example, exceptionally complete specimens of a large smooth trimerophyte, collected by D. C. McGregor in 1960, display opposite-decussate

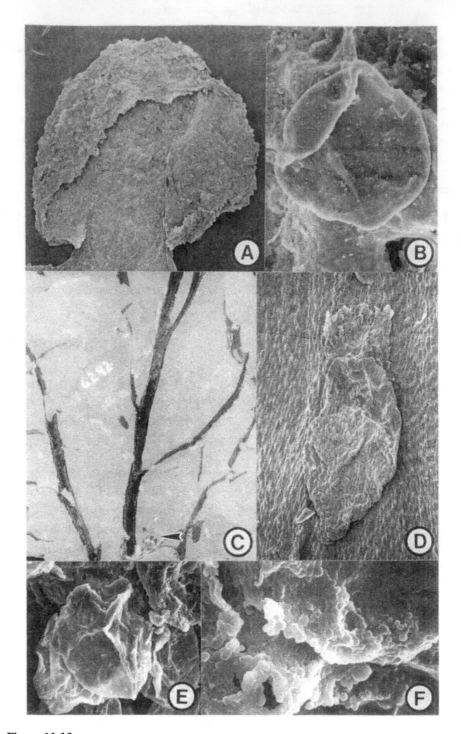

Figure 11.16.

A: *Crenaticaulis verruculosus* sporangium with spores, portion of fertile axis macerated from matrix. Cap-aux-Os Member, Association 1, Seal Rock Landing, SRL93-O17. Whole sporangium (SEM). Note valves of unequal size and papillate cells typical of *C. verruculosus*. × 22. **B:** *Crenaticaulis verruculosus*. Proximal view of *in situ* spore from (A) (SEM). Note curvaturate ridges visible near equator (*arrow*). × 1,000. **C:** New genus E (zosterophyll). Cap-aux-Os Member, Association 1, Seal Rock. Compressions of axes. Note isotomous branching and circinate bud (*arrow*). GSC 6292. × 0.6. **D:** New genus E. Axillary tubercle located laterally and just below dichotomizing axis (SEM). Note chevron pattern of papillate cellular thickenings. SRW95-P3. × 35. **E:** Proximal view of spore from spore mass isolated from monotypic assemblage of new genus E (SEM). Note long trilete ridges and hints of *curvaturae perfectae*. SRL93-D. × 1,000. **F:** Detail of ornamentation of several other spores from the spore mass in (E). Note irregular gemmae. × 3,300.

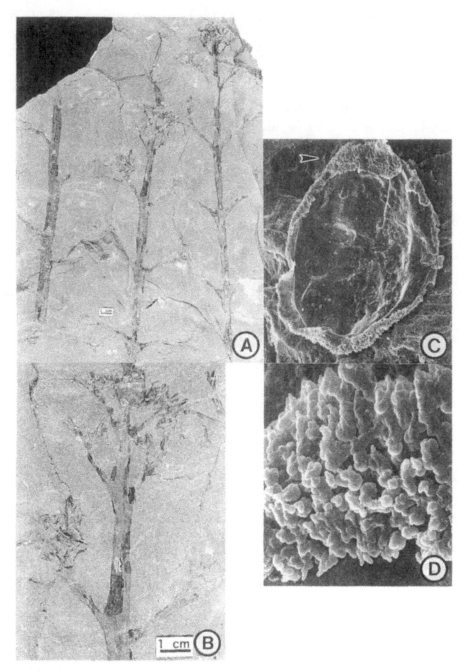

Figure 11.17.
A: Large trimerophyte from probable type locality of *"Psilophyton" robustius* Dawson 1871 (further details in appendix). Cap-aux-Os Member, Association 2, Fort Péninsule, GSC 5586. Slab with compressions of three axes, fragment of a once 1.5-m-long slab bearing complete axes, bases curved as though attached to a rhizome (scale, 1 cm). Note alternate lateral branches on specimen to left, opposite-decussate branching on specimens center and right, and sporangia terminating apex as well as lateral branches. × 0.3. **B:** Terminal portion of central axis in (A) (scale, 1 cm). Sporangia borne on lateral and terminal branches appear to be at the same stage of development. Object at top of photograph is an unrelated axis fragment. × 0.8. **C:** Proximal view of spore extracted from sporangium of axis from *Trimerophyton robustius* type locality (SEM). Note partially detached exoexine (which extends across entire distal portion of grain), exposing smooth endoexine and poorly defined trilete aperture beneath. × 650. **D:** Detail of ornamentation of spore in (C) (*arrow*). Note exoexine comprising more or less conate biform elements seated on very irregular reticulum. × 5,000.

and alternate branching side by side on the same slab (figure 11.17A). The site from which these specimens were collected is apparently the same site from which axes of "*Psilophyton*" *robustius* were collected (Dawson 1871: figure 138; also illustrated in Kräusel and Weyland 1961: pl. 4, figure 1). Opposite-decussate, alternate, and "trichotomous" branching may also be observed on different specimens of the same slab of *Pertica quadrifaria* (paratype, USNM 169002A) and in different specimens of *Pertica dalhousii* (GSC 54741, 54742). In the course of this study, we have maintained the generic distinction, applying the name *Trimerophyton* to those large trimerophytes with smooth axes and relatively lax fertile branchlets (figure 11.17A,B). This appears to be true of the specimens described by Hopping (1956), as well as the specimen in figure 138 of Dawson (1871), the upper portion of which is in the possession of the Redpath Museum (RM 12.529). We have applied the name *Pertica* to those specimens with sparse to densely scattered, small, acicular trichomes and tightly clustered fertile branches (where present). Slight differences in spore ornamentation visible at the SEM level (figure 11.17C,D) may also distinguish *Trimerophyton* from *Pertica*; however, many more specimens must be examined to determine whether these differences are consistent within genera. Nonetheless, specimens are difficult to assign to one genus or the other, and we believe that these genera require thorough revision to determine whether, in fact, they are distinct, and what characters may be useful to consistently distinguish species.

12

Effects of the Middle to Late Devonian Spread of Vascular Land Plants on Weathering Regimes, Marine Biotas, and Global Climate

Thomas J. Algeo, Stephen E. Scheckler, and J. Barry Maynard

The Middle to Late Devonian was an interval of major changes in both the terrestrial and marine biospheres. In the terrestrial realm, evolutionary innovations among early vascular land plants resulted in a far-reaching transformation of the structure and composition of terrestrial floras, heralded by the appearance of the first trees and forests and the spread of plants into harsher upland habitats. Increases in the size and geographic distribution of land plants at this time led to transient elevation of chemical and physical weathering rates, rapid increases in soil volume, and long-term changes in the hydrologic cycle, sediment fluxes, and landform stabilization. In the marine realm, a major biotic crisis (sometimes termed the Frasnian–Famennian mass extinction in reference to its stratigraphic peak, but consisting of eight to ten separate events) resulted in the extermination of about 70 to 80 percent of shallow-marine species and to the disappearance of the Middle Paleozoic reef-building community of stromatoporoids and tabulate corals. In the aftermath of extinction peaks, opportunistic taxa experienced short-lived acmes, and cold-water-adapted taxa radiated into vacated warm-water niches. Many of the extinction peaks were correlative with deposition of extensive organic-rich shale horizons in marine environments, reflecting widespread

development of anoxic bottomwater conditions in epicontinental seas, and elevated rates of organic carbon burial. These conditions (in addition to the intensified pedogenic weathering of silicates) led to rapid drawdown of atmospheric CO_2 levels, global climatic cooling, and the onset of continental glaciation toward the end of the Devonian period.

A variety of causal mechanisms have been postulated for these Middle and Late Devonian global events, including sea-level changes (Johnson 1974), oceanic overturn (Wilde and Berry 1984), tectonoclimatic factors (Copper 1986), bolide impacts (McLaren 1982), and environmental disturbances associated with the spread of vascular land plants (Algeo et al. 1995; Algeo and Scheckler 1998). All of these hypotheses account for the mass extinction event and for aspects of climate change linked to the global carbon cycle, but few of them satisfactorily explain unusual features of the biotic crisis (e.g., its duration and episodicity) or address the question of potential links with coeval developments in the terrestrial realm. In this regard, the "Devonian plant hypothesis" of Algeo et al. (1995) and Algeo and Scheckler (1998) is a holistic model that links coeval developments in the terrestrial and marine realms by focusing on the critical role of weathering, especially the function of soils as a geochemical interface between the atmosphere/hydrosphere and lithosphere, that of plants as a mediator of pedogenic weathering intensities and fluxes, and the effects of variable weathering fluxes on environmental conditions in epicontinental seas. The primary goals of this chapter are (1) to review key evolutionary developments among Middle and Late Devonian vascular land plants, (2) to consider the potential impact of these developments on weathering processes, soil production, and the sediment and water cycles, and (3) to evaluate the "Devonian plant hypothesis" vis-à-vis alternative models invoking eustatic, oceanic, tectonoclimatic, or extraterrestrial mechanisms as an explanation for Middle and Late Devonian global events.

EVOLUTIONARY INNOVATIONS AMONG EARLY VASCULAR LAND PLANTS

Colonization of land surfaces by terrestrial vegetation was a protracted event, spanning the Ordovician to Carboniferous at a minimum. The pace of this event was dictated by the rate of appearance of evolutionary innovations in response to environmental stresses—for example, cutins to limit water loss, rhizoids and tracheids for water transport and nutrient uptake, cellulose plus lignin for structural support, and spores and seeds for dispersal (Chaloner and Sheerin 1979; Knoll et al. 1984; Gensel and Andrews 1987; Thomas and Spicer 1987; Selden and Edwards 1989; Beerbower et al. 1992). The earliest land plants (Early?–Middle Ordovician to Early Silurian) were small, nonvascular, and possibly thalloid (i.e., lacking morphological differentiation into roots, stems, and leaves), representing a bryophytic grade of evolution (Thomas and Spicer 1987; Beerbower et al. 1992). Vascular (i.e., tracheid-bearing) plants evolved and diversified during the Late Silurian and Early Devonian, but these plants remained small and ecologically restricted to moist lowland habitats during this interval (figure 12.1A; Gensel and Andrews 1984, 1987; Gray 1985, 1993; Edwards and Berry 1991; Strother et al.

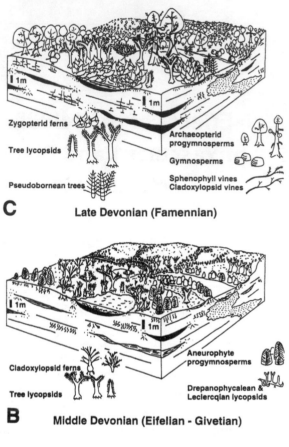

C   Late Devonian (Famennian)

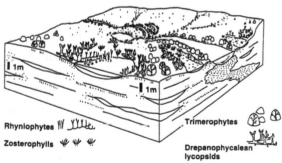

B   Middle Devonian (Eifelian - Givetian)

A   Early Devonian (Siegenian/Pragian - Emsian)

Figure 12.1.
Paleobotanical and paleoecological reconstructions of an Early Devonian (Siegenian/Pragian—Emsian) upland floodplain (**A**), a Middle Devonian (Eifelian—Givetian) upland floodplain (**B**), and a Late Devonian (Famennian) upland floodplain (**C**). Data from Scheckler (1986a,b and unpublished) and Gensel and Andrews (1984, 1987). Common vegetation types, inferred relative sizes, densities, and heterogeneity of spacing, as well as the effects of organic carbon burial and of their roots on soil formation and texture are depicted.

1996). However, competitive pressures for sunlight favored taller plants, and between the Pragian and the Givetian, secondary tissues evolved (i.e., wood, phloem, and bark with cork), providing the structural support for vascular plants to attain larger sizes (arborescence) (figure 12.1B; Banks 1980; Niklas 1985; Thomas and Spicer 1987; Beerbower et al. 1992). Finally, reproductive innovations led to the appearance of seeds in the Famennian, freeing plants from dependence on moist lowland habitats and permitting colonization of drier, harsher inland areas (figure 12.1C; Chaloner and Sheerin 1979).

The role of land plants in soil development and landscape stabilization increased in direct relation to their size and geographic distribution. Early land plants had relatively little effect on their physical environment because of their small size, shallow root systems, and ecological restriction to lowland habitats. The significance of the appearance of secondary supporting tissues and the seed habit was that these innovations permitted substantial increases in the size and geographic coverage of land plants. This increased plant biomass resulted in increases in soil volume and soil production rates. For this reason, the discussion of evolutionary innovations among early vascular land plants will focus on the development of arborescence, root systems, and the seed habit.

Arborescence

The maximum size of vascular land plants increased greatly during the Pragian to Givetian (late Early to late Middle Devonian) (figures 12.1, 12.2; Chaloner and Sheerin 1979; Edwards and Berry 1991; Beerbower et al. 1992; Algeo et al. 1995). By the late Givetian, woody shrubs and medium-sized trees had appeared independently in several clades—for example, cladoxylopsid ferns such as *Pseudosporochnus*, lepidosigillarioid lycopsids such as *Eospermatopteris*, aneurophyte progymnosperms such as *Rellimia* and *Tetraxylopteris*, and archaeopterid progymnosperms such as *Svalbardia* (Banks 1980; Gensel and Andrews 1984; Mosbrugger 1990). Among the early arborescent plants, archaeopterids established dominance during the Late Devonian (figure 12.1; Beck 1981; Scheckler 1986a,c; Edwards and Berry 1991). Some late Devonian archaeopterids were large having trunks thicker than 1.5 m in diameter and maximum heights in excess of 30 m (figure 12.2; Thomas and Spicer 1987; Snigirevskaya 1988, 1995; Trivett 1993; Meyer-

Berthaud et al. 1999, 2000). Further, archaeopterids (especially *Archaeopteris*) were among the most diverse and abundant arborescent plants of that age, ranging from tropical to boreal paleolatitudes (Petrosyan 1968; Beck 1981; Cross 1983; Gensel and Andrews 1984). High concentrations of archaeopterid remains are known from many Late Devonian (especially middle to late Frasnian) terrestrial and marginal marine successions, and monospecific archaeopterid forests may have been common in floodplain habitats at that time (Beck 1964, 1981; Stubblefield et al. 1985; Fairon-Demaret 1986; Scheckler 1986c; Thomas and Spicer 1987). After peaking in abundance between the mid Frasnian and mid Famennian, *Archaeopteris* declined precipitously during the mid to late Famennian, although still locally abundant [e.g., Bear Island (Nathorst 1902)], and disappeared by the early Tournaisian (figure 12.2; Beck 1981; Scheckler 1986a,c; Beerbower et al. 1992).

The reasons for the success of archaeopterid progymnosperms are not known for certain but may have been linked to their ability to produce multiple new crown branches (Meyer-Berthaud et al. 1999, 2000) and to create and maintain favorable microenvironments. Mass shedding of their leafy lateral branches (Beck 1964; Scheckler 1978; Meyer-Berthaud et al. 1999, 2000) would have created a deep mat of litter that not only would have affected soil moisture, pH, and humic content, but also might have favored their own free-sporing, heterosporous reproduction to the potential detriment of other plants, such as aneurophyte progymnosperms, which did not shed branch parts. Aneurophyte reproduction was probably homosporous (e.g., Taylor and Scheckler 1996), and, since the progymnosperm sporophytes show no evidence for clonal growth, it is very likely that aneurophyte gametophytes grew in the same well-lit, open habitats favored by sporophytes. Archaeopterids produced shaded, litter-strewn forests, and the vulnerability of aneurophytes to shading may have contributed to their rapid decline in diversity and abundance as archaeopterids spread during the early to mid Frasnian (Scheckler and Banks 1971a,b; Scheckler 1986a,c; Beerbower et al. 1992; Hill et al. 1997). The reason for the precipitous decline of *Archaeopteris* in the mid to late Famennian is

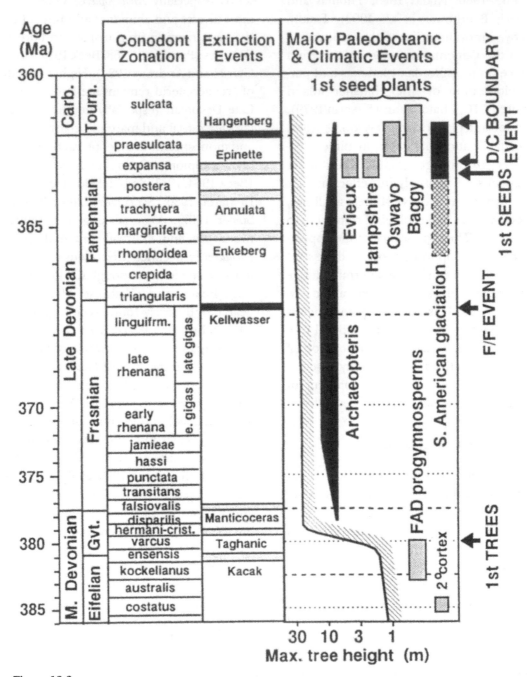

MIDDLE - LATE DEVONIAN PALEOBOTANICAL EVENTS

Figure 12.2.

Middle and Late Devonian botanical, marine, and climatic events. **A:** Trees and roots: first appearances for secondary cortex (periderm) and progymnosperms (Scheckler and Banks 1974; Chaloner and Sheerin 1979; Banks 1980; Gensel and Andrews 1984; Thomas and Spicer 1987; Stewart and Rothwell 1993; Taylor and Taylor 1993), maximum tree heights (Chaloner and Sheerin 1979; Gensel and Andrews 1984; Mosbrugger 1990), and peak abundance of *Archaeopteris* (Beck 1964; Scheckler 1986a,b; Beerbower et al. 1992). **B:** Seed plants: first seeds in Evieux, Hampshire, Oswayo, and Baggy formations (Fairon-Demaret 1986; Rothwell and Scheckler 1988); also shown is late Famennian continental glaciation in South America (Caputo 1985; Caputo and Crowell 1985), which is age constrained by miospore data (Streel 1986, 1992). Conodont zonation scheme from Ziegler and Sandberg (1984, 1990); extinction events are shown for comparison. Timescale is from Harland et al. (1990), although recent radiometric age studies (e.g., Claoué-Long et al. 1992) cast unknown. Because of differences in habitat and reproductive strategy, direct competition with early seed plants is not likely a factor in its demise (Scheckler 1986a,b; Rothwell and Scheckler 1988).

Root Systems

Increases in the size of vascular land plants during the Middle and Late Devonian were undoubtedly accompanied by increases in the size of root systems. Among modern higher plants, a strong positive correlation exists between shoot mass and root mass, with roots accounting for 15 to 30 percent of total floral biomass in conifer forests and up to 75 percent in grasslands (Russell 1977). With regard to primary productivity, root growth is even more important, accounting for 40 to 85 percent of annual biomass turnover in modern forest, shrub, and grassland ecosystems (Fogel 1985; Comeau and Kimmins 1989). Furthermore, in harsh environments an even larger proportion of total net production is put into rootlets (e.g., Keyes and Grier 1981), suggesting that as early land plants began to colonize marginal new habitats, an important physiological adaptation may have been larger root systems.

Increases in the biomass and penetration depth of root systems coincident with the spread of arborescent vegetation during the Middle and Late Devonian are evidenced by root traces and by permineralized and compressed fossil roots in contemporaneous paleosols (e.g., see figure 12.4; Driese et al. 1997). The earliest root traces associated with vascular plants appeared in the late Pragian (mid-Early Devonian) in the form of short (centimeter-long), forked roots produced by herbaceous lycopods (e.g., *Asteroxylon, Drepanophycus*) and trimerophytes (e.g., *Psilophyton*). The appearance of deeply penetrating root systems coincided with the development of arborescent vegetation in the late Middle and Late Devonian (figure 12.3; Beerbower et al. 1992). In the Eifelian and Givetian, the root systems of archaeopterid and aneurophyte progymnosperms were generally shallow (<20 cm) (Scheckler 1986a, and unpubl. data; Beerbower et al. 1992), but by the Frasnian and Famennian, dominant genera such as *Archaeopteris* had massive root systems with penetration depths in excess of 100 cm (figures 12.3, 12.4; e.g., Beck 1953, 1967; Walker and Harms 1971; Snigirevskaya 1984a,b, 1988, 1995; Retallack 1985; LeJeune 1986; Driese et al. 1997; Scheckler, unpubl. data). Later gymnosperms possessed much the same type of root system as these progymnosperms and were probably equally effective with regard to pedoturbation (Scheckler 1995). Some nonprogymnosperms also developed extensive root systems during the Late Devonian (e.g., the tree lycopods *Cyclostigma* and *Lepidodendropsis/Protostigmaria*, the cladoxylalean fern *Pseudosporochnus*, and the zygopterid fern *Rhacophyton*, but these taxa were commonly restricted to wetland habitats and their root systems were generally shallower and unbranched or much less branched than those of contemporaneous progymnosperms (figure 12.3; Leclercq and Banks 1962; Schweitzer 1969; Scheckler 1974, 1986a,b; Jennings et al. 1983). The greater pedoturbational effectiveness of progymnosperms and gymnosperms relative to other clades was a result of differences in root structure: (1) the presence in both groups of a vascular cambium that allowed continuous, perennial growth with storage potential (i.e., wood and ray parenchyma), and (2) internal (i.e., endogenous and adventitious) production of lateral rootlets that could repeatedly arise and repenetrate a given volume of soil (figure 12.4; Scheckler, in press).

Seed Habit

Innovations in reproductive systems played a major role in the greening of land masses. The earliest land plants had a pteridophytic

Figure 12.2. (*continued*)
some aspects of this timescale in doubt. Famennian epoch is also supported by a recent radiometric study that yielded a date for the Devonian–Carboniferous boundary some 8 m.y. younger than that of the Harland et al. timescale (354.2±4.0 Ma vs. 362.4 Ma). Hence, substantial revision of the Devonian timescale is likely in the future.

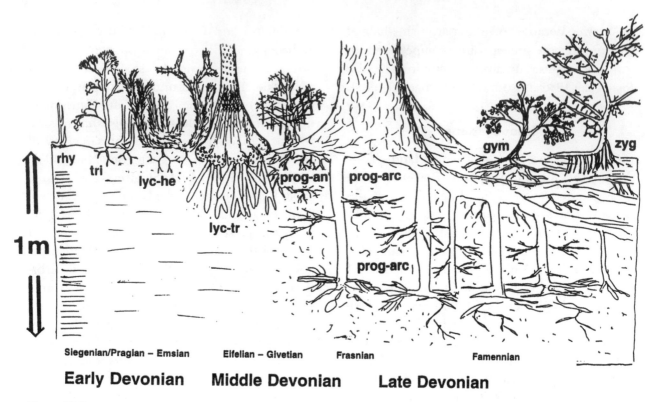

Figure 12.3.
Diagram of the relative sizes, morphologies, and penetration depths of selected Early, Middle, and Late Devonian plants. *rhy*, rhyniophytes such as *Aglaophyton* or *Horneophyton*; *tri*, trimerophytes such as *Psilophyton*; *lyc-he*, early herbaceous lycopods such as *Asteroxylon* or *Drepanophycus*; *lyc-tr*, early tree lycopods such as *Lepidosigillaria* or *Cyclostigma*; *prog-an*, aneurophyte progymnosperms such as *Tetraxylopteris*; *prog-arc*, *Archaeopteris* progymnosperms; *zyg*, zygopterid ferns such as *Rhacophyton*; *gym*, early gymnosperms such as *Elkinsia* or *Moresnetia*. Bar scale = 1 m.

reproductive mode—that is, free-living gametophytes grew from dispersed spores and produced sperm that swam over moist plant or soil surfaces to egg-bearing archegonia of the same or other gametophytes (Kenrick 1994; Remy et al. 1993; Remy and Hass 1996). By the Middle Devonian, two different sizes of spores containing reduced internal gametophytes (a condition known as heterospory) had developed in several clades including archaeopterid progymnosperms. Further differentiation of spores led, by the late Famennian, to the appearance of seeds, in which a single spore matures into a megagametophyte within a nutrient-bearing and protective ovule, and which is fertilized by small gametes producing microgametophytes (i.e., pollen) (Chaloner and Sheerin 1979; Rothwell and Scheckler 1988; DiMichele et al. 1989; Edwards and Berry 1991; Bateman and DiMichele 1994). The development of seeds freed vascular land plants from

dependence on aqueous sperm dispersal and moist lowland environs (e.g., figure 12.1A), permitting occupation of previously closed habitats such as drier upland areas (e.g., figure 12.1C). Ovules also reduced the risk of gametophyte desiccation, increased the rate of successful sexual unions via pollination, and provided greater adaptability to diverse ecological conditions, allowing rapid colonization of harsh primary successional stages in floodplain and delta environments and rapid early growth of seedlings under unfavorable conditions such as in shaded forest understory environments (Scheckler 1986b; Thomas and Spicer 1987; Rothwell and Scheckler 1988; Edwards and Berry 1991). Thus, the seed habit was an important factor in the rapid spread and diversification of gymnosperms during the latest Devonian and Early Carboniferous (e.g., Chaloner et al. 1977; Gillespie et al. 1981; Fairon-Demaret 1986, 1996; Fairon-Demaret and

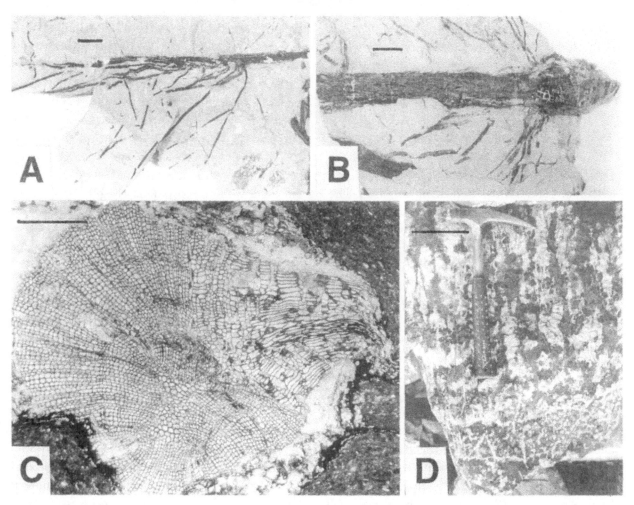

Figure 12.4.
Late Devonian root systems. **A:** Compressed young root system (*Callixylon petryi*) of progymnosperm tree *Archaeopteris* that shows numerous lateral rootlets. Parts of this specimen (UAPC S4630) are also petrified and demonstrate that the lateral rootlets were endogenous. **B:** Compressed older root system of *Archaeopteris* to show clusters of adventitious lateral rootlets (specimen UAPC S4631b). Specimens in (A) and (B) are from the lower Frasnian Yahatinda Formation of Alberta, Canada. Scale bars in (A) and (B) = 1 cm. **C:** Cross-section of a petrified aneurophyte progymnosperm root that shows the secondary xylem and secondary phloem produced by its perennial bifacial vascular cambium (n.b., asymmetry is common in roots that press against objects in the soil). Also visible on the right is a longitudinal section of an endogenous lateral rootlet; note cell arrangement at its point of attachment to the tetrarch main root. This specimen (CUPC 1990.4-9R) is from the lower Frasnian Oneonta Formation of New York. Scale bar = 1 mm. **D:** Vertical profile of a paleosol horizon (vertisol) showing in its upper part wedge-shaped peds and large calcrete tubes typical of those formed around major roots of *Archaeopteris* in semiarid environments, and in its lower part small root traces typical of the much-branched root systems of either young progymnosperms (e.g., *Archaeopteris*) or early gymnosperms (e.g., *Moresnetia* or *Dorinnotheca*), all of which are present in this flora. The paleosol is from the mid Famennian Evieux Formation of Belgium. Scale bar = 10 cm.

Scheckler 1987; Rothwell and Scheckler 1988; Rothwell et al. 1989; Beerbower et al. 1992; Hilton and Edwards 1996).

The appearance of seeds can be dated rather precisely to the mid to late Famennian (figure 12.2). The earliest known seeds are *Moresnetia zalesskyi* and *Dorinnotheca streelii* (Evieux Formation, Belgium) and *Elkinsia polymorpha* (Hamp-

shire Formation, West Virginia), which date to the lower VCo spore zone (Famennian 2c). These were followed by the slightly younger seeds *Archaeosperma arnoldii* (Oswayo Formation, Pennsylvania), *Aglosperma quadripartita* [Quartz Conglomerate Group (below the Tongwynlais Formation) of the Upper Old Red Sandstone at Taffs Well, Wales], *Spermolithus devonicus*

(Kiltorcan Formation, Eire), *Kerryia mattenii* (Coomhola Formation, Eire), and *Xenotheca devonica* (Baggy Formation, England), which date to the LV-LN spore zones (Fa2d-basal Tn1b) of the latest Devonian (Chaloner et al. 1977; Gillespie et al. 1981, Fairon-Demaret 1986, 1996; Fairon-Demaret and Scheckler 1987; Rothwell and Scheckler 1988; Rothwell et al. 1989; Rothwell and Wight 1989; Streel and Scheckler 1990; Hilton and Edwards 1996). The Fa2c and Fa2d (lower VCo-LV) spore zones correlate with the upper *postera*–lower *expansa* and upper *expansa*– *praesulcata* conodont zones, respectively (Richardson and Ahmed 1988). Seed plants were not common, however, until after the latest Devonian collapse of the progymnosperm *Archaeopteris* forests, when gymnosperms diversified and filled ecological niches in Early Carboniferous floodplain and upland forest environments (Scheckler 1986a; Rothwell and Scheckler 1988; Beerbower et al. 1992). Moreover, competition with seed plants does not seem to have been a major factor in the demise of *Archaeopteris*, since they occupied different floodplain habitats and exploited different features of successful reproductive biologies (Scheckler 1986a,b; Rothwell and Scheckler 1988; Beerbower et al. 1992).

EFFECTS OF EARLY VASCULAR LAND PLANTS ON WEATHERING PROCESSES, PEDOGENESIS, AND THE HYDROLOGIC CYCLE

Weathering Processes

The spread of forests and associated, deeply rooted soils during the late Middle and Late Devonian is likely to have caused major changes in weathering processes and pedogenesis (soil formation). With regard to chemical weathering in the soil environment, higher land plants play a major role in the production of (1) organic acids, which are released through root mycorrhizae, (2) humic and fulvic acids, which are produced through bacterial decay of complex organic compounds, and (3) carbonic acid, which is generated through oxidation of

organic matter in soils (Knoll and James 1987; Johnsson 1993; Drever 1994). Substantially greater quantities of acid are produced per unit land area in vascular than in nonvascular plant ecosystems because of the greater gross primary productivity of the former—for example, 600 to 6,000 g m^{-2} yr^{-1} in forests and grasslands versus less than 1,000 g m^{-2} yr^{-1} in nonvascular bryophytic and less than 100 g m^{-2} yr^{-1} in algal–lichen ecosystems (Fogel 1985; Longton 1988; Comeau and Kimmins 1989). This accounts for observations of higher rates of chemical weathering in vascular plant ecosystems (e.g., Cawley et al. 1969; Berner 1994), which are estimated to exceed those of nonvascular ecosystems by a factor of about seven on the basis of rock-weathering studies (Cochran and Berner 1993; Moulton and Berner 1998) and soil and groundwater analyses (Drever and Zobrist 1992; Cochran and Berner 1996). In addition to acid production, the greater weathering effectiveness of higher plants is also caused by the three-dimensional (i.e., volumetric) contact area of their rootlets and roothairs compared to the largely two-dimensional (i.e., surficial) contact area of algae and lichens with weathering substrates (e.g., Russell 1977; Viles and Pentecost 1994). These observations cast doubt on proposals that pre-Silurian microfloras consisting of bacteria, algae, and lichens could have brought about an intensity of chemical weathering *per unit land area* comparable to that of higher land plants (e.g., Schwartzman and Volk 1989; Keller and Wood 1993; Horodyski and Knauth 1994).

With regard to physical weathering processes, the spread of higher land plants may have changed the residence time of material in the soil environment and the texture and mineralogy of pedogenic weathering products (figure 5; Drever 1994). Vegetation retards the transport of sediment from hillslopes, leading to a transport-limited as opposed to a weathering-limited soil regime (Stallard 1985; Johnsson 1993). In the absence of a densely interwoven mat of plant roots, weathering products are rapidly swept from hillslopes, accumulating as immature sediments in alluvial fans and

HYDROLOGIC CYCLE & PEDOGENIC WEATHERING

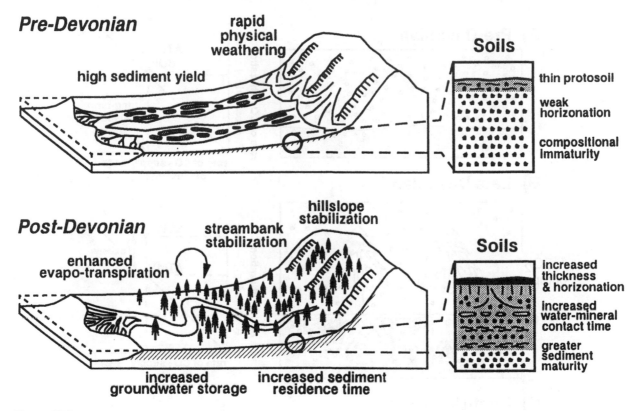

Figure 12.5.
Changes in the hydrologic cycle and pedogenic weathering processes as a consequence of the Middle-to-Late Devonian spread of vascular land plants (see text for discussion).

braided streams. Thus, development of a higher land-plant cover may have stabilized land surfaces, allowing more time for weathering of surface rocks to a finer-grained, compositionally more mature product (Schumm 1968; Johnsson 1993; Drever 1994). This process may account for apparent secular changes in sediment maturity, such as an increase in the clay content and compositional maturity of Silurian fluvial clastics relative to equivalent Cambro-Ordovician facies (Cotter 1978; Feakes and Retallack 1988; Retallack 1985, 1990). Secular changes in compositional maturity also may have come about through selective weathering of minerals containing nutrients essential for plant growth. For example, pre-Silurian sandstones are richer in chemically labile K-feldspars, and lower Paleozoic clastic sediments contain higher K_2O/Na_2O ratios than equivalent Mesozoic–Cenozoic facies, suggesting more complete decomposition of K-bearing minerals

from the mid Paleozoic onward because of the potassium demand of higher land plants (Basu 1981; Maynard et al. 1982; Holland 1984; Nesbitt et al. 1996).

With regard to rates of chemical and physical weathering, it is important to note that changes in these parameters associated with the spread of vascular land plants in the Middle and Late Devonian were transient rather than permanent (figure 12.6). This transience is dictated by the fact that, at geologic timescales (e.g., $> 10^7$ yr), fluctuations in chemical weathering rates are dampened by negative feedbacks involving (1) longer time constants for the sources of atmospheric CO_2 (i.e., volcanic and metamorphic degassing) than for its sinks (i.e., silicate weathering and organic carbon fixation), and (2) temperature dependence of chemical weathering rates [i.e., cooling as a consequence of atmospheric CO_2 drawdown would reduce the rate of CO_2 utilization via silicate weathering

ATMOSPHERIC CO$_2$ DRAWDOWN

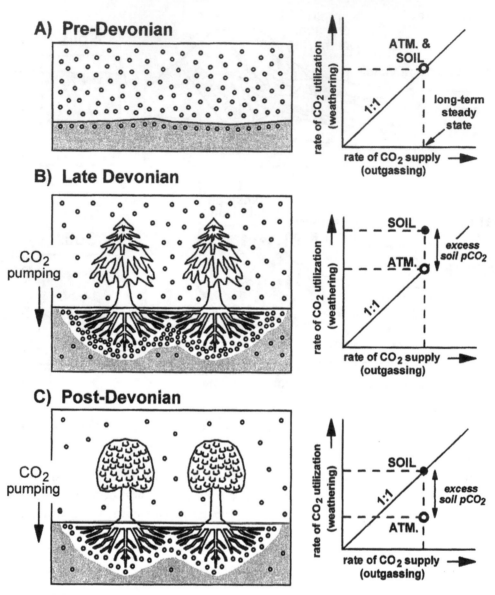

Figure 12.6.
Changes in atmospheric and soil CO$_2$ levels as a consequence of the Middle-to-Late Devonian spread of vascular land plants. **A:** In the largely unvegetated pre-Devonian world, atmosphere and soil CO$_2$ levels were similar. **B:** During the Middle and Late Devonian, active pumping of carbon into soils by land plants initiated atmospheric pCO$_2$ drawdown. **C:** in the post-Devonian, terrestrial floras maintain artificially low atmospheric CO$_2$ and high soil CO$_2$ levels via intensive carbon pumping. Because of a long-term steady state in CO$_2$ supply via volcanic and metamorphic degassing and the requirement that geochemical sources and sinks balance, CO$_2$ levels in the soil environment (where silicate mineral weathering takes place) must have been similar both before and after the spread of land plants. However, during the transient interval in which CO$_2$ uptake by terrestrial floras first intensified but atmospheric CO$_2$ levels remained high, soil CO$_2$ levels are likely to have been much higher than at present, resulting in enhanced pedogenic weathering.

(Berner and Rye 1992; Caldeira 1992; Berner 1994)]. However, significant short-term perturbations in global weathering rates (figure 12.6B) are possible because the rate of supply of CO$_2$ via degassing is effectively constant at the timescales associated with pedogenic weathering processes (e.g., $<< 10^7$ yr). Thus, a transient increase in weathering intensity at these timescales (e.g., driven by land-plant evolution) would have to be counteracted by (1) a decrease in atmospheric pCO_2, (2) a decrease in global temperature, or (3) both. Eventually, this would yield a new equilibrium state (figure 12.6C) in which global weathering intensity is comparable to that prior to the transient disturbance (figure 12.6A), and in which atmospheric pCO_2, global temperature, or both, remain permanently at levels lower than those prior to the disturbance. Because of the stronger dependence of silicate weathering rates on temperature than on soil pCO_2 (Velbel 1993; Gwiazda and Broecker 1994), the dominant factor in producing a negative feedback on transient increases in weathering intensity is likely to be global cooling rather than atmospheric CO_2 drawdown. These considerations suggest that any increases in chemical and physical weathering rates associated with the spread of vascular land plants would have been a transient phenomenon, which is consistent with a transient sedimentation rate anomaly reported for the Middle and Late Devonian (Dineley 1984; Gregor 1985; Algeo et al. 1995; Retallack 1997; Algeo and Scheckler 1998).

Pedogenesis

The spread of vascular land plants resulted in profound changes in soil types, the thickness and geographic extent of soils, and rates of pedogenesis on a global scale (Retallack 1990, 1997; Algeo et al. 1995). Prior to the advent of vascular plants, land surfaces may have consisted largely of barren rock and thin microbial protosoils (Retallack 1985, 1986, 1990, 1992; Beerbower et al. 1992), similar to modern "desert crusts" formed of cyanobacterial mats (Campbell 1979; Whitford and Freckman 1988; Watson 1992). The advent of vascular land plants appears to have been closely linked to global changes in soil types, as most of the major types of the modern world first appeared during the Devono-Carboniferous (figure 12.7). For example, the oldest known histosol is the Rhynie Chert from the Early Devonian, and the earliest probable alfisols, ulti-

sols, and spodosols are all of Late Devonian or Early Carboniferous age (Retallack 1986; Mack and James 1992) (but note that classification of paleosols using modern soil nomenclature is tentative because of the lack of information on pH and base status). These soil types, which form today mainly beneath temperate-zone forests, probably reflect contemporaneous afforestation of land surfaces. Vertisols, although present in pre-Devonian successions, became increasingly common during the Devonian (e.g., Allen 1986; Driese and Mora 1993) and, especially, during the Carboniferous (e.g., Goldhammer and Elmore 1984; Wright 1987; Wright and Robinson 1988; Ettensohn, Dever, and Grow 1988; Goldstein 1991; Vanstone 1991; Muchez et al. 1993; Driese et al. 1994; Retallack and Germán-Heins 1994; Gill and Yemane 1996). Vertisols (see figure 12.4D) are characterized by wedge-shaped peds bounded by slickensided surfaces that result from high concentrations of swelling clays (e.g., smectite), and such clays are generated in large quantities in the type of seasonally wet climate that is associated with temperate-zone vegetation (Wilding and Tessier 1988; Mack and James 1992).

A fundamental change in soil weathering processes may be indicated also by a major shift in the clay-mineral composition of shales during the Devono-Carboniferous from illite- and chlorite-dominated assemblages to smectite- (or mixed-layer illite/smectite-) and kaolinite-dominated assemblages (Weaver 1967, 1989; Algeo et al. 1995). Smectite and kaolinite are produced in large amounts in the soil environment, principally through alteration of biotite and plagioclase feldspar precursors. Production of these minerals is closely associated with soil leaching: smectites and related expandable clay minerals are the major weathering products of many parent rock types subject to moderate leaching in temperate to semiarid climates, whereas kaolinite is favored under conditions of strong leaching in humid tropical climates (Velde 1985; Singer and Munns 1991). Although smectite-rich paleosols are known to have developed in warm, seasonally wet paleoclimates in the absence

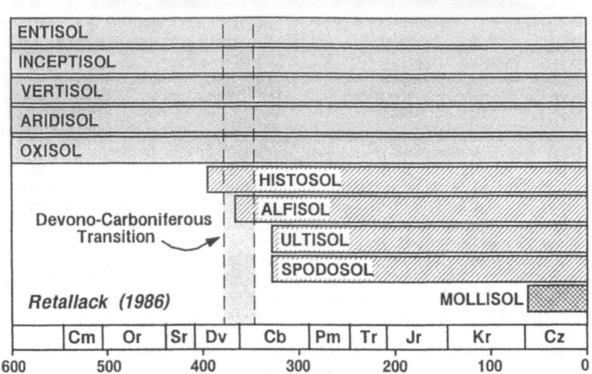

Figure 12.7.
Phanerozoic age ranges of major soil types (Retallack 1986, 1990). The dominant soil types of modern swamps (histosols) and forests (alfisols, ultisols, and spodosols) appeared during the Devonian and Carboniferous.

of terrestrial vegetation (e.g., Nesbitt and Young 1982; Driese and Foreman 1992; Driese et al. 1992), the range of environments favoring smectite formation is likely to have broadened considerably with the advent of a terrestrial plant cover. Other clay minerals and soil types produced largely or wholly through pedogenic processes first appear or exhibit significant increases in abundance during the Devono-Carboniferous, such as (1) palygorskite and sepiolite (Weaver and Beck 1977; Rateev and Timofeev 1979), (2) bauxite and laterite, which are weathering products of intense chemical leaching in perennially wet tropical and seasonally wet temperate climates, respectively (Nicholas and Bildgen 1979; Retallack 1986, 1990, 1992). These trends, although not well documented, are consistent with intensified chemical weathering and enhanced pedogenic production of clay minerals from the Late Devonian onward.

Soil thicknesses appear to have increased

rapidly in association with the spread of arborescent vegetation in the Middle and Late Devonian (figures 12.3, 12.5). Although vegetation is only one of several factors influencing soil formation, comparison of paleosols of different ages that developed under otherwise similar conditions of climate, parent materials, and slope suggests that vegetation plays a rather important role in pedogenesis. For example, paleosols of Ordovician to Carboniferous age that developed on subtropical alluvial outwash fans exhibit progressive increases in clay content, soil structuring (i.e., ped formation), and profile maturity (i.e., horizonation) with decreasing age (Retallack 1985). These changes are correlated with increases in root penetration depths—0 cm for Late Ordovician, about 5 to 25 cm for Late Silurian, and about 100 cm for Late Devonian paleosols. Further, translocation of clays and iron to form a pale leached A horizon and a purple illuviated B horizon, which is associated with temperate-zone forest soils today, is first encountered in wet lowland soils of the early Frasnian and in dry upland soils of the Early Carboniferous (Retallack 1985). The timing of these pedogenic changes is consistent with the timing of important terrestrial floral developments likely to have influenced soil development—for example, widespread afforestation of lowlands in the Frasnian and the spread of gymnosperms into upland areas in the Early Carboniferous.

Rates of pedogenesis are likely to have varied considerably in association with the changes in soil types and thicknesses documented here. Rapid increases in soil thickness during the Middle and Late Devonian are *prima facie* evidence for a (transient) elevation of chemical weathering rates, because modern soils exhibit a strong positive correlation between rates of breakdown of soil mineral constituents and the density and rate of growth of roots and rootlets [which are linked to rates of production of soil acids (Newman and Andrews 1973; Russell 1977)]. This inferred acceleration of soil formation rates, although primarily due to increases in terrestrial floral biomass, also may have been promoted by positive feedbacks between (1) evolving root systems and soil development, via a shift from a weakly rooted rhizomatous mode to a deeply rooted free-standing mode (figures 12.3, 12.4) as soils increased in thickness and compositional maturity (Banks 1985), and (2) accumulating soil humus and floral biomass, via enhanced rates of recycling of organic detritus by soil fungi and bacteria, permitting greater standing floral crops (Thomas and Spicer 1987; Klepper 1987; Retallack 1990). Whereas any increase in soil formation rates during the Middle and Late Devonian would have been transient, changes in soil depth (and structure and type) (figure 12.5) were permanent, reflecting a new equilibrium maintained dynamically by the continuous presence of a large terrestrial floral biomass since the Late Devonian.

Increases of organic carbon burial occurred not only in marine black shales of this time, but also in coeval terrestrial coal beds. Partly this was the result of the huge expansion of terrestrial primary productivity. But the evolution of large, wetland-tolerant plants (tree lycopsids and

some zygopterid ferns) formed increasingly common and thicker coal beds throughout the Givetian, Frasnian, and Famennian (figure 12.1; Schweitzer 1969; Goodarzi et al. 1989, 1994; Goodarzi and Goodbody 1990; Scheckler 1986a,b,c; Beerbower at al. 1992). By the latest Devonian to Early Carboniferous, arborescent lycopsids (e.g., *Cyclostigma, Protolepidodendropsis, Lepidodendropsis, Sublepidodendron*) were fully wetland tolerant and formed extensive coal swamps (Scheckler 1986a; Beerbower et al. 1992), a clade-dependent ecosystem that would continue through most of the Carboniferous (DiMichele and Phillips 1996).

The Hydrologic Cycle

The spread of terrestrial vegetation is likely to have altered the global hydrologic cycle (figure 12.5). With regard to recirculation of atmospheric water, forests promote evapotranspiration and increase precipitation through changes in land surface albedo and atmospheric turbulence (Shukla and Mintz 1982; Sud et al. 1993). With regard to surface runoff, a dense vegetative cover reduces runoff, total discharge, and peak discharge during floods *for a given level of precipitation* (Schumm 1968, 1977; Faulkner 1990). Even though the rate of evapotranspirative recirculation (and, hence, precipitation) would have increased in a vegetated world over that of an earlier nonvegetated world, surface runoff nonetheless would have been reduced in the former as a result of the greater storage capacity of the thick soils developed in association with vegetation.

The reduction in runoff associated with vegetation also would have led to lower landscape erosion rates and sediment yields. Sediment yields in the modern world peak in semiarid environments with an annual precipitation of about 250 to 350 mm yr^{-1} (figure 12.8A,B; Summerfield 1991), but in the absence of a higher plant cover, sediment yields may have covaried positively with increases in precipitation to much higher precipitation rates (figure 12.8C,D; Schumm 1968, 1977). The reduction in modern sediment yields at precipitation rates greater than 250 to 350 mm yr^{-1}

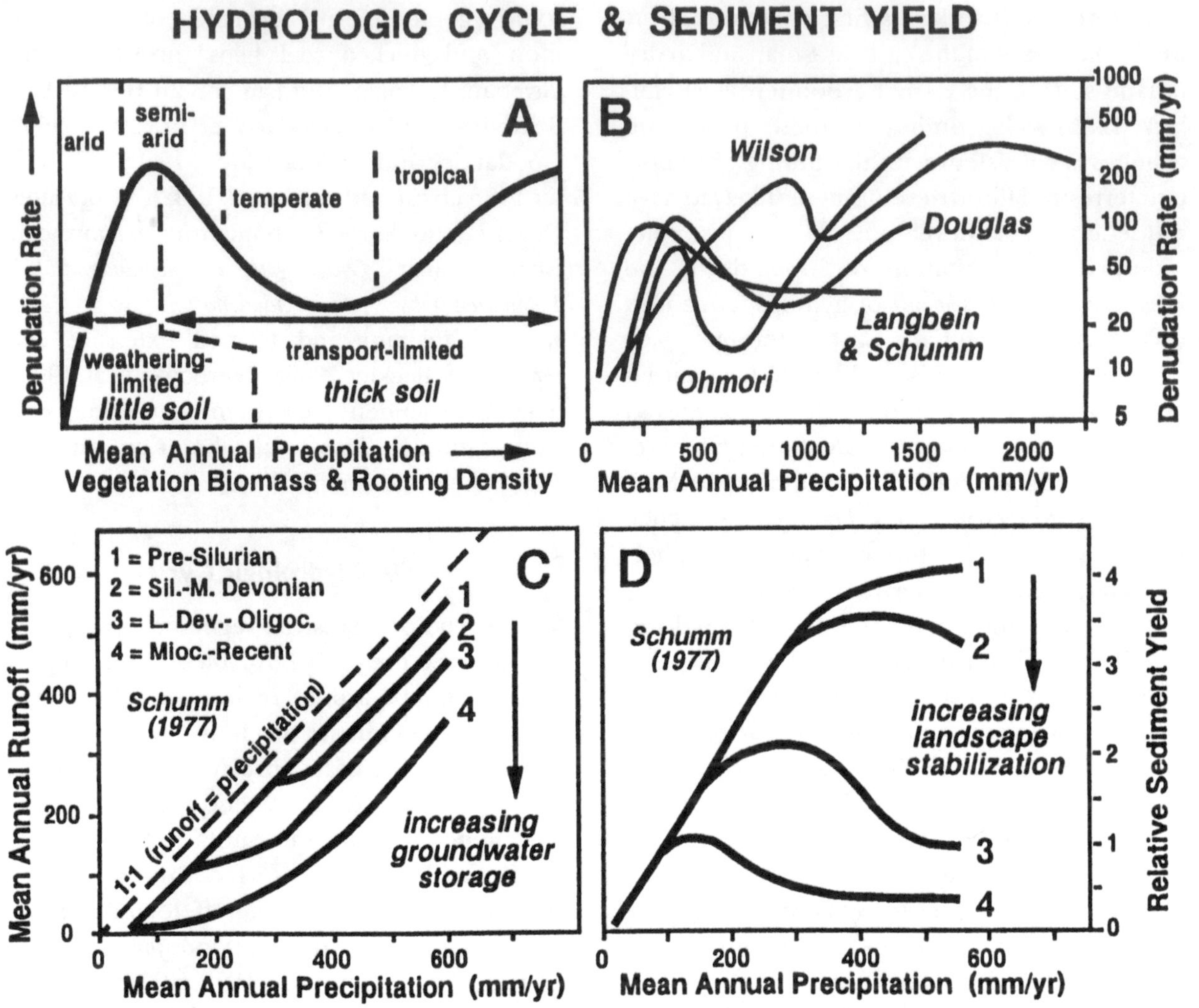

HYDROLOGIC CYCLE & SEDIMENT YIELD

Figure 12.8.
Changes in the hydrologic cycle and sediment yields associated with the Middle-to-Late Devonian spread of vascular land plants. **A:** Theoretical relation between precipitation and landscape denudation rates, and associated climate zones, weathering regimes, and soil thicknesses (Stallard 1985); because floral biomass correlates strongly with precipitation, a similar relation between biomass and landscape denudation rates may be inferred. **B:** Four estimates of present-day precipitation—denudation rate relations (Summerfield 1991). **C,D:** Theoretical relations between precipitation and mean annual runoff (C) and precipitation and sediment yield (D) for four geologic epochs (Schumm 1968, 1977); because floral biomass correlates strongly with precipitation, similar relations between biomass and runoff (C) and biomass and sediment yield (D) may be inferred. See text for further discussion.

(figure 12.8B) is the result of physical stabilization of land surfaces by vegetation in more humid environments—for example, higher-plant root systems serve to bind soil particles, and the plants themselves serve as windbreaks (Moore 1984). These stabilizing effects may be evidenced by secular changes in fluvial morphology: a reported shift within fluvial facies of the Appalachian Basin from dominantly braided character in the pre-Silurian to mixed braided-meandering character

in the Silurian and later (Cotter 1978). This pattern is consistent with the predicted effects of a higher land-plant cover, which would cause an increase in annual fluvial discharge (related to intensification of the hydrologic cycle) and a reduction in sediment yield, favoring meandering streams over braided streams (figure 12.5; Schumm 1968, 1977). The fact that this shift occurred in the Silurian rather than the Devonian may reflect preferential colonization of floodplains by the earliest vascular land plants (Thorne 1990; Beerbower et al. 1992) and points out the potential for diachronous effects as a consequence of the progressive spread of land plants from the Ordovician to the Carboniferous.

MIDDLE AND LATE DEVONIAN MARINE EVENTS AND GLOBAL CLIMATE CHANGE

The Middle and Late Devonian is characterized by important events not only in the terrestrial but also in the marine realm. In the latter, a protracted biotic crisis eliminated the Middle Paleozoic reef community and resulted in a far-reaching turnover among tropical-marine benthic groups. The biotic crisis was probably related in some manner to contemporaneous development of widespread anoxia in shallow epicontinental seas, and bottomwater anoxia contributed to high rates of organic carbon burial and to C- and S-isotopic anomalies in marine sediments of this age. Changes in the global carbon cycle associated with massive burial of organic carbon (as well as with enhanced pedogenic weathering of silicates) were the primary factor leading to coeval changes in global climate, such as rapid drawdown of atmospheric CO_2, climatic cooling, and the onset of continental glaciation. Whether events in the marine realm and changes in global climate during the Middle and Late Devonian were precipitated by or otherwise linked to the evolutionary development of vascular land plants (discussed previously) is open to question, although a strong circumstantial case can be made in favor of such a scenario.

Marine Biotic Crisis

The Middle and Late Devonian biotic crisis, one of the "Big Five" mass extinctions of the Phanerozoic (Sepkoski 1986; Raup and Boyajian 1988; McGhee 1996), resulted in the disappearance of about 20 to 25 percent of families, 45 to 60 percent of genera, and 70 to 82 percent of species of marine biota (Sepkoski 1986, 1996; Jablonski 1991). This biotic crisis differed from other well-studied mass extinctions in three important respects: (1) duration, spanning an interval of about 20 to 25 million years (Sepkoski 1996; McGhee 1996), (2) episodicity, comprising at least eight to ten separate events (House 1985; Bayer and McGhee 1986; Scrutton 1988; Becker 1993), and (3) selectivity, being largely restricted to tropical marine biota (Bambach 1985; Sepkoski 1986; Scheckler 1986a).

With regard to episodicity, this crisis was characterized by fluctuations in extinction rates producing a "stepwise" extinction pattern with peak intensities at intervals of a few million years (Walliser 1996a,b). Many biotic groups exhibit this stepwise extinction, including rugose corals, brachiopods, ammonoids, and benthic foraminifera (House 1985; Copper 1986; Bayer and McGhee 1986; Kalvoda 1986, 1990; Scrutton 1988; Boucot 1990; Simakov 1993; Schindler 1993; Becker 1993; Talent et al. 1993; McGhee 1996). The most profound extinctions were associated with (1) the Frasnian–Famennian (F/F) boundary (or Upper Kellwasser horizon), and (2) the Devonian–Carboniferous (D/C) boundary [or Hangenberg horizon (Sepkoski 1986, 1996)]. The duration of these extinction peaks is uncertain, but estimates based on conodont zones suggest events of less than 20 ky duration (Sandberg, Ziegler, et al. 1988).

With regard to selectivity, the Middle and Late Devonian biotic crisis was particularly severe for tropical marine organisms, especially benthic groups. The Middle Paleozoic reef community, dominated by stromatoporoids and tabulate corals, was under stress as early as the Eifelian, although massive reefs continued to be constructed into the late Givetian and early Frasnian (Burchette 1981; Sepkoski 1986; James 1983; Bambach 1985). These large reefs declined rapidly during the mid to late Frasnian and disappeared at the F/F boundary, to be replaced during the Famennian by small stromatolitic structures built by cyanobacteria (Krebs 1974; Playford et al. 1976, 1984). A few marine biotic groups thrived during the biotic crisis, especially high-latitude and cold-water taxa [e.g., siliceous sponges (hexactinellids and desmosponges)] (Rigby 1979; Racki 1990; Geldsetzer, Goodfellow and McLaren 1993). In taxonomic groups experiencing widespread mortality among shallow-water species, deep-water members survived disproportionately and in some cases subsequently radiated into vacated warm-water niches (e.g., spiriferid and rhynchonellid brachiopods, tornoceratid ammonoids, labechiid stromatoporoids, and nondissepimentate rugose corals) (Pedder 1982; Copper 1986; Stearn 1987; Kalvoda 1990). Other groups exhibited short-lived acmes following extinction maxima—for example, certain phytoplankton (e.g., tasmanitids and styliolinids) and the reef-building calcareous algal association of Epiphytales and *Renalcis-Shuguria* that had flourished previously only in the Cambrian (Chuvashov and Riding 1984). These patterns suggest opportunistic expansion of preadapted survivors and crisis taxa following a major environmental disturbance.

Black Shales and Marine Anoxic Events

Many of the Middle and Late Devonian extinction episodes just discussed were correlative with deposition of extensive organic-rich shale horizons in marine environments, reflecting widespread development of anoxic bottomwater conditions in epicontinental seas at that time (e.g., Sandberg, Poole, and Johnson 1988; Sandberg, Ziegler, et al. 1988; Joachimski and Buggisch 1993) (note that whether anoxic conditions existed coevally in the global ocean is unknown). In eastern North America, black shales range in age from the late Eifelian through the earliest Mississippian, with the greatest concentration in the Frasnian and Famennian—for example, Ohio-Sunbury and Chattanooga-Maury shales (Appalachian Basin), New Albany Shale (Illinois Basin), Antrim Shale (Michigan Basin), and Kettle Point and Long Rapids formations (Ontario) (Kepferle et al. 1982; Kepferle and Pollock 1983; Robl and Barron 1988; Russell and Barker 1983; Kluessendorf et al. 1988; Telford 1988). In western North America, deposition of black shales was limited to deep basinal environments in the Givetian and early Frasnian (e.g., the Pilot

Basin) but spread over wide areas from the mid Frasnian through the Famennian (e.g., Morrow and Geldsetzer 1988; Sandberg, Poole, and Johnson 1988). Outside North America, organic-rich facies are common in Middle–Upper Devonian marine successions—for example, in western and central Europe (Feist 1990; Schindler 1990, 1993; Buggisch 1991; Schönlaub et al. 1992; Hladil and Kalvoda 1993; Racki et al. 1993; Joachimski and Buggisch 1993), Morocco (Wendt and Belka 1991), South China (Hou et al. 1988; Bai and Ning 1988), Brazil (Popp and Barcellos-Popp 1986; Russell 1990), and the former U.S.S.R. (Avkhimovitch et al. 1988). Interregional correlation of these black shale horizons implies synchroneity of changes in marine environmental conditions over millions of square kilometers, but the degree of correlativity of these horizons at an intercontinental scale is uncertain. Also noteworthy is that many Middle and Late Devonian lacustrine successions contain organic-rich facies and record mass mortality events associated with deep-water anoxia—for example, the Achanarras, Sandwick, and Edderton "fish beds" of the Orcadian Basin of Scotland (Mykura 1983; Trewin 1986).

Marine Geochemical Anomalies

Widespread deposition of Devono-Carboniferous black shales (and, later, Carboniferous coals) led to a substantial increase in the burial rate of organic carbon on a global scale, and to associated changes in the C- and S-isotopic composition of many reservoirs in the Earth's exogenic system. A quantitative measure of changes in organic carbon burial rates is provided by the marine carbonate C-isotopic record, in which higher $\delta^{13}C$ values are linked to enhanced burial of ^{13}C-depleted organic matter and to increases in the $^{13}C/^{12}C$ ratio of the ocean-surface dissolved inorganic carbon (ΣCO_2) reservoir and of carbonates precipitated therefrom (Kump 1991; Joachimski and Buggisch 1993). The Phanerozoic marine carbonate C-isotopic record exhibits a net +4‰ shift during the Devono-Carboniferous, from

+0.5‰ PDB in the Middle Devonian to +4.5‰ PDB in the Late Carboniferous (Veizer et al. 1986; Popp et al. 1986; Lohmann and Walker 1989; Berner 1989). This shift in $\delta^{13}C$ appears to have occurred in a stepwise manner—for example, with abrupt increases (+2‰ to +3‰) associated with individual episodes of widespread organic carbon burial and followed by decreases of somewhat lesser magnitude. For those C-isotopic excursions that have been well studied (e.g., the F/F boundary Upper Kellwasser horizon; Playford et al. 1984; McGhee et al. 1986; Buggisch 1991; Halas et al. 1992; Joachimski and Buggisch 1993; Joachimski et al. 1994; Wang et al. 1996), the shifts appear to have been synchronous and of uniform magnitude on a global scale. The synchroneity, uniformity, and facies-independence of these C-isotopic excursions are strong evidence that they represent global shifts in $\delta^{13}C$ of the oceanic surface ΣCO_2 reservoir (Joachimski and Buggisch 1993). Reports of negative carbonate $\delta^{13}C$ excursions at major boundaries (e.g., F/F) (Xu et al. 1986; Wang et al. 1991; Geldsetzer et al. 1993; Yan et al. 1993; Wang et al. 1996) are likely to represent either geochemical changes in water masses isolated from the global ocean or contamination of carbonates by ^{13}C-depleted organic matter (e.g., Wang et al. 1996).

In marine environments, high rates of organic carbon burial are generally associated with increased production and burial of reduced sulfur, mainly as iron sulfides (Berner 1978; Berner and Raiswell 1983; Berner and Canfield 1989). Coburial of reduced (organic) C and reduced (sulfide) S during the Late Devonian is evidenced locally by (1) positive covariance of total organic carbon and total sulfur (a proxy for reduced S) in black shales of that age (e.g., Leventhal 1987), and (2) association of positive sulfide S-isotopic anomalies with organic-rich horizons (e.g., Geldsetzer et al. 1987; Wang et al. 1996), and globally by (3) positive covariance of marine carbonate $\delta^{13}C$ and marine evaporite $\delta^{34}S$ values (Holser et al. 1989). A major excursion in marine evaporite $\delta^{34}S$ values (+8‰ to +10‰) between the Middle Devonian and mid Mississippian is likewise

attributable to enhanced burial of ^{34}S-depleted reduced S (Holser et al. 1989). Coburial of large quantities of reduced C and S during the Late Devonian would have released large quantities of molecular oxygen (O_2) to the atmosphere (Kump and Garrels 1986; Berner and Canfield 1989), which is evidenced by coeval increases in the abundance of fossil charcoal, reflecting a greater frequency of forest fires in an oxygen-rich atmosphere (Chaloner 1989; Jones and Chaloner 1991), and by gigantism of terrestrial invertebrates, reflecting an enhanced ability to take up oxygen in tissues through diffusion (Graham et al. 1995).

Global Climate Change

Atmospheric CO_2 levels dropped precipitously during the Devono-Carboniferous as a consequence of elevated silicate weathering rates and enhanced burial of organic carbon (Berner 1992, 1997, and pers. comm., 1995). Atmospheric pCO_2 estimates based on global carbon cycle models suggest a decrease from pre-Devonian concentrations of 4 to 20 PAL (present atmospheric level, generally taken as a preindustrial value of 270 ppmV) to mid-Carboniferous concentrations of about 1 PAL (Berner 1991, 1993, 1994). Imprecision in these estimates is largely the result of uncertainty regarding relative rates of chemical weathering by nonvascular algal-lichen floras versus higher land plants (Berner 1994), and recent studies supporting low weathering efficiency ratios (about 0.15) (e.g., Drever and Zobrist 1992; Cochran and Berner 1996; Moulton and Berner 1998) would favor higher atmospheric pCO_2 estimates for the pre-Devonian (about 12 to 20 PAL). The latter values are independently supported by C-isotopic analysis of soil carbonates (Mora et al. 1991, 1996; Yapp and Poths 1992, 1996) and by analysis of stomatal densities in contemporaneous terrestrial plants (McElwain and Chaloner 1995).

Rapid drawdown of atmospheric CO_2 levels is likely to have been the main cause of coeval climatic cooling, culminating in a brief episode of continental glaciation in the latest Devonian.

Evidence for climatic cooling during the Devono-Carboniferous is provided by a net +4‰ shift in the marine carbonate $\delta^{18}O$ record—that is, from −5‰ PDB (or less) to −1‰ PDB (Lohmann and Walker 1989), most of which occurred within a time interval of about 7 to 15 m.y., spanning the Devono-Carboniferous boundary (Popp et al. 1986; Dunn 1988; Brand 1989). The rapid, ubiquitous nature of this O-isotopic shift argues against both a diagenetic mechanism (e.g., Degens and Epstein 1962) and a secular change in the $\delta^{18}O$ composition of seawater (e.g., Holland 1984). Rather, the shift is likely to reflect a significant cooling of tropical sea surface temperatures (SSTs) (i.e., by about 16°) (cf. Craig 1965), or, if coeval growth of Gondwanan icesheets caused a +0.5‰ ^{18}O enrichment of seawater (Crowley and Baum 1991), by about 14°C. Thus, assuming no other changes in the $\delta^{18}O$ composition of seawater, tropical SSTs may have cooled from about 40°C in the Devonian to near-modern temperatures of about 24° to 26°C in the Carboniferous (Veizer et al. 1986). Although these estimated temperature changes are large, they may not be unrealistic in view of recent studies demonstrating temperature changes of 3° to 8°C in tropical SSTs at glacial timescales (i.e., <<1 m.y.) during the Pleistocene (Emiliani and Ericson 1991; Rostek et al. 1993; Anderson and Webb 1994). The strongest argument in favor of temperature control of Devono-Carboniferous marine carbonate $\delta^{18}O$ values is the link it provides to coeval changes in organic carbon burial rates, atmospheric CO_2 levels, and global climate.

Rapid global cooling is likely to have been responsible for a brief episode of continental glaciation during the latest Devonian. Glacigenic deposits of probable Famennian age, including diamictites with striated and polished pebbles, rhythmites with dropstones, erratic boulders, and striated pavements, are widespread in northern Brazil, and potentially correlative deposits are present in North Africa and Argentina, implying glaciation over an area more than 3,500 km long (Caputo 1985; Caputo and Crowell 1985; Frakes et al. 1992). Caputo (1985) assigned this event a "mid-Famennian"

age, but Streel (1986, 1992) redated it as latest Famennian ("Strunian" = middle *praesulcata* conodont zone). The latter age is more consistent with Late Devonian eustatic trends—a late Famennian sea-level fall (Johnson et al. 1985) may be the eustatic signature of continental icesheet growth.

CAUSES OF MIDDLE AND LATE DEVONIAN MARINE EVENTS AND GLOBAL CLIMATE CHANGE

There is, at present, no consensus regarding the causes of Middle and Late Devonian marine anoxic and extinction events and coeval changes in global climate. A variety of mechanisms have been postulated, including plate tectonics, eustasy, oceanic overturn, and bolide impacts, all of which contain plausible elements. In discussing causality, however, one should bear in mind that most events in Earth history are a response not to a single factor but to complex interactions among a set of factors, and that it is necessary to distinguish between proximate and ultimate causes for a given event. Furthermore, to achieve robustness, a model for global events of a given age must account for *all* coeval phenomena through a set of interrelated causal factors (i.e., an appeal to separate mechanisms for different phenomena would represent special pleading for temporal coincidences). With regard to the Middle and Late Devonian, valid hypotheses must account for unique aspects of the marine biotic crisis (i.e., its duration, episodicity, and selectivity) and its relationship to geochemical anomalies in coeval marine strata. Existing models for Middle and Late Devonian global events will be evaluated here and compared with the model of Algeo et al. (1995), which focuses on the potential role of vascular land plant evolution.

Existing Models for Marine Events and Global Climate Change

Tectonism and climate change can have far-reaching consequences on the evolution and extinction of contemporaneous biotas. The Middle-to-Late Devonian biotic crisis has been attributed to cooling of tropical seasurface waters as a consequence of changes in ocean circulation or plate geometry (e.g., closure of a warm paleo-Tethyan Ocean and deflection of cold high-latitude currents into equatorial areas) (Copper 1986; Stanley 1988; Schindler 1990, 1993). If climatic cooling did play a direct role in the biotic crisis, then it must have occurred episodically and in concert with marine anoxic events, bcause of the correlativity of such events with extinction horizons (Algeo et al. 1995). Although multiple factors may have contributed to climate change at a range of timescales, stresses on marine biotas are more likely to have been caused by short-term cooling associated with episodes of increased organic carbon burial rather than to long-term effects associated with changes in ocean circulation. Global warming has also been proposed as a mechanism for the Devonian biotic crisis (Thompson and Newton 1988; Brand 1989), but this hypothesis is difficult to reconcile with evidence of contemporaneous continental glaciation (Caputo 1985; Caputo and Crowell 1985), drawdown of atmospheric pCO_2 (Berner 1991, 1994), and disproportionate destruction of tropical rather than boreal fauna (McGhee 1996).

Tectonic factors may have been important in either triggering or enhancing individual episodes of marine anoxia and organic carbon burial (e.g., through thrust loading of foreland basins and areal expansion of mid-depth oxygen-minimum zones) (Ettensohn, Miller, et al. 1988). In support of this idea are (1) early initiation of anoxia in areas proximal to active orogens, as evidenced, for example, by black shales of Pragian age in the northern Appalachian Basin, adjacent to the Acadian Orogen (Woodrow et al. 1988), and (2) migration of anoxic areas cratonward in response to foreland basin thrust loading, as in the Late Devonian central Appalachian Basin (Ettensohn, Miller, et al. 1988). Although tectonic factors were important at a minimum in enhancing marine anoxia in periorogenic areas, their significance for widespread deposition of organic-rich facies in tectonically stable epicratonic regions is unclear.

Global sea-level changes (eustasy) have been widely cited as a possible mechanism for both marine anoxic and extinction events. The Middle and Late Devonian biotic crisis has been attributed both to sea-level rises (transgressions), which would promote bottomwater anoxia through increased water depths and enhanced water-column stratification (Berry and Wilde 1978; Johnson et al. 1985), and to sea-level falls (regressions), which would cause habitat loss within epicontinental seas (Johnson 1974; Sandberg, Poole, and Johnson 1988). Although most black shale horizons of that age are thought to be correlative with eustatic transgressions (e.g., Sandberg, Poole, and Johnson 1988; Sandberg, Ziegler, et al. 1988), a role for sea-level rises in precipitating episodes of extinction is doubtful because eustatic cycles of this age are similar in frequency and amplitude to others throughout the Phanerozoic, few of which are associated with mass mortality (Johnson et al. 1985; Haq et al. 1987). Further, although eustatic rises during the Middle and Late Devonian may have predisposed epicontinental seas to low-oxygen conditions, additional factors must have been operative at this time to have produced intense anoxia on an intercontinental scale. On the other hand, models invoking marine regressions are fundamentally improbable because (1) Middle and Late Devonian sea-level elevations were generally high and stable or rising (e.g., Johnson et al. 1985), and (2) much larger regressions in association with Pleistocene continental icesheet growth have had little effect on marine biotas.

Overturn of sulfidic open-ocean bottomwaters, as a consequence either of long-term surfacewater cooling and thermocline breakdown (Wilde and Berry 1984) or of a bolide impact (Geldsetzer et al. 1987), has been proposed to account for both the marine biotic crisis and widespread anoxia in epicontinental seas during the Middle and Late Devonian. A fundamental prediction of this model is a catastrophic decline in marine primary productivity following overturn (i.e., a Strangelove Ocean) (Hsü and McKenzie 1985; Kump 1991), resulting in (1) a reduced flux of labile organic matter from marine phytoplankton to the seafloor, (2) lower bacterial sulfate reduction rates (because of the paucity of labile organic matter) and, hence, lower marine sulfide $\delta^{34}S$ values, and (3) lower marine carbonate $\delta^{13}C$ values as a consequence of a reduction in the C-isotopic fractionation between shallow- and deep-water masses that is dynamically maintained by phytoplankton. However, existing petrographic and isotopic data are inconsistent with this model: (1) organic matter in Middle–Upper Devonian black shales is overwhelmingly of marine algal origin (Maynard 1981; Jaminski et al. 1998), (2) early diagenetic sulfides in black shales exhibit significant ^{34}S enrichments (Maynard 1980; Goodfellow and Jonasson 1984; Geldsetzer et al. 1987; Halas et al. 1992), suggesting elevated rates of bacterial sulfate reduction owing to the greater availability of labile marine organic matter, and (3) marine anoxic event horizons correlate with positive carbonate $\delta^{13}C$ excursions, suggesting *enhanced* rather than diminished photosynthetic fixation and burial of ^{13}C-depleted organic matter (Joachimski and Buggisch 1993).

A bolide impact was postulated as a mechanism for the Frasnian–Famennian mass extinction and associated marine anoxic event (McLaren 1982), and intensive research has generated some evidence for one or more bolide impacts at or stratigraphically close to the F/F boundary, including (1) iridium anomalies in Australia, China, and Canada (Playford et al. 1984; Hurley and Van der Voo 1990; Geldsetzer et al. 1987; Wallace et al. 1991; Wang et al. 1991, 1993, 1994), (2) microtektites in China and Belgium (Wang 1992; Wang and Chatterton 1993; Claeys et al. 1992; Claeys and Casier 1994), and (3) a probable impact site in Sweden, the 55-km-diameter Siljan Ring structure (368 ± 1 Ma) (Hodge 1994; McGhee 1996). However, the strength of the geochemical evidence for an impact is questionable, because (1) concentrations of iridium and other platinum group elements are generally low compared to those at the Creta-

ceous–Tertiary boundary, (2) some elemental ratios are not compatible with a meteoritic origin, and (3) enrichment of platinum group metals in some sections is probably the result of other factors, such as slow sedimentation rates, changes in redox conditions, or cyanobacterial concentration (Playford et al. 1984; Geldsetzer et al. 1987; Dyer et al. 1989; Orth et al. 1990; Wang et al. 1993). Furthermore, not all of the reported iridium anomalies and microtektite horizons coincide stratigraphically with the F/F boundary, and a number of boundary sections proximal to the presumed impact site lack physical or geochemical evidence of such an event (McGhee et al. 1986; Hurley and Van der Voo 1990; Claeys et al. 1992; Nicoll and Playford 1993; Wang et al. 1994; McGhee 1996).

Important issues need to be addressed with regard both to the evidence for a Late Devonian bolide impact and to the significance of such an event for contemporaneous shallow-marine ecosystems. The intense search for physical and chemical evidence within a narrow stratigraphic window centered on the F/F boundary raises the possibility of sampling bias and suggests the need for a control test—a search for evidence of extraterrestrial inputs in "background" strata that are not proximal to the F/F boundary. Also unresolved is whether impact layers are more prevalent than average within Upper Devonian strata because of a preservation bias associated with anoxic marine environments (e.g., exclusion of burrowing benthos would preclude disruption of thin impact ejecta blankets that would otherwise be bioturbated out of existence). Such a preservation bias might account for the fact that evidence of bolides has been found at several horizons within the Upper Devonian, some of which are not correlated with extinction episodes. Finally, even if one or more impacts did indeed occur during the Late Devonian, it would be necessary to consider (1) whether these played any significant role in the Middle and Late Devonian biotic crisis, and (2) if so, whether that role was amplified relative to the size of the impactor because of preexisting destabilizing stresses within Late Devonian shallow-marine ecosystems. Thus, the most important questions turn not on the existence of an impact but on its consequences.

The "Devonian Plant Hypothesis"

Whereas the models just discussed take little or no consideration of terrestrial developments, the "Devonian plant hypothesis" of Algeo et al. (1995) and Algeo and Scheckler (1998) focuses on potential links between coeval events in the terrestrial and marine realms. Critical in establishing such links is the role of soils as a geochemical interface between the atmosphere/hydrosphere and lithosphere, and the role of plants as a mediator of weathering intensities and geochemical fluxes (figure 12.9). Specifically, enhanced pedogenesis associated with the rapid spread of vascular land plants during the Middle and Late Devonian is likely to have resulted in elevated fluxes of soil solutes (especially biolimiting nutrients) as a consequence of (1) enhanced mineral leaching, (2) fixation of nitrogen by symbiotic root microbes, and (3) shedding of plant-derived detrital carbon compounds. The large flux of terrestrial plant litter is likely to have promoted development in riparian and paralic habitats of complex food webs of detritivores, bacteria, and fungi, which were capable of yielding soluble growth factors that enhanced the productivity of contemporaneous marine phytoplankton. Thus, elevated riverborne nutrient fluxes may have promoted eutrophication of semi-restricted epicontinental seas and stimulated algal blooms (figure 12.9). Circumstantial evidence for high marine primary productivity during the Middle and Late Devonian is (1) the large proportion of marine algal matter in coeval black shales (e.g., Maynard 1981; Jaminski et al. 1998), and (2) the wide geographic but restricted stratigraphic distribution of enigmatic fossils such as *Protosalvinia*, which may have been the product of such algal blooms (Schopf and Schwietering 1970). Analogous links are known from modern coastal waters and restricted marine basins, in which natural or anthropogenic nutrient loading has led to eutrophication and algal blooms

Figure 12.9.
Flowchart model linking the development of arborescence and seeds among vascular land plants to Middle and Late Devonian marine biotic and global sedimentological, geochemical, and climatic events. Events are arrayed by relative duration with transient effects on the left and long-term effects on the right. *Solid outlines* indicate documented geologic records; *dashed outlines* indicate processes inferred to link records (see text for discussion).

(e.g., Fong et al. 1993; Lyons et al. 1993; Turner and Rabalais 1994). However, although the repeated temporal coincidence of marine extinction events and organic-rich horizons supports a link of both phenomena to episodic development of widespread bottomwater anoxia, it is uncertain whether oxygen deprivation itself or some related factor (e.g., climate cooling via enhanced organic carbon burial) was the principal agent of biotic destruction.

Support for this model may be provided by temporal relations between events in the terrestrial and marine realms during the Middle and Late Devonian (see figure 12.2). Potentially significant temporal coincidences include (1) a rapid increase in the size of vascular land plants (see figures 12.1, 12.3) and the onset of marine anoxia in many areas during the Eifelian and Givetian, (2) diversification and spread of progymnosperms and an increase in the frequency and intensity of marine anoxic events in the late Givetian, (3) the mid to late Frasnian spread of *Archaeopteris* forests and the Frasnian–Famennian boundary event, and (4) the mid to late Famennian appearance of seed plants and the Devonian–Carboniferous boundary event. Of these temporal coincidences, only the last one is reasonably tightly constrained, i.e., the appearance of seed plants is known to predate the D/C boundary (Hangenberg) extinction event by about two conodont zones, probably representing no more than 1 m.y. (Sandberg, Ziegler, et al. 1988). This lag could be imputed to the temporally sigmoidal expansion characteristic of most biotic radiations (Sepkoski 1979), and, in this regard, it is important to bear in mind that the effects of plants on their physical environment depend less on first appearances than on actual increases in their abundance, biomass, and geographic distribution. In the case of seed plants, this expansion did not begin until just before and after the D/C boundary (Scheckler 1986a; Rothwell and Scheckler 1988; Beerbower et al. 1992).

Although the "Devonian plant hypothesis" is robust in accounting for most observed biotic and geochemical patterns associated with the Middle and Late Devonian, it is important to note that land plants were not evolving in a vacuum, and that the environmental effects associated with their evolution were mediated by coeval tectonic, eustatic, and atmospheric developments. Thus, any accurate representation of causation of these events must depend on a complex set of interactions among multiple factors (e.g., figure 12.9). In this regard, an important question is whether plant-induced changes in weathering processes represented merely a background factor predisposing epicontinental seas to low-oxygen conditions, in which case Middle and Late Devonian marine anoxic episodes must have been triggered by other factors (e.g., eustatic transgressions), or whether specific paleobotanical developments (e.g.,

rapid spread of a newly evolved taxon) could have precipitated episodes of marine anoxia and biotic extinction lasting no longer than 1 million years. The answer to this question has significant implications for rates and patterns of land plant evolution and terrestrial ecosystem development. A second question is whether changes in global climate and geochemical fluxes, related to the spread of land plants, produced stresses in marine ecosystems that increased their susceptibility to other agents of destruction. Specifically, if marine ecosystems were already under severe stress because of weathering-related changes in geochemical fluxes, then a relatively small bolide or drop in global seawater temperature might have been sufficient to cause widespread ecological disruption. Since the frequency of bolide impacts rises sharply with decreasing diameter (Shoemaker et al. 1990), temporal coincidence of Middle and Late Devonian marine anoxic events with bolide impacts might not have been statistically improbable under such conditions. Additionally, global temperature, including that of seawater, was declining at this time (Berner 1997, 1998) presumably as a result of atmospheric CO_2 depletion.

TESTING THE "DEVONIAN PLANT HYPOTHESIS"

Support for the "Devonian plant hypothesis" of Algeo et al. (1995) and Algeo and Scheckler (1998) rests primarily on two considerations: (1) temporal correlations between important developments in the terrestrial and marine realms during the Middle and Late Devonian, such as the appearance of arborescence and seeds and peak episodes of marine anoxia and biotic extinction (figure 12.2), and (2) a plausible mechanism to link these developments without appeal to temporal coincidences or multiple causalities (figure 12.9). Although coevality of events neither demonstrates causality nor reveals its directionality, a model that successfully links a wide range of coeval events to a single ultimate cause or conjunction of factors is implicitly self-supporting. However, the critical test of any hypothesis is its ability to make testable predictions (cf. Alvarez et al. 1980), and, in this context, the "Devonian plant hypothesis" offers a variety of testable predictions that could be used to test its validity. Facets of the hypothesis that might serve as a focus for future research efforts include (1) relative timing of Middle and Late Devonian paleobotanical developments and marine anoxic and extinction events, (2) secular changes in the composition and structure of paleosols (e.g., depth of rooting and degree of textural and chemical horizonation), (3) secular changes in the bulk mineralogy, geochemistry, and texture of clastic sediments, (4) secular changes in the provenance and geochemistry of sedimentary organic matter, and (5) secular changes in the C-, S-, and O-isotopic composition of marine precipitates as a record of geochemical changes in the oceanic/atmospheric system. It should be noted that strong spatial heterogeneity in most of the relevant paleobotanical, sedimentological, and geochemical parameters is likely to make assessment of the "Devonian plant hypothesis" inherently more difficult than the search for physical and chemical evidence of bolides, and, hence, a far larger database will be needed to provide a fair test of the hypothesis. Furthermore, no single dataset is likely to resolve the issue, and acceptance or rejection of the hypothesis ultimately will depend on a better understanding of interrelations among diverse phenomena.

CONCLUSIONS

The evolutionary development and spread of vascular land plants during the Middle and Late Devonian may have been an important factor in coeval marine anoxic and extinction events. Although taxonomically diverse, pre-Middle Devonian terrestrial floras had little effect on their physical environment because of shallow root systems and narrow habitat ranges. According to the "Devonian plant hypothesis," the influence of land plants on pedogenic weathering processes and global geochemical cycles

increased substantially during the Middle and Late Devonian with the advent of arborescence (tree-sized stature), which increased depths of rooting and pedogenesis, and the seed habit, which freed plants from reproductive dependence on moist lowland habitats and allowed colonization of drier upland areas. These developments resulted in a transient increase in rates of pedogenesis and in large permanent increases in the thickness and areal extent of deeply weathered soil profiles. Accelerated pedogenesis led to increased sediment yields through episodic disturbance of developing soils and to increased riverine nutrient fluxes through enhanced rates of pedogenic chemical weathering and soil microbial nitrogen fixation. Elevated nutrient fluxes and bacterial or fungal breakdown of plant detritus during aqueous transport produced eutrophic conditions in semirestricted epicontinental seaways, stimulating algal blooms and causing development of widespread anoxia. Enhanced rates of silicate mineral weathering and organic carbon burial drew down atmospheric pCO_2, resulting in changes in marine carbonate equilibria, the C- and O-isotopic composition of the oceanic ΣCO_2 reservoir and marine carbonates, and global climate.

Temporal correlations provide circumstantial evidence in favor of terrestrial–marine teleconnections: (1) rapid increases in the maximum size of vascular land plants and onset of episodic marine anoxia during the Eifelian–Givetian, (2) advent of large, deeply rooted progymnosperms and increases in the frequency and intensity of marine anoxic events during the late Givetian, (3) a peak in the abundance of archaeopterid forests during the mid Frasnian to mid Famennian, which may or may not have been related to the Frasnian–Famennian boundary Kellwasser Event, and (4) the appearance and rapid diversification of seed plants during the late Famennian, probably less than 1 million years prior to the Devonian–Carboniferous boundary Hangenberg Event. Correlativity of Middle and Late Devonian black shales with extinction horizons implicates bottomwater anoxia or a related environmental parameter (e.g., increased turbidity or seawater cooling) as the proximate cause of the biotic crisis among tropical-marine invertebrates. The Middle-to-Late Devonian spread of vascular land plants and consequent changes in pedogenic weathering processes and global geochemical fluxes may have been the ultimate cause for coeval marine anoxic and extinction events and for coeval changes in global climate.

ACKNOWLEDGMENTS

Thanks to Robert Berner and two anonymous readers for thorough reviews of the manuscript, which also benefited from discussions with Elso Barghoorn, William Chaloner, Geoffrey Creber, Patricia Gensel, Herman Pfefferkorn, Greg Retallack, and Jacek Jaminski. Research support for study of black shales and global geochemical cycles is acknowledged from the University of Cincinnati Research Council and National Research Council (TJA) and the U.S. Department of Energy (JBM). Research support for study of early terrestrial ecosystems (SES) is acknowledged from the National Science Foundation (BSR 83-15254 and IBN 97-28719), Harvard University (Bullard Fellowship 1999), National Geographic Society (NGS 2409-81), Virginia Center for Coal and Energy Research (1984-87), and the Polar Continental Shelf Project, Canada (1989).

13

Diversification of Siluro-Devonian Plant Traces in Paleosols and Influence on Estimates of Paleoatmospheric CO_2 Levels

Steven G. Driese and Claudia I. Mora

Plant rhizome, plant root, and animal traces are abundant in paleosols (fossil soils) preserved in Paleozoic terrigenous successions within the Appalachian Basin. This "pedological signature" of vascular land plant evolution and diversification has been largely ignored by paleobotanists. Paleosols crop out extensively in the Appalachian region of eastern North America, from the Canadian Maritime Provinces southward to the Tennessee—Alabama border along the western side of the Appalachian Orogen (figure 13.1). The paleosols occur primarily in terrigenous clastic redbed deposits ranging in age from Late Ordovician to Early Permian (Mora and Driese 1999), which encompasses a time interval characterized by rapid evolution and diversification of terrestrial ecosystems (table 13.1). Because the paleosols formed under relatively constant sediment source area and pedogenic environments (table 13.1), they share generally uniform physical and chemical properties and are thus suitable for investigating evolutionary advances in the gross morphology of plant roots and rhizomes preserved as traces. Citing specific examples that occur within Appalachian redbed paleosols, we report distinct secular variation in root trace morphology during the Sil-

Figure 13.1.
Map showing distribution of Ordovician to Devonian redbed paleosols in Appalachian region of eastern United States and maritime Canada. Paleosol locality numbers are keyed to Table 13.1.

urian and Devonian periods that records important changes in plant rhizome and root systems. We consider the importance of these evolutionary advances in effecting changes in paleosol morphology and chemistry, especially the stable

Table 13.1　Stratigraphic occurrences of Ordovician to Devonian redbed paleosols, and the botanical and biological advances that affected their morphology and chemistry

System	Stage	Marine-parented coastal paleosols	Non-marine-parented inland alluvial paleosols	Botanical advances[9]	Biological advances[10]
D E V O N I A N	Famennian			*Archaeopteris* progymnosperms; seed reproduction (seed ferns); megaphyllous leaves; **deep roots** (10–15 cm diam., 1.0–1.5 m long)	
	Frasnian		Catskill Formation (PA,NY)[8]		Spiders
	Givetian	Catskill Formation (NY)[7]	Malbaie Formation (Q)[6]	*Eospermatopteris* cladoxylaleans?; **first trees** (arborescence); **shallow roots** (2–3 cm diam., 10–20 cm long); secondary xylem; progymnosperms; heterospory	
	Eifelian				
	Emsian		Battery Point Formation (Q)[5]		Insects
	Pragian			Trimerophytes; lycophytes; first leaves (microphyllous); zosterophyllophytes; **first roots** (1–3 mm diam., 1–3 cm long) **Major diversification of land plants**	
	Lochkovian				
S I L U R I A N	Pridoli			Rhyniophytes	Centipedes
	Ludlow	Moydart Fm. (NS)[4] Bloomsburg Fm. (PA)[3]		**Roots absent;** rhizomatous mats (0.5–1.5 mm diam., 0.5–1.5 cm long); **first vascular plants (tracheids)**	Millipedes
	Wenlock				**First soil animals** (trigonotarbids)
	Llandovery				
ORD	Ashgill	Juniata Formation (TN, VA)	Juniata Formation (PA)[2]	**First nonvascular land plants?;** homospory	Soil animal traces?

Area examined in this study: Appalachian Basin in the United States and Canada.

U.S. states: NY, New York; PA, Pennsylvania; TN, Tennessee; VA, Virginia. Canadian provinces: NS, Nova Scotia; Q, Québec.

References: Banks et al. (1985)[9]; Beerbower (1985)[9,10]; Boucot et al. (1974)[4]; Bridge and Gordon (1985)[8]; Bridge and Willis (1994)[7]; Cant and Walker (1976)[5]; Diemer (1992)[8]; DiMichele and Hook (1992)[9,10]; Dineley (1963)[4]; Driese and Foreman (1991, 1992)[1]; Driese and Mora (1993)[8,9]; Driese et al. (1992)[3,9]; Driese et al. (1997)[7–9]; Elick et al. (1996, 1998a)[5,9]; Feakes and Retallack (1988)[2,9,10]; Gensel and Andrews (1984, 1987)[9]; Gordon (1988)[10]; Gordon and Bridge (1987)[8]; Gray (1985)[9]; Gray and Shear (1992)[10]; Hoskins (1961)[3]; Kidston and Lang (1917, 1920a,b, 1921a)[9]; Lawrence and Rust (1988)[5,6]; Mägdefrau (1952)[9]; Mora et al. (1991)[3,8,9]; Mora et al. (1996)[3,7,9]; Mora and Driese (1999)[1–9]; Rahmanian (1979)[8]; Retallack (1985, 1986; 1992)[2,3,8–10]; Retallack (1993)[2,9,10]; Retallack and Feakes (1987)[2,10]; Rickard (1969)[3]; Rust (1984)[6]; Sevon (1985)[8]; Sevon and Woodrow (1985)[8]; Shear (1991)[10]; Stewart and Rothwell (1993)[9]; Strother (1988)[3]; Thompson (1970)[1]; Walker (1971)[8]; Walker and Harms (1971)[8]; Willis and Bridge (1988)[7,8]; Woodrow et al. (1973)[7,8]; Yeakel (1962)[2].

isotope chemistry of pedogenic carbonate and subsequent paleoatmospheric pCO_2 estimations.

GENERAL ASPECTS OF PALEOZOIC REDBED PALEOSOLS

For nearly 200 Ma, until the Appalachian Basin was deformed by compression during the Alleghanian orogeny (Pennsylvanian—Permian), a depositional pattern was established that persisted through the development of three major Paleozoic clastic wedges (Colton 1970; Meckel 1970). From the foothills of linear highland uplifts, extending west and northwest into the Appalachian Basin, piedmont alluvial fans graded downslope to a broad alluvial plain, which in turn led to low-gradient delta-plain and coastal mud-flat environments at the interface with a shallow-marine system (Yeakel 1962; Meckel 1970; Thompson 1970; Thompson and Sevon 1982; Sevon et al. 1988; Cotter and Driese 1998). Proximal, higher-gradient alluvial facies are largely coarser-grained, light-colored sandstones and conglomerates. Redbed deposits with paleosols, consisting largely of fining-upward sequences of red channel sandstone overlaid by red shale and siltstone, were deposited lower on the alluvial and deltaic plain, and in coastal-margin mudflat environments (table 13.1). As a result of the relative constancy of depositional processes, Appalachian paleosol-bearing deposits are all redbeds with grossly similar physical and chemical attributes (Mora and Driese 1999).

Paleogeographical reconstructions place the Appalachian Basin region at about 20° to 30° south paleolatitude during Silurian time (Ziegler et al. 1979; Scotese et al. 1979; Scotese and McKerrow 1990; Scotese and Golonka 1992); by Late Devonian to Mississippian time, the region was located at about 4° to 20° south paleolatitude (Van der Voo et al. 1979; Kent 1985; Kent and Miller 1988; Van der Voo 1988; Scotese and McKerrow 1990; Scotese and Golonka 1992). Paleoclimate models for Silurian time predict warm, moist winters and hot, dry summers (Ziegler et al. 1977; Golonka et al. 1994). The Devonian paleoclimate was subtropical to tropical and strongly controlled by the orographic effects of the Acadian orogen, which would have blocked southeasterly trade winds, resulting in a seasonally wet and dry (monsoonal) pattern of precipitation (Woodrow et al. 1973; Woodrow 1985; Witzke 1990; Golonka et al. 1994).

Morphological and mineralogical features of Appalachian redbed claystone paleosols support the paleoclimate models. The abundance of vertic (shrink/swell) features preserved in most of the paleosols (cf. Driese and Foreman 1992; Driese et al. 1992; Driese and Mora 1993; Caudill et al. 1996; Mora and Driese 1999) is consistent with the seasonally moist (four to eight dry months each year) and the typically tropical to warm temperate climate in which many Holocene vertisols develop (Ahmad 1983; Dudal and Eswaran 1988; Coulombe, Dixon, and Wilding 1996; Coulombe, Wilding, and Dixon 1996). Warm tropical to subtropical paleoclimates and seasonal moisture deficit are also suggested by the occurrence of pedogenic car-

bonate in nearly all of the redbed paleosols. In Quaternary soils, pedogenic carbonate forms under conditions of low mean annual precipitation (<50 cm/yr) or under higher precipitation where there is a significant moisture deficit because of high evaporation or evapotranspiration (Goudie 1983; Wright and Tucker 1991). Such a paleoclimate can be inferred for the Appalachian Basin region throughout the Paleozoic era, based on the widespread distribution of redbed paleosols with vertic features (figure 13.1).

METHODOLOGY

Paleosols were first examined in the field and described using the methodology of Retallack (1988). Strata enclosing the paleosols were described and logged using standard field techniques. Root and rhizome traces were identified and interpreted in the field according to criteria outlined in Sarjeant (1975), Pfefferkorn and Fuchs (1991), and Bockelie (1994). Selected hand-specimens of paleosols containing root and rhizomatous traces and pedogenic carbonate were impregnated with epoxy and slabbed; standard petrographic thin sections were prepared from these samples and treated with a dual stain of Alizarin red S and potassium ferricyanide (Dickson 1965, 1966) to identify carbonate mineral phases present. Some larger (>0.5 cm in diameter) root and rhizomatous traces were repeatedly thin sectioned to evaluate their three-dimensional morphology. Micromorphological analysis was conducted on paleosols using standard soil petrographic techniques (Brewer 1976; Wright 1990; FitzPatrick 1993). Thin sections containing pedogenic carbonate were examined under cathodoluminescence (CL; 10 to 12 kv and 150 to 200 μa beam current) using a Citl Cold Cathode Luminescence 8200 mk3 microscope to evaluate recrystallization and growth zonation (Solomon and Walkden 1985; Machel and Burton 1991). X-ray radiography on 1- to 2-cm-thick rock slabs was utilized in an attempt to better interpret three-dimensional morphology, but this was generally unsuccessful because of the lack of strong contrast in material properties between root traces and enclosing paleosol matrix.

Pedogenic micrite (calcite) was identified by transmitted light and CL microscopy, sampled using a dental drill, roasted at 375°C for 1 hr to remove organic material, and reacted with phosphoric acid, following the method of McCrea (1950). Bulk organic matter was concentrated to 0.5 to 3 weight-percent (Mora et al. 1996), mixed with CuO, and combusted at 1,050°C for 15 minutes to produce CO_2. Oxygen and carbon isotope ratios were measured on the VG903 mass spectrometer at the University of Tennessee and are reported in standard δ-permil notation relative to PDB, with a precision of ±0.1‰ for carbonates and ±0.2‰ for organic carbon.

BIOLOGICAL FEATURES IN EARLY PALEOZOIC PALEOSOLS

Background Information

Major diversification and adaptive radiation of land plants and concurrent evolution of herbivorous land animals occurred during the Paleozoic era, marked, in particular, by the rise of vascular plants during the Early and Middle Devonian (table 13.1; see also Edwards 1980; Gensel and Andrews 1984, 1987; Gray and Shear 1992; Edwards and Selden 1993). The appearance of soil macroflora and macrobiota is manifested by changes in soil morphology and chemistry (Mora and Driese 1999). Silurian and Early Devonian plant communities probably occupied wet coastal-margin environments because of their shallow root systems (cf. Mägdefrau 1952; Gensel and Andrews 1984, 1987; Remy et al. 1997), although the recent discovery of late Early Devonian plants with large roots does cast some doubts on this generalization (Elick et al. 1996, 1998a).

The oldest fossilized plant roots with cell structure reported are those of the early lycophyte *Asteroxylon*, reported by Kidston and Lang (1920a, 1921a) from the Rhynie Chert (Lower Devonian, Pragian) of Scotland, which were 1 to 5 mm in diameter and may have been adventi-

tious. Associated smaller plants (*Rhynia, Horneophyton*) in the Rhynie Chert apparently anchored in the soil by horizontal, 1- to 3-mm-diameter rhizomes that represented below-soil extensions of the aerial portions of the plants; very fine (<100 μm in diameter) hairlike rhizoids extended outward from the rhizomes into the soil (Kidston and Lang 1917, 1920b, 1921a). Remy et al. (1997) proposed two major groups of plant growth and life forms in the Rhynie Chert. The first group consisted of plants with creeping rhizomatous parts (e.g., *Rhynia gwynnevaughanii* and *Aglaophyton major*), whereas the second included plants with true subterranean organs (rhizomes *sensu stricto*; e.g., *Asteroxylon mackiei* and *Horneophyton lignieri*).

By mid-Devonian time, plants probably occupied drier parts of the coastal-alluvial plain between the river courses, and by the Late Devonian, relatively diverse communities of shrubs and trees, particularly vascular plants with well-developed root systems, had begun to spread into continental interiors (Banks et al. 1985; Beerbower 1985; Gensel and Andrews 1984, 1987; DiMichele and Hook 1992). Although concentrated on levees and near standing bodies of water, plants had expanded into drier interfluve areas (Banks et al. 1985), as evidenced by the abundance of root traces in paleosols formed in floodplain and overbank deposits (Driese and Mora 1993; Driese et al. 1997; Elick et al. 1998a; Mora and Driese 1999). The impact of these paleobotanical changes on weathering and global geochemical cycles has been recently discussed by Algeo et al. (1995).

Criteria for Distinguishing Animal from Plant Root Traces

Differentiation between animal and plant root traces can be a formidable problem in stratigraphic sequences that include both terrestrial and marine deposits (see, e.g., chapter 11), or in terrestrial paleosols that have been chemically and biologically overprinted by marine processes associated with flooding and transgression (e.g., Driese and Foreman 1991, 1992; Driese et al. 1992). Even more difficult is the taxonomic assignment of fossil roots without direct evidence for attachment to the aerial portions of plants. Our approach in what follows is to describe the morphology of probable plant traces without attempting any taxonomic assignment. Although we cannot rule out the possibility that some of the traces might be of animal origin, we offer these generalized criteria for differentiation between plant and animal traces, which represent modifications of criteria presented by Sarjeant (1975), Klappa (1980), Pfefferkorn and Fuchs (1991), and Bockelie (1994):

1. Animal traces are more likely to maintain a constant diameter along their entire length (proportional to the body size of the animal), whereas root traces may decrease in diameter progressively away from a primary axis (or axes) and also decrease with each bifurcation.

2. Plant root traces may branch and bifurcate either downward or upward (except horizontal rhizomatous systems), but more commonly branch downward, whereas only a few animal traces (e.g., *Chondrites*) show a strongly expressed downward-branching pattern.

3. Animal traces commonly have a meniscate terrigenous sediment infilling, produced during backfilling by the actions of the organism's appendages, whereas plant root traces do not.

4. Plant root traces may have illuviated clays that exhibit birefringence fabric under cross-polarized light; this colloidal clay was deposited as a hypocoating that lined the margin of the root while it was alive and after death and decay, and it may contain organic carbon remains in the center derived from degraded plant tissues.

5. Plant root traces are more commonly calcretized (carbonate-cemented) than are animal burrows, and they typically consist of a micritic sheath with carbon isotope compositions that are ^{13}C depleted relative to other carbonate phases in the paleosols as a consequence of micrite precipitation from soil solutions that include a significant amount of isotopically light, root-respired CO_2.

Animal Traces

Animal traces are the most common macroscale biological features of Ordovician and Silurian redbed paleosols (see table 13.1). Retallack and Feakes (1987) reported clay-, sand-, and silt-filled burrows up to 1 cm in diameter and with prominent meniscate structures occurring in alluvial paleosols in the Juniata Formation (Upper Ordovician) of central Pennsylvania; they interpreted these traces as the earliest evidence of dry soil animals. Driese and Foreman (1991, 1992) described vertical, sand- and silt-filled animal burrows 1 to 2 mm in diameter and up to 15 cm long occurring in pedogenically modified tidal-flat deposits in the Juniata Formation (Upper Ordovician) of eastern Tennessee; these burrows were associated with a marine flooding (transgressive) surface and formed by marine invertebrates utilizing the paleosol as a firmground substrate. Large clay-, sand-, and silt-filled animal burrows, commonly 1 to 3 cm in diameter and up to 30 cm long, with prominent meniscate infillings and localized iron reduction in their centers, are abundant in paleosols formed in more proximal parts of the Bloomsburg Formation (Upper Silurian) redbed succession in central Pennsylvania and southeastern New York (see Retallack 1986). *Chondrites* burrows occur within vertic paleosols in the more distal parts of the Bloomsburg Formation (Upper Silurian) in central Pennsylvania, commonly at the tops of paleosols where they are truncated by marine flooding surfaces and are extensively gleyed (Driese et al. 1992).

Animal burrows are less important features in Devonian and younger paleosols, probably because of the increasing dominance of plant root traces, although animal traces remain common in other, nonpedogenic parts of some sedimentary successions. For example, in Middle and Upper Devonian Catskill Magnafacies rocks of New York, common animal traces include meniscate, ribbed burrows of the freshwater bivalve *Archanodon, Beaconites*, and other types of arthropod traces, millipede and myriapod traces, and dipnoan (lungfish) aestivation burrows (Bridge and Droser 1985; Bridge and Gordon 1985; Bridge and Titus 1986; Gordon 1988). Problematic calcified structures occur in the Cap-aux-Os Member of the Battery Point Formation (Lower Devonian) along the north shore of Gaspé Bay at D'Aiguillon, in northeastern Québec (figure 13.2). These traces are 2 to 5 mm in diameter and up to 30 cm long and consist of vertical to subvertical, nonbranching micritic tubes, which occur in depositional sequences interpreted by Hotton et al. (chapter 11) as dominantly coastal-marine or estuarine, but intercalated with paleosols. The probable animal traces are characterized by a cloudy micritic calcite sheath with a sharply defined outer margin and coarsely crystalline, sparry ferroan calcite infilling or meniscate terrigenous clastic sand and silt infilling, typical of animal traces (figure 13.2).

Puzzling Rhizomatous Traces

The earliest microscale terrestrial plant traces occur in redbed paleosols of the Appalachian Basin within the upper part of the Bloomsburg Formation (Upper Silurian), at several localities in central Pennsylvania. The traces are preserved as networks of highly anastomosing, micrite-lined, 0.5- to 1.5-mm-diameter tubes (figure 13.3). The rhizomatous networks have a dominantly horizontal, matlike aspect; networks also occur within macroscale carbonate pseudonodules, where they were engulfed by later pedogenic carbonate precipitation. The tubes generally are defined by a thin micritic sheath consisting of one or more generations of cloudy micrite and/or microspar, with central pore spaces occluded by one or more generations of clear, equant calcite spar cement (figure 13.3B,C; see also Driese et al. 1992: their figure 7A,B). The networks extend downward less than 1 cm, as seen in thin sections cut perpendicular to bedding, and their anastomosing pattern is especially striking in thin sections cut parallel to bedding. Fine, 50- to 100-μm-diameter calcified filaments extend laterally away from the larger

Figure 13.2.
Photomicrograph photomosaic (plane-polarized light) of micritic vertical tube interpreted as estuarine animal burrow from Battery Point Formation (Lower Devonian), north shore of Gaspé Bay (D'Aiguillon section), Québec, showing meniscate sediment fill (*s*), hematitic outer rind (*h*), cloudy micritic sheath (*m*) and partial pore-filling, and ferroan calcite spar cement pore-filling (*c*).

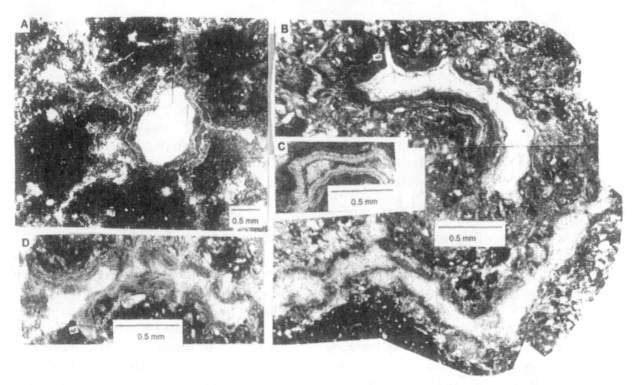

Figure 13.3.
Photomicrograph photomosaics (plane-polarized light) showing examples of Late Silurian rhizomatous traces (or calcified whole plants?) occurring in Appalachian redbed paleosols. **A:** Central hollow pore filled with clear, equant calcite spar cement (*white*), with five cloudy micrite to microspar branches (*gray*) radiating outward from the pore into the surrounding paleosol matrix (*opaque*); thin section cut parallel to bedding. Bloomsburg Formation (Upper Silurian), Danville, PA. **B:** Filamentous micritic tubes (*gray, laminated*) filled with clear, equant calcite spar cement (*white*); thin section cut perpendicular to bedding. Note smaller nodes (*arrows*) that branch off primary tubes. **C:** Enlargement showing details of filamentous, branching, and tapering fine micritic tubes, thin section cut perpendicular to bedding. Note that one or more layers of micrite and microspar comprise outer sheath of filaments, whereas centers are occluded by sparry calcite cement. **D:** Filamentous to feathery micritic tubes (*gray, laminated*) filled with clear, equant calcite spar cement (*white*); thin section cut perpendicular to bedding. Note smaller branches (*arrows*) that branch off primary tubes. Figure parts (B), (C), and (D) from Bloomsburg Formation (Upper Silurian), Port Clinton, PA.

tubes with about a 90° angle of intersection (figure 13.3B); these are much larger than rhizoids, attached to larger rhizomes, reported by Kidston and Lang (1917, 1920a) from early vascular plants preserved in the Rhynie Chert (Lower Devonian, Pragian). The Bloomsburg calcified filaments are also larger than structures interpreted by Klappa (1979a) as calcified organic filaments of Quaternary soil fungi, algae, actinomycetes, and root hairs of vascular plants. Morphologically similar rhizomatous structures occur within large micrite nodules found in paleosols in the Moydart Formation (Upper Silurian, Ludlovian—Pridolian) of Nova Scotia.

These Late Silurian rhizomatous structures may represent calcified whole plants, either bryophyte or rhyniophyte grade, which were 0.5 to 3 mm thick, or perhaps calcified fungal mycelia or possibly lichen rhizines (cf. Klappa 1979b; C. Hotton, pers. comm., 1997). The plants or fungi were living in moist, pedogenically modified deposits of coastal mudflat and marsh environments (Driese et al. 1992). Alternatively, the traces may even be calcified thalli of *Nematothallus*, a nonvascular nematophyte that was reported by Strother (1988) from an organic carbon-rich lens in the Bloomsburg Formation exposure near Port Clinton, Pennsylvania, the section that has yielded the best-preserved plant remains. Because no data exist on the rhizome systems for early (*Cooksonia*-type) rhyniophytes, and because of the morphological (diameter, length) similarity between the calcified rhizomatous features described here and bunched strands of *Nematothallus*, we cannot discount the possibility that these are simply mats of calcified thalli (P. Gensel, pers. comm., 1996).

Earliest Macroscale Root Traces

Macroscale root traces occurring in paleosols of the Battery Point Formation (upper Lower Devonian, Emsian) of Québec are considerably larger than the rhizomatous features noted in Upper Silurian paleosols. Fine, 1 to 3 mm in diameter, clay-lined root traces up to 15 cm long occur in red claystone paleosols of the Cap-aux-Os Member along the north shore of Gaspé Bay between Penouille and Fort Peninsule; they are the earliest examples, in the Appalachian Basin succession, of root traces exhibiting a dominantly fibrous, vertical disposition. Plants producing these traces may have included trimerophytes such as *Psilophyton* and *Trimerophyton*, based on their size and association with plant macrofossils in sub- and superjacent layers.

One or more varieties of exceptionally large, clay-lined, and sand-filled root traces and attached aerial portions of stems occur in the overlying Prével Member (latest Emsian) along the south shore of Gaspé Bay (Elick et al. 1996, 1998a). These traces have primary axes that are 0.5 to 3 cm in diameter and show a rhizomatous habit in which primary axes generally follow horizontal clay seams (figure 13.4A; see also Elick et al. 1998a: their figures 2, 3A). However, the traces also bifurcate, branch, and taper downward (figure 13.4B,C; see also Elick et al. 1998a: their figures 2, 3E,F) in a manner reminiscent of Schweitzer's (1980b) reconstruction of the root system of the early lycophyte *Drepanophycus spinaeformis* (cf. Rayner 1984; *D. qujingensis* in Li and Edwards 1995). Upward dichotomous H-branching, clay-lined, and sand-filled traces attached to the horizontal rhizomes were interpreted by Elick et al. (1996, 1998a) as casts of the above-ground portions of the plants (figure 13.4D; see also Elick et al. 1998a: their figure 3B,C), which may have been larger (1 to 2 m tall) trimerophytes such as *Pertica* (P. Gensel, pers. comm., 1996), or perhaps a large lycophyte such as *Drepanophycus*, based on their gross morphology and the rare preservation of

spinelike protuberances (cf. Banks and Grierson 1968; Rayner 1984; Li and Edwards 1995).

Micritic rhizoliths occur in red silty paleosols of the Cap-aux-Os Member exposed on the south shore of Gaspé Bay, near East Sandy Beach (Cap Haldimand), and comprise an outer micritic sheath, an inner clay or iron oxide lining, and a calcite cement pore-filling. Although these traces superficially resemble micritic tubes interpreted as marine animal traces seen in the Cap-aux-Os Member along the north shore of Gaspé Bay at D'Aiguillon (figure 13.2), they are most certainly plant in origin because of their downward-tapering and branching, and their association with illuviated clay and iron oxides filling interpedal pores and noncalcretized root traces. The central pore-filling cement in traces from East Sandy Beach is nonferroan calcite of probable meteoric (freshwater) origin, whereas ferroan calcite of probable burial origin occludes the pores of the traces from D'Aiguillon. The plant(s) producing these traces from East Sandy Beach is/are uncertain.

Root Traces of the Earliest Trees

Middle Devonian (Givetian) paleosols exposed near Gilboa, New York, contain large (up to 55 cm in diameter and 10 m tall) *in situ* stumps of probable arborescent cladoxylaleans (*Eospermatopteris*) (see Goldring 1924, 1927; Willis and Bridge 1988: their figures 6D, 7D; Bridge and Willis 1994: their figure 10A; Boyer and Matten 1996; Driese et al. 1997: their figure 4). Root traces outlined by carbon films are attached to the strongly flared stump casts, which comprised an aerial root mantle; the attached root traces exhibit a distinctive strap-like morphology, up to 1 to 3 cm in diameter and at least 15 cm long (see Willis and Bridge 1988: their figure 7E; Driese et al. 1997: their figure 6A,B). Root traces preserved in the enclosing tidal/estuarine sandstone (Bridge and Willis 1994: their figure 10B; Driese et al. 1997: their figures 6C,D) probably represent aerial roots produced by the stem that grew down and extended horizontally (C. Hotton, pers. comm., 1997). In the subjacent paleosol, the strap-like roots are identifi-

Figure 13.4.
Examples of Early Devonian root traces from Appalachian redbed paleosols. Battery Point Formation, Prével Member (Lower Devonian), south shore of Gaspé Bay (Ft. Prével section), Québec. 15 cm scale card in (B), (C), and (D). **A:** Medium to large, clay-lined, and sand-filled root traces exhibiting both strongly rhizomatous and downward-branching and tapering morphology within sandstone paleosol. Hammer handle is 20 cm long. **B:** Downward-branching and tapering, clay-lined root traces attached to horizontal rhizome traces (*a*) and larger-diameter root trace (*b*) that is nearly 60 cm long. **C:** Finer clay-lined root traces that branch and bifurcate downward. **D:** Upward-branching, clay-lined, and sand-filled trace (*arrow*) that may represent preserved above-ground portion of plant.

able by their higher sand and silt content relative to the clayey, siltstone paleosol (see Driese et al. 1997: their figure 13.6B). Driese et al. (1997) interpreted the dominantly horizontal root trace morphology and the strongly flared morphology of the stump casts as compatible with a soil environment that was waterlogged and poorly oxygenated. This interpretation is consistent with numerous characteristics of the paleosol, including drab gray-green coloration, high organic carbon and pyrite content, and absence of pedogenic carbonate.

By Late Devonian (Famennian) time, large (up to 18 m tall) arborescent progymnosperms (*Archaeopteris*) occupied well-drained, upland Catskill alluvial environments characterized by seasonal moisture variations (Banks et al. 1985;

Gensel and Andrews 1987). Driese et al. (1997) described probable *Archaeopteris* root traces 10 to 15 cm in diameter and up to 1.5 m deep attached to stump casts, which exhibit a "modern" morphology characterized by deep, vertical taproots and large secondary roots that branch from primary taproots; they occur in channel-margin sandstone deposits of the Duncannon Member of the Catskill Formation (Famennian) near Trout Run, Pennsylvania (figure 13.5; see also Driese et al. 1997: their figures 7, 8, 10). The external three-dimensional morphology of these root and stump casts resembles that of transported, permineralized stumps and attached roots of *Callixylon trifilevii* described by Snigirevskaya (1984b, 1988) from the Upper Devonian of the Donetz Basin in

Figure 13.5.
Examples of root traces from early trees, Appalachian Basin paleosols. **A,B:** *Archaeopteris?* sandstone stump casts (*s*) with attached deep taproot traces (*arrows*), rooted in silty sandstone paleosol formed from channel margin deposits. Centimeter scale on ruler. Catskill Formation, Duncannon Member (Upper Devonian), Trout Run, PA.

Russia, in which the stumps have a moderately flared base and taproots attach to the stump at about a 60° angle. The Catskill tree root traces are preserved as endocortical casts in sandstone, with evidence for multiple stages of sediment infilling accompanying progressive decay of organic tissues (see Driese et al. 1997: their figures 12A–C). The 10- to 30-cm-diameter stump casts display a crude reticulate structure of uncertain origin around the periphery (see Driese et al. 1997: their figure 12D). The corm-like morphology of the stump casts, as well as the downward branching and tapering of the attached root traces (figure 13.5; see also Driese et al. 1997: their figures 8, 10), precludes their interpretation as dipnoan aestivation burrows, such as are described elsewhere in Catskill deposits (Woodrow and Fletcher 1968). Gordon (1988: her figure 6) reported possible upright(?) tree trunk casts, preserving a concentric internal structure and ranging from 5 to 15 cm in diameter and up to 1 m long, in overbank deposits of the Oneonta Formation (Famennian) near West Kaaterskill Falls, New York.

Casts and rhizoconcretions of root traces of possible juvenile *Archaeopteris* (e.g., *Eddya?*) are common in the Upper Devonian Catskill succession and consist of a single, central taproot ranging from 2 to 5 cm in diameter and up to 50 cm in length, with numerous secondary and tertiary lateral roots (see Driese and Mora 1993: their figure 4B). The dominantly horizontal to subhori-

zontal, 1- to 3-mm-diameter rhizoliths could also have been produced by shrubby aneurophyte progymnosperms; such traces are exceedingly common in Middle and Late Devonian claystone paleosols; they consist of a micritic outer cortex and an inner calcite spar cement pore-filling (see Driese and Mora 1993: their figure 4B,C,F; Driese et al. 1997: their figure 13). These smaller root traces are more common in fine-grained floodplain and overbank deposits, whereas the larger tree stump casts and root traces occur principally in channel-margin sandstone deposits. This preservational bias reflects either (1) a greater preference of the arborescent vegetation for wetter near-channel environments, (2) higher sedimentation rates near active channels, allowing for more rapid burial and better preservation potential, or (3) selective destruction of large root traces in floodplain deposits by pedoturbation processes (Driese et al. 1997).

INTERPRETATION OF BIOLOGICAL FEATURES

Evolutionary History of Plant Traces

The case examples presented (figures 13.2–13.6) indicate that there are significant changes in the morphology of root traces and related organic features in Upper Ordovician to Upper Devonian paleosols. The absence of root or rhizomatous features in Upper Ordovician paleosols is consistent with what little is known of the flora of this time. Although diminutive rhizomatous traces that are probably attributable to land plants occur in Upper Silurian paleosols, the majority of biological features seen in Ordovician and Silurian paleosols are probably animal in origin. Lower Devonian paleosols are the first to contain abundant root traces with modern aspects such as downward branching and bifurcation, and progressive decreases in diameter. As plants increased in stature and arborescence was achieved, rooting depth increased substantially, to in excess of 1 m by Late Devonian time. Many plant traces in Appalachian redbed paleosols are associated with pedogenic carbonate, and the morphology of the carbonate parallels land plant and root trace evolution and diversification (see table 13.1)—that is, relatively simple pedogenic carbonate morphology occurs in Ordovician and Silurian paleosols, and there is an increase in the diversity of forms of pedogenic carbonate in the Middle to Late Paleozoic.

Effects of Pedogenic Paleoenvironment on Root Trace Morphology

In addition to secular changes, differences in root trace morphology may be related to the sedimentary paleoenvironment that the plants inhabited. Although it is generally assumed that early (Ordovician–Silurian) land plant communities occurred dominantly in coastal-margin settings (e.g., Gensel and Andrews 1984, 1987; Beerbower 1985; DiMichele and Hook 1992; Gray and Shear 1992), niche-partitioning of Early Devonian taxa by sedimentary environment has been recently documented (chapter 11), and by Late Devonian time, land plants inhabited a range of environments from coastal margin to well-drained, alluvial uplands. Changes in paleosol and root trace morphology along a coastal-margin to alluvial-plain transect were evaluated for Late Devonian (late Frasnian to middle Famennian) exposures of the Irish Valley and Sherman Creek Members (Catskill Formation) near Selinsgrove, Pennsylvania (Capelle and Driese 1995; Cotter and Driese 1998). Along this transect, which includes some 1,100 m of con-

tinuous outcrop, three general types of root traces occur: (1) small, clay-lined, and fibrous traces, 1 to 3 mm in diameter and up to 5 cm long, occurring in dense clusters (figure 13.6A), (2) slender, dichotomously branching, clay-lined or calcretized root traces up to 30 cm long (figure 13.6B), and (3) thicker root traces, 0.5 to 2 cm in diameter and greater than 30 cm long, with distinctive lateral branching off primary roots; these traces are sand-filled, clay-lined, and rarely calcretized (figure 13.6C).

The Irish Valley Member of the Catskill Formation begins in the Upper Frasnian and extends to the Frasnian–Famennian boundary (Berg et al. 1983). Irish Valley Member paleosols formed in muddy coastal-margin environments characterized by repeated transgressive-regressive cycles (Walker 1971; Walker and Harms 1971, 1975; Cotter and Driese 1998) and generally contain vertical and horizontal examples of the small, clay-lined root traces (figure 13.6A). Many of the Irish Valley Member paleosols show significant retention of primary depositional fabrics, because of either low density of plant coverage or frequent depositional events that terminated pedogenesis. Pedogenic carbonate is very sparse and thin, comprised of dolomitic lenses with tubular-fenestral morphology resembling the rhizomatous networks of the Upper Silurian Bloomsburg Formation paleosols, suggesting precipitation in association with matlike vegetation of diminutive stature.

The Sherman Creek Member of the Catskill Formation extends from the Frasnian–Famennian boundary into the Famennian (Berg et al. 1983). Sherman Creek Member paleosols developed in a lower to middle alluvial plain setting, characterized by sluggish, low-gradient rivers, and poorly to moderately drained floodplains that were subjected to occasional marine incursions (Rahmanian 1979; Sevon 1985; Cotter and Driese 1998). Pedogenesis was more extensive, as indicated by the greater thicknesses of the paleosol profiles and their differentiation into distinct "horizons," especially pedogenic carbonate (B_k) horizons, probably because of the longer time available for pedo-

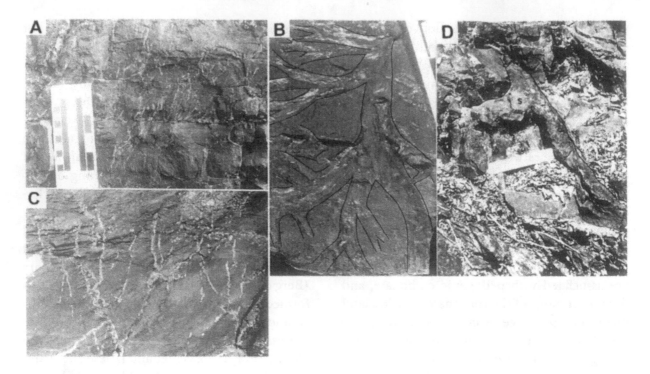

Figure 13.6.

Examples of Late Devonian root traces, in a shoreline-to-landward paleoenvironmental transect. Catskill Formation, Irish Valley and Sherman Creek Members (Upper Devonian), Selinsgrove, PA. **A:** Small, clay-lined root traces typical of coastal-margin setting preserved in red, clayey siltstone paleosol, upper Irish Valley Member. Scale card is 15 cm. **B:** Slender, elongate, clay-lined, branching and tapering root traces characteristic of lower alluvial plain setting preserved in red, silty claystone paleosol, lower Sherman Creek Member. Centimeter divisions on scale card. **C:** Large calcretized root traces typical of mid to upper alluvial plain setting preserved in red siltstone paleosol, middle Sherman Creek Member. Part of ruler shown is 5 cm long. **D:** Large *Stigmaria*-like sandstone root casts (*s*) preserved in alluvial channel-margin sandstone paleosol, middle Sherman Creek Member. Ruler is 15 cm long.

genesis between depositional events, compared to Irish Valley Member sediments (Capelle and Driese 1995; Cotter and Driese 1998). The Sherman Creek paleosols exhibit an upward increase in root trace size, from small, to slender and long, to the robust, thick, long traces described previously (figure 13.6B,C). Large, drab-haloed root traces are common, suggesting the presence of significant organic material in the root trace; drab haloes result from localized iron reduction caused by organic decay (Retallack 1990). In addition, rare *Stigmaria*-like root traces in a fluvial sandstone body in the upper Sherman Creek Member (figure 13.6D) indicate the existence of some type of arborescent vegetation, perhaps large lycophytes or progymnosperms such as *Archaeopteris*. Pedogenic carbonate in the Sherman Creek paleosols is morphologically complex but is dominated by the rhizolith morphology (figure 13.6C), with

variable replacement of micrite and calcite by dolomite; nodular pedogenic carbonate morphologies, typical of the vertic claystone paleosols in the overlying Duncannon Member, are less common.

These systematic changes in size, depth, and density of root traces, abundance and thickness of pedogenic carbonate, and paleosol morphology are, in part, a consequence of the progressively larger and deeper rooting of land plants found higher on alluvial plain, as compared to those inhabiting coastal-margin environments (Capelle and Driese 1995). These observations suggest that comparisons of early and mid-Paleozoic plant ecosystems are best restricted to those developed in comparable paleoenvironments; environmental factors such as soil drainage and salinity, as well as geomorphic stability, can exert a strong influence on plant distribution.

EARLY LAND PLANT DEVELOPMENT AND ESTIMATES OF SILURO-DEVONIAN ATMOSPHERIC CO₂ LEVELS

Early land plant evolution occurred under conditions of elevated atmospheric carbon dioxide levels (Yapp and Poths 1992; Berner 1994; McElwain and Chaloner 1995; Mora et al. 1996). Indeed, the rise of terrestrial vascular plants may have accelerated the drawdown of atmospheric CO_2 from very high levels in the Silurian to near modern-day levels during the late Paleozoic (Berner 1994, 1997; Retallack 1997). The soil carbonate carbon isotope CO_2 paleobarometer of Cerling (1991) is based on studies of production and diffusion of CO_2 in modern soils (cf. Cerling 1984), and has been successfully applied in studies of paleosols of Late Devonian and younger age (Cerling 1991; Mora et al. 1991, 1996; Ghosh et al. 1995; Ekart and Cerling 1996). Its successful application to ancient paleosols results in part from the similarities in the composition of soil organic matter and average rate and depth of CO_2 production in ancient and modern soils (Cerling 1991; Mora et al. 1996; Retallack 1997). The importance of other factors affecting soil CO_2 production and diffusion, such as soil permeability and porosity, can be minimized by restricting studies to paleosols that share sedimentological and morphological similarities (Mora et al. 1996; Mora and Driese 1999). Siluro-Devonian paleosols pose a number of special problems for the soil CO_2 paleobarometer, which will be briefly discussed here. At present, quantifying these factors is very difficult and the errors in CO_2 estimates for this time period must be considered correspondingly large (Mora et al. 1996).

Stable Isotope Composition of Pedogenic Carbonate in Siluro-Devonian Paleosols

Pedogenic carbonate is present in many Appalachian redbed paleosols, occurring with two general morphologies, micritic rhizoliths or nodules (cf. Driese and Mora 1993; Mora et al. 1996; Mora and Driese 1999). Stable carbon and oxygen isotope compositions of pedogenic carbonate from selected Siluro-Devonian paleosols are shown in figure 13.7. Pedogenic carbonate generally preserves a pedogenic carbon isotope signature; oxygen isotopes have been reset by exchange with water-rich (but carbon-poor) fluids during burial diagenesis. The carbon isotope compositions tend towards a constant, minimum value in time-equivalent paleosols and show the marked influence of isotopically light, soil-respired CO_2. Detailed interpretation and discussion of the pedogenic carbonate morphology and stable isotope geochemistry of these profiles are described elsewhere (cf. Mora et al. 1991, 1993, 1996; Driese and Mora 1993; Mora and Driese 1999).

Depth and Rate of CO₂ Production

Sparse and shallowly rooted primitive plants (see figures 13.2, 13.3; table 13.1; see also Stewart and Rothwell 1993; Gensel and Andrews 1984, 1987) suggest low rates of soil respiration and shallow mean production of CO_2 in Silurian and Early Devonian soils. Both factors result in lower estimates of atmospheric CO_2 for a given isotopic composition of pedogenic carbonate (figure 13.8A; see also Cerling 1991). Possible mediating factors include incorporation of

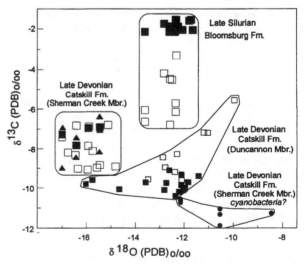

Figure 13.7.
Stable carbon and oxygen isotope compositions of pedogenic carbonate in Siluro-Devonian Appalachian redbed paleosols; micrite in rhizoliths (*closed symbols*), nodules (*open symbols*), spherulitic microspar and spar (*triangles*), and densely clustered spherulites (*circles*).

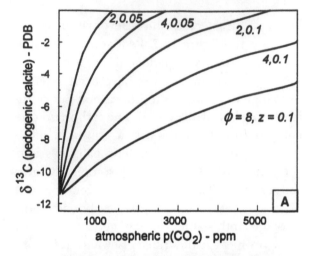

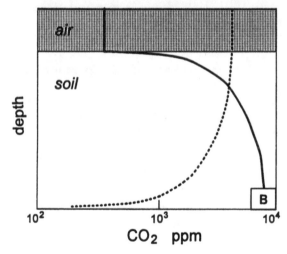

Figure 13.8.

A: Calculation of the isotopic composition of soil carbonate varying the respiration rate (N, mmol/m²/hr) and average depth of CO_2 production (z, *m*) using the soil CO_2 model of Cerling (1991); other details of the calculation can be found in Mora et al. (1993). Calculated pCO_2 is sensitive to shallow CO_2 production or low respiration rates, both of which are likely for mid-Paleozoic soils; these factors may be mediated by pedogenic processes, as described in text. **B:** Schematic depth profile for soil pCO_2 calculated for instances where soil pCO_2 >> atmospheric pCO_2 (*solid line*) and where atmospheric pCO_2 >> soil pCO_2 (*dashed line*). Although the profiles are quite different, in both cases, carbonate precipitated in equilibrium with the isotopically lightest soil CO_2 would yield minimum estimates of atmospheric pCO_2.

organic matter deeper in the soils by pedoturbation, the physical mixing that occurs as a result of shrinking and swelling of the soil in response to seasonal moisture, or bioturbation. Rapid sedimentation and burial of paleosols in

geomorphically unstable environments, or development of compound paleosols would have similar results. Translocation of organic matter deeper in Siluro-Devonian soils would effectively increase both the rate and the depth of CO_2 production, making them more comparable to younger paleosols, despite their significantly different soil ecology.

Preferential plant colonization of moister sedimentary environments, such as the coastal margin, would effectively decrease the soil porosity and permeability of Siluro-Devonian paleosols compared to younger redbed paleosols. This effectively increases the organically derived proportion of soil CO_2 (Mora et al. 1996), thereby decreasing estimates of atmospheric CO_2.

Modern rates of soil respiration and soil CO_2 concentrations result in a typical depth profile for soil CO_2, shown in figure 13.8b. For Ordovician–Silurian soils, low rates of soil respiration, combined with high levels of atmospheric CO_2, may have resulted in an opposite profile (figure 13.8b). Isotope depth profiles like this are observed in the Late Silurian Bloomsburg Formation (Driese et al. 1992). In this case, minimum pCO_2 estimates are made from the isotopically lightest samples.

Composition of Soil Organic Matter

The isotopic composition of soil organic matter is perhaps the most critical variable in the soil carbonate paleobarometer. The likely predominance of nonvascular plant types such as bryophytes, fungi, and lichens in Silurian–early Devonian soils suggest that the composition of soil organic matter may be variable. The carbon isotope composition of lichens, for example, varies over more than 10 permil, largely dependent on the nature of the photobiont (Lange et al. 1988; Máguas et al. 1995). By the Late Devonian, soil organic matter was likely dominated by vascular plant material. The carbon isotope composition of terrestrial organic matter shows secular variations that correspond to variations in $\delta^{13}C$ of atmospheric CO_2 and marine carbonate (Mora et al. 1996). The carbon isotope composition of soil organic matter reported by Mora et al. (1996) shows a shift of approximately +4 permil between the Late Silurian and the early Permian, slightly less than the +6 permil shift determined for the $\delta^{13}C$ of atmospheric CO_2. Isotopic discrimination in C_3 vascular plants is controlled in part by stomatal conductance (Farquhar et al. 1989), and it is interesting to speculate whether the late Paleozoic isotopic trend was somewhat dampened by plant stomatal control.

Estimates of Siluro-Devonian pCO₂

We have previously used the isotopic compositions of pedogenic carbonate from redbed vertic paleosols in nine stratigraphic successions, ranging in age from Late Silurian to early Permian, to estimate paleoatmospheric levels of CO_2; these data are tabulated in Mora et al. (1996). The estimates are summarized in figure 13.9; they include preliminary results for the Early Devonian, based on micrite nodules sampled from the Prével Member of the Battery Point Formation (late Emsian) from the south shore of Gaspé Bay near Fort Prével (Elick et al. 1996). Estimates of atmospheric CO_2 levels from geographically separated, time-equivalent paleosols are consistent, suggesting that a coherent record of changing atmospheric chemistry is preserved in the ancient soil record. Our results suggest that atmospheric CO_2 levels declined from 10 to 18 × present atmospheric levels (PAL) in the late Silurian to approximately 1 × PAL in the early Permian, closely following a decline predicted by the theoretical carbon mass–balance model of Berner (1994). The most significant decrease, between the Late Silurian and Late Devonian, coincides with a period of rapid evolution and diversification of the terrestrial ecosystem (cf. tables 13.1, 13.2), and the new data from Gaspé suggest that the Early Devonian may have been the time of initiation of the steep decline in pCO₂ (figure 13.9).

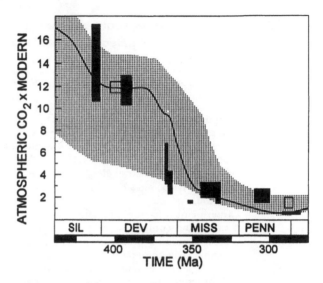

Figure 13.9.
Estimates of Siluro-Devonian Paleozoic atmospheric CO_2 levels from Appalachian paleosols (*closed symbols*), calculated using the soil carbonate carbon isotope paleobarometer of Cerling (1991), are consistent with levels calculated from stomatal density (*open symbols*; McElwain and Chaloner 1995) and a carbon cycle model (*line and shaded error envelope*; Berner 1997). Atmospheric CO_2 levels dipped sharply between the late Silurian and early Permian, approximately coincident with afforestation of the land surface; the steepest decline occurs in Early Devonian time, corresponding to development of true root systems and increased root trace density in paleosols.

SUMMARY AND CONCLUSIONS

Detailed macro- and micromorphological and isotopic characterization of Siluro-Devonian paleosols and pedogenic carbonate yield a pedological signature of vascular land plant evolution and diversification. Root trace morphology suggests progressive increases in root size and depth of rooting, and possibly colonization of drier, more upland environments. Late Ordovician paleosols lack evidence for plant roots or rhizomes. Rhizomatous traces observed in Late Silurian (Ludlovian–Pridolian) paleosols in Pennsylvania and Nova Scotia are diminutive (0.5 to 2 mm in diameter, 0.5 to 2 cm long) and provide evidence of plants living in coastal mudflat and marsh environments. Early Devonian (Emsian) paleosols from the Battery Point Formation of Gaspé, Québec, exhibit larger (2 to 3 mm in diameter, 15 cm long), downward-branching, calcretized or clay-lined

Table 13.2   Hypotheses about the generalized effects of Silurian–Devonian plant evolution on terrestrial ecosystems, paleosol and soil carbonate morphology, the carbon isotope composition of soil carbonate, and calculated atmospheric CO2 levels

Age	Characteristics of the paleovegetation/ecology	Effect on paleosol and soil carbonate morphology	Effect on soil carbonate $\delta 13C$ and calculated pCO_2
Mid to Late Silurian	Diminutive plants (cm to tens of cm tall), confined to coastal marsh and floodplain environments due to primitive reproductive strategies. No true roots, but small rhizomatous systems.	Sparse and diminutive vegetation exerts only a weak influence on weathering rates. Soils are thin, poorly structured, and poorly horizonated; no coals deposited. Small, sparse nodules or calcretized rhizomes.	Shallow production of soil CO_2 and low rates of respiration results in soil carbonate compositions dominated by isotopically heavy atmospheric CO_2. Calculated pCO_2 very dependent on model parameters.
Early to Mid Devonian	Larger plants (up to 1–2 m tall), still mainly confined to wet areas bordering stream channels, ephemeral floodponds, and coastal marshes. Root systems larger; dominated by shallow (<10 cm), lateral rhizomes.	Weathering rates are accelerated. Soils are thicker, better structured, and better horizonated. Very rare coals, possibly dominated by nonvascular vegetation. Carbonate nodules and small rhizoliths.	Variable $\delta^{13}C$, depending mostly on depth of calcite precipitation; deeper carbonates show more significant biological influence and yield more consistent estimates of pCO_2.
Post-Middle to Late Devonian	Large plants (up to 18 m tall). More advanced reproductive strategies permit more complete colonization of terrestrial environments. Arborescent component dominantly confined to wet areas bordering stream channels; shrubby component occupying drier interfluve areas. Meter-deep root systems with tap and lateral morphologies.	Weathering rates greatly accelerated. Soils are thick, well structured, well horizonated. Rare thin coals. Carbonate micromorphology more distinctly biogenic; large rhizoliths that may show several cement generations.	Soil CO_2 dominantly controlled by soil organic matter. Many model parameters for estimation of atmospheric pCO_2 similar to geologically younger examples.

References: Algeo et al. (1995); Banks et al. (1985); Beerbower (1985); Cerling (1991); DiMichele and Hook (1992); Driese et al. (1997); Elick et al. (1998a); Gensel and Andrews (1984, 1987); Gray (1985); Gray and Shear (1992); Mora and Driese (1999); Mora et al. (1991, 1996); Retallack (1985, 1986, 1992); Shear (1991); Stewart and Rothwell (1993).

root traces, as well as exceptionally large (2 to 3 cm in diameter, up to 1 m deep) root traces produced by plants of larger stature. Strap-like, dominantly horizontal root traces of the first large trees (*Eospermatopteris*) from Middle Devonian (Givetian) paleosols are small relative to the size of the plant, and this may be because of the wet, marsh-like setting that these plants occupied. By Late Devonian (Famennian) time, plants occupied well-drained, upland alluvial environments, and the corresponding root systems are large (10 to 15 cm in diameter, up to 1.5 m deep) and exhibit a morphology characterized by deep vertical taproots and large secondary roots that branch from primary taproots.

The stable carbon isotope compositions of pedogenic carbonate in paleosols can be used to estimate paleoatmospheric CO_2 levels (Cerling 1991). Our studies suggest a steep mid to late Paleozoic decline in paleoatmospheric CO_2 levels ranging from 10 to 18 × PAL in the Late Silurian, to 2 to 5 × PAL in the Late Devonian, and to 1 × PAL in the early Permian (Mora et al. 1996); very similar decreases are predicted by the long-term carbon mass balance model of Berner (1994). Shallow depths of rooting and relatively low soil productivity in Silurian soils pose a special challenge to application of the paleobarometer; however, these problems may be mediated by colonization of moister soils and/or translocation of organic matter to greater depths by pedoturbation.

First-order controls on paleosol morphology and stable isotope chemistry therefore include (1) soil development in the presence or absence of land plants with significant root systems, and (2) variations in geomorphic, sedimentary, and pedogenic environments. The morphology and stable isotope chemistry of Ordovician and Silurian paleosols differ substantially from those of late Paleozoic paleosols, in part because of the impact of vascular plant development.

ACKNOWLEDGMENTS

This research was supported by PRF Grant 25678-AC8-C-SF94, and by NSF Grants EAR-9206540 and EAR-9418183 awarded to Mora and Driese. We appreciate the comments of reviewers Paul Wright (Cardiff), Vicky Beck (Bristol), as well as those of coeditor Dianne Edwards. We gratefully acknowledge the cooperation of the Canadian Parks Service, as well as helpful discussions with Pat Gensel (North Carolina) and Carol Hotton (Smithsonian). Tom Algeo (Cincinnati) and Jennifer Elick (Tennessee–Knoxville) graciously provided field and laboratory assistance. Ed Cotter (Bucknell) and John Bridge (Binghamton) provided important outcrop locality information. Dick Beerbower (Binghamton) provided a preliminary Battery Point Formation sample set from Gaspé, and Paul Strother (Boston) provided sample material from the Bloomsburg Formation in Pennsylvania and the Moydart Formation from Arisaig.

REFERENCES

Abrahamson, W. G., ed. 1989. *Plant–Animal Interactions.* New York: McGraw Hill.

Ahlberg, P. 1991. Tetrapod or near-tetrapod fossils from the Upper Devonian of Scotland. *Nature* 354: 298–301.

———. 1995. *Elginerpeton pancheni* and the earliest tetrapod clades. *Nature* 373: 420–425.

Ahlberg, P., and A. R. Milner. 1994. The origin and early diversification of tetrapods. *Nature* 368: 507–514.

Ahmad, N. 1983. Vertisols. In L. P. Wilding, N. E. Smeck, and G. F. Hall, eds. *Pedogenesis and Soil Taxonomy. II. The Soil Orders*, pp. 91–123. New York: Elsevier.

Algeo, T. J., R. A. Berner, J. B. Maynard, and S. E. Scheckler. 1995. Late Devonian oceanic anoxic events and biotic crises: "Rooted" in the evolution of vascular land plants? *Geol. Soc. Am. Today* 5(45): 64–66.

Algeo, T. J., and S. E. Scheckler. 1998. Terrestrial-marine teleconnections in the Devonian: Links between the evolution of land plants, weathering processes, and marine anoxic events. *Phil. Trans. R. Soc. Lond.* B353: 113–130.

Allen, J. R. L. 1986. Pedogenic calcretes in the Old Red Sandstone facies (late Silurian–early Carboniferous) of the Anglo-Welsh area, southern Britain. In V. P. Wright, ed. *Paleosols: Their Recognition and Interpretation*, pp. 58–86. Oxford: Blackwell.

Allen, K. C. 1980. A review of *in situ* Late Silurian and Devonian spores. *Rev. Palaeobot. Palynol.* 29: 253–270.

Allen, K. C., and D. L. Dineley. 1988. Mid-Devonian to mid-Permian floral and faunal regions and provinces. In A. L. Harris and D. J. Fettes, eds. *The Caledonian-Appalachian Orogen.* (*Geol. Soc. Lond. Spec. Publ.*) 38: 531–548.

Almond, J. E. 1985. The Silurian-Devonian fossil record of the Myriapoda. *Phil. Trans. R. Soc. Lond.* B309: 227–237.

———. 1986. Studies on Palaeozoic Arthopoda. Unpublished doctoral thesis, University of Cambridge, U.K.

Alvarez, L. W., W. Alvarez, F. Asaro, and H. V. Michel. 1980. Extraterrestrial cause for the Cretaceous-Tertiary extinction. *Science* 208: 1095–1108.

Ananiev, A. R. 1960. Studies in the Middle Devonian flora of the Altai-Sayan mountain region. *Botanicheskii*

Zhurnal (Academy of Sciences, USSR) 45: 649–666 (in Russian with English summary).

Anderson, D. M., and R. S. Webb. 1994. Ice-age tropics revisited. *Nature* 367: 23–24.

Anderson, J. M., and H. M. Anderson. 1985. *The Palaeoflora of Southern Africa: Prodromus of Southern African Megafloras, Devonian to Lower Cretaceous.* Rotterdam: A. A. Balkema.

Anderson, J. M., and P. Ineson. 1983. Interactions between soil arthropods and microbial populations in carbon, nitrogen and mineral nutrient fluxes from decomposing leaf litter. In J. Lee and S. McNeill, eds. *Nitrogen as an Ecological Factor*, pp. 413–432. Oxford: Blackwell Scientific.

Andrews, H. N., P. G. Gensel, and W. H. Forbes. 1974. An apparently heterosporous plant from the Middle Devonian of New Brunswick. *Palaeontology* 17: 387–408.

Andrews, H. N., P. G. Gensel, and A. E. Kasper. 1975. A new fossil plant of probable intermediate affinities (trimerophyte-progymnosperm). *Can. J. Bot.* 53: 1719–1728.

Andrews, H. N., A. E. Kasper, W. H. Forbes, P. G. Gensel, and W. G. Chaloner. 1977. Early Devonian flora of the Trout Valley Formation of northern Maine. *Rev. Palaeobot. Palynol.* 23: 255–285.

Andrews, H. N., A. E. Kasper, and E. Mencher. 1968. *Psilophyton forbesii*, a new Devonian plant from northern Maine. *Bull. Torrey Bot. Club* 95: 1–11.

Appel, H. M. 1993. Phenolics in ecological interactions: The importance of oxidation. *J. Chem. Ecol.* 19: 1521–1552.

Arms, K., P. Feeney, and R. C. Lederhouse. 1974. Sodium: Stimulus for puddling behaviour by tiger swallowtail butterflies, *Papilio glaucus. Science* 185: 372–374.

Arnold, C. A. 1941. Observations on fossil plants from the Devonian of eastern North America. V. *Hyenia banksii* sp. nov. *Contrib. Mus. Paleontol. Univ. Mich.* 6: 53–57.

Arthur, M. A., and T. J. Fahey. 1993. Controls on soil solution chemistry in a subalpine forest in North-Central Colorado. *Soil Sci. Soc. Am. J.* 57: 1123–1130.

Avkhimovitch, V. I., N. S. Nekryata, and T. G. Obuhovskaya. 1988. Devonian palynostratigraphy of the Pripyat Depression, Byelorussia. In N. J. McMillan, A. J. Embry, and D. J. Glass, eds. *Devonian of the World*, vol. 3: *Paleontology, Paleoecology and Biostratigraphy*, pp. 559–566. Calgary, Alberta: Canadian Society of Petroleum Geologists, Mem. 14.

Bai, S., and Z. Ning. 1988. Faunal change and events across the Devonian-Carboniferous boundary of Huangmao section, Guangxi, south China. In N. J. McMillan, A. J. Embry, and D. J. Glass, eds. *Devonian of the World*, vol. 3: *Paleontology, Paleoecology and Biostratigraphy*, pp. 147–157. Calgary, Alberta: Canadian Society of Petroleum Geologists, Mem. 14.

Bai, Z. Q., and S. L. Bai. 1990. Palaeogeographic position of the South China Plate during the Lower-Middle Devonian. *Acta Geol. Sin.* 3: 199–205.

Baldauf, S. L., J. R. Manhart, and J. D. Palmer. 1990. Different fates of the chloroplast *tuf*A gene following its transfer to the nucleus in green-algae. *Proc. Nat. Acad. Sci., USA* 87: 5317–5321.

Bambach, R. K. 1985. Phanerozoic marine communities. In D. M. Raup and D. Jablonski, eds. *Patterns and Processes in the History of Life*, pp. 407–428. Berlin: Springer.

Banks, H. P. 1944. A new Devonian lycopod genus from southern New York. *Am. J. Bot.* 31: 649–659.

———. 1966. Devonian flora of New York State. *Empire State Geogram* 4: 10–24.

———. 1968. The early history of land plants. In E. T. Drake, ed. *Evolution and Environment: A Symposium Presented on the One Hundredth Anniversary of the Foundation of the Peabody Museum of Natural History at Yale University*, pp. 73–107. New Haven, CT: Yale University Press.

———. 1973. Occurrence of *Cooksonia*, the oldest vascular land plant macrofossil, in the Upper Silurian of New York State. *J. Indian Bot. Soc.* 50A: 227–235.

———. 1975a. The oldest vascular land plants: A note of caution. *Rev. Palaeobot. Palynol.* 20: 13–25.

———. 1975b. Early vascular land plants: Proof and conjecture. *BioScience* 25: 730–737.

———. 1975c. Reclassification of Psilophyta. *Taxon* 24: 401–413.

———. 1980. Floral assemblages in the Siluro-Devonian. In D. L. Dilcher and T. N. Taylor, eds. *Biostratigraphy of Fossil Plants*, pp. 1–24. Stroudsburg, PA: Dowden, Hutchinson and Ross.

———. 1981. Peridermal activity (wound repair) in an early Devonian (Emsian) trimerophyte from the Gaspé Peninsula, Canada. *Paleobotanist* 28–29: 20–25.

———. 1985. Early land plants. *Phil. Trans. R. Soc. Lond.* B309: 197–200.

———. 1992. The classification of early land plants—revisited. *Palaeobotanist* 41: 36–50.

Banks, H. P., P. M. Bonamo, and J. D. Grierson. 1972. *Leclercqia complexa* gen. et sp. nov., a new lycopod from the late Middle Devonian of eastern New York. *Rev. Palaeobot. Palynol.* 14: 19–40.

Banks, H. P., and B. J. Colthart. 1993. Plant-animal-fungal interactions in early Devonian trimerophytes from the Gaspé Peninsula, Canada. *Am. J. Bot.* 80: 992–1001.

Banks, H. P., and M. R. Davis. 1969. *Crenaticaulis*, a new genus of Devonian plants allied to *Zosterophyllum*, and its bearing on the classification of early land plants. *Am. J. Bot.* 56: 436–449.

Banks, H. P., and J. D. Grierson. 1968. *Drepanophycus spinaeformis* Göppert in the early Upper Devonian of New York State. *Palaeontographica B* 123: 113–120.

Banks, H. P., J. D. Grierson, and P. M. Bonamo. 1985. The flora of the Catskill clastic wedge. In D. L. Woodrow and W. D. Sevon, eds. *The Catskill Delta*. Geological Society of America Spec. Pap. 201: 125–141.

Banks, H. P., S. Leclercq, and F. M. Hueber. 1975. Anatomy and morphology of *Psilophyton dawsonii*, sp. n. from the late Lower Devonian of Quebec (Gaspé) and Ontario, Canada. *Palaeontographica Americana* 8: 77–127.

Bär, P., and W. Riegel. 1980. Mikrofloren des höchsten Ordovizium bis tiefen Silurs aus der Unteren Sekondi-Serie von Ghana (Westafrika) und ihre Beziehung zu den Itaim-Schichten des Maranhao-Beckens in NE-Brasilien. *Neues Jahrbuch für Geologie und Paläontologie Abhandlungen* 160: 42–60.

Barclay, W. J., P. A. Rathbone, D. E. White, and J. B. Richardson. 1994. Brackish water faunas from the St. Maughans Formation: The Old Red Sandstone section at Ammons Hill, Hereford and Worcester, UK, re-examined. *Geol. J.* 29: 369–379.

Bartels, D., A. Furini, J. Ingram, and F. Salamini. 1996. Responses of plants to dehydration stress: A molecular analysis. *Plant Growth Regul.* 20: 111–118.

Bartels, D., K. Schneider, G. Terstappen, D. Piatkonski, and F. Salamini. 1990. Molecular cloning of abscisic acid-modulated genes which are induced during desiccation of the resurrection plant *Craterostigma plantagineum*. *Planta* 181: 27–34.

Baschnagel, R. A. 1966. New fossil algae from the Middle Devonian of New York. *Trans. Am. Microsc. Soc.* 85: 297–302.

Basinger, J. F., M. E. Kotyk, and P. G. Gensel. 1996. Early land plants from the Late Silurian-Early Devonian of Bathurst Island, Canadian Arctic Archipelago. *Geol. Surv. Canada, Cur. Res.* 1996B: 51–60.

Bassett, M. G., J. D. Lawson, and D. E. White. 1982. The Downton Series as the fourth series of the Silurian System. *Lethaia* 15: 1–24.

Basu, A. 1981. Weathering before the advent of land plants: Evidence from unaltered detrital K-feldspars in Cambrian-Ordovician arenites. *Geology* 9: 132–133.

Bateman, R. M. 1991. Palaeoecology. In C. J. Cleal, ed. *Plant Fossils in Geological Investigation: The Palaeozoic*, pp. 34–116. New York: Ellis Harwood.

———. 1996. An overview of lycophyte phylogeny. In J. M. Camus, M. Gibby, and R. J. Johns, eds. *Pteridology in perspective*, pp. 405–415. Kew: Royal Botanic Gardens.

Bateman, R. M., and W. A. DiMichele. 1994. Heterospory: The most-iterative key innovation in the evolutionary history of the plant kingdom. *Biol. Rev.* 69: 345–417.

Bayer, U., and G. R. McGhee, Jr. 1986. Cyclic patterns in the Paleozoic and Mesozoic: Implications for time scale calibrations. *Paleoceanography* 1: 383–402.

Beck, C. B. 1953. A new root species of *Callixylon*. *Am. J. Bot.* 40: 226–233.

———. 1960. The identity of *Archaeopteris* and *Callixylon*. *Brittonia* 72: 351–368.

———. 1962. Reconstruction of *Archaeopteris* and further consideration of its phylogenetic position. *Am. J. Bot.* 49: 373–382.

———. 1964. Predominance of *Archaeopteris* in Upper Devonian flora of western Catskills and adjacent Pennsylvania. *Bot. Gaz.* 125: 126–128.

———. 1967. *Eddya sullivanensis*, gen. et sp. nov., a plant of gymnospermic morphology from the Upper Devonian of New York. *Palaeontographica* 121B: 1–22.

———. 1981. *Archaeopteris* and its role in vascular plant evolution. In K. J. Niklas, ed. *Paleobotany, Paleoecology, and Evolution*, vol. 1, pp. 193–230. New York: Praeger.

Beck, C. B., and W. Stein. 1993. *Crossia virginiana* gen. et sp. nov., a new member of the Stenokoleales from the Middle Devonian of southwestern Virginia. *Palaeontographica* 229B: 115–134.

Beck, C. B., and D. C. Wight. 1988. Progymnosperms. In C. B. Beck, ed. *Origin and Evolution of Gymnosperms*, pp. 1–84. New York: Columbia University Press.

Becker, R. T. 1993. Anoxia, eustatic changes, and Upper Devonian to lowermost Carboniferous global ammonoid diversity. In M. R. House, ed. *The Ammonoidea: Environment, Ecology, and Evolutionary Change*, pp. 115–163. Oxford: Clarendon, Oxford, Syst. Assoc. Spec. vol. 47.

Beerbower, J. R. 1985. Early development of continental ecosystems. In B. H. Tiffney, ed. *Geological Factors and the Evolution of Plants*, pp. 47–92. New Haven, CT: Yale University Press.

Beerbower, J. R., J. A. Boy, W. A. DiMichele, R. A. Gastaldo, R. Hook, N. Hotton, III, T. L. Phillips, S. E. Scheckler, and W. A. Shear. 1992. Paleozoic terrestrial ecosystems. In A. K. Behrensmeyer, J. D. Damuth, W. A. DiMichele, R. Potts, H.-D. Sues, and S. L. Wing, eds. *Terrestrial Ecosystems through Time*, pp. 205–325. Chicago: University of Chicago Press.

Behrensmeyer, A. K., J. D. Damuth, W. A. DiMichele, R. Potts, H.-D. Sues, and S. L. Wing. 1992. *Terrestrial Ecosystems through Time*. Chicago: University of Chicago Press.

Belk, D., and G. A. Cole. 1975. Adaptational biology of desert temporary-pond inhabitants. In N. F. Hadley, ed. *Environmental Physiology of Desert Organisms*, pp. 207–226. Stroudsburg, PA: Dowden, Hutchinson and Ross.

Bell, A. D. 1991. *Plant Form: An Illustrated Guide to Flowering Plant Morphology*. Oxford: Oxford University Press.

Bell, W. A. 1922. A new genus of Characeae and new Merostomata from the Coal Measures of Nova Scotia. *Proc. Trans. R. Soc. Can.*, Sect. IV, Ser. III, 16: 159–167.

Bendix-Almgreen, S., J. A. Clack, and H. Olsen. 1990. Upper Devonian tetrapod palaeoecology in the light of new discoveries in East Greenland. *Terra Nova* 2: 131–137.

Benfy, P. N. 1999. Is the shoot a root with a view? *Current Opinion in Plant Biology* 2: 39–43.

Berg, T. M., M. K. McInerney, J. H. Way, and D. B. MacLachlan. 1983. Stratigraphic correlation chart of Pennsylvania. *PA Geol. Surv. General Geology Report 75* (chart).

Bernays, E., G. A. Cooper-Driver, and M. Bilgener. 1989. Herbivores and plant tannins. *Adv. Ecol. Res.* 19: 263–303.

Berner, R. A. 1978. Sulfate reduction and the rate of deposition of marine sediments. *Earth Planet. Sci. Lett.* 37: 492–498.

———. 1989. Biogeochemical cycles of carbon and sulfur and their effect on atmospheric oxygen over Phanerozoic time. *Palaeogeog. Palaeoclim. Palaeoecol.* 75: 97–122.

———. 1991. A model for atmospheric CO_2 over Phanerozoic time. *Am. J. Sci.* 291: 339–376.

———. 1992. Weathering, plants, and the long-term carbon cycle. *Geochim. Cosmochim. Acta* 56: 3225–3231.

———. 1993. Paleozoic atmospheric CO_2: Importance of solar radiation and plant evolution. *Science* 261: 68–70.

———. 1994. GEOCARB II: A revised model of atmospheric CO_2 over Phanerozoic time. *Am. J. Sci.* 294: 56–91.

———. 1997. The rise of plants and their effect on weathering and atmospheric CO_2. *Science* 276: 544–546.

———. 1998. The carbon cycle and CO_2 over Phanerozoic time: The role of land plants. *Phil. Trans. R. Soc. Lond.* B353: 75–82.

Berner, R. A, and K. Caldeira. 1997. The need for mass balance and feedback in the geochemical carbon cycle. *Geology* 25: 955–956.

Berner, R. A, and D. E. Canfield. 1989. A new model of atmospheric oxygen over Phanerozoic time. *Am. J. Sci.* 289: 333–361.

Berner, R. A., and R. Raiswell. 1983. Burial of organic carbon and pyrite sulfur in sediments over Phanerozoic time: A new theory. *Geochim. Cosmochim. Acta* 47: 855–862.

Berner, R. A., and D. M. Rye. 1992. Calculation of the Phanerozoic strontium isotope record of the oceans from a carbon cycle model. *Am. J. Sci.* 292: 136–148.

Berry, C. M. 1993. Devonian plant assemblages from Venezuela. Unpublished Ph.D. thesis, Cardiff: University of Wales.

———. 1994. First record of the Devonian lycophyte *Leclercqia* from South America. *Geol. Mag.* 131: 269–272.

———. 1997. Diversity and distribution of Devonian Protolepidodendrales. *Palaeobotanist* 45: 209–216.

Berry, C. M., J. E. Casas, and J. M. Moody. 1993. Diverse Devonian plant assemblages from Venezuela. *Doc. Lab. Geol. Fac. Sci. Lyon* 125: 29–42.

Berry, C. M., J. Cordi, and W. E. Stein. 1997. Morphological models of Devonian Iridopteridales. *Am. J. Bot* 84 (Supplement): 129.

Berry, C. M., and D. Edwards. 1994. New data on the morphology and anatomy of the Devonian zosterophyll *Serrulacaulis* Hueber and Banks from Venezuela. *Rev. Palaeobot. Palynol.* 81: 141–150.

———. 1995. New species of the Devonian lycophyte *Colpodexylon* from the Devonian of Venezuela. *Palaeontographica B* 137: 59–74.

———. 1996a. The herbaceous lycophyte genus *Haskinsia* Grierson and Banks from the Devonian of western Venezuela, with observations on the leaf morphology and fertile specimens. *Bot. J. Linn. Soc.* 122: 103–122.

———. 1996b. *Anapaulia moodyi* gen et sp nov, a probable iridopteridalean compression fossil from the Devonian of Venezuela. *Rev. Palaeobot. Palynol.* 93: 127–145.

———. 1997. A new species of the lycopsid *Gilboaphyton* Arnold from the Devonian of Venezuela and New York State with a revision of the closely related genus *Archaeosigillaria* Kidston. *Rev. Palaeobot. Palynol.* 96: 47–70.

Berry, C. M., and M. Fairon-Demaret. 1997. A reinvestigation of the cladoxylopsid *Pseudosporochnus nodosus* Leclercq et Banks from the Middle Devonian of Goé, Belgium. *Int. J. Plant Sci.* 158: 350–372.

Berry, C. M., and W. E. Stein. 2000. A new genus of anatomically and morphologically preserved iridopteridalean from the Devonian of western Venezuela. *Int. J. Plant Sci.* 161: 807–827.

Berry, W. B. N., and P. Wilde. 1978. Progressive ventilation of the oceans—an explanation for the distribution of the lower Paleozoic black shales. *Am. J. Sci.* 278: 257–275.

Bertrand, P. 1913. Étude du stipe de *l'Asteropteris noveboracensis. Compt. Rend. XII Congr. Geol. Int.*, Ottawa.

Bewley, J. D. 1979. Physiological aspects of desiccation tolerance. *Ann. Rev. Plant Physiol.* 30: 195–238.

———. 1995. Physiological aspects of desiccation tolerance—a retrospect. *Int. J. Plant Sci.* 156: 393–403.

Bewley, J. D., and M. Black. 1994. *Seeds: Physiology of Development and Germination*, 2nd ed. New York: Plenum Press.

Bewley, J. D., and J. E. Krochko. 1982. Desiccation tolerance. In O. L. Lange, P. S. Nobel, C. B. Osmond, and H. Ziegler, eds. *Physiological Plant Ecology II. Encyclopedia of Plant Physiology*, N.S., vol. 12B., pp. 325–378. Berlin: Springer-Verlag.

Bewley, J. D., and M. J. Oliver. 1992. Desiccation tolerance in vegetative plant tissues and seeds: Protein synthesis in relation to desiccation and a potential role for protection and repair mechanisms. In G. N. Somero, C. B. Osmond, and C. L. Bolis, eds. *Water and Life. Comparative Analysis of Water Relationships at the Organismic, Cellular, and Molecular Levels*, pp 141–160. Berlin: Springer-Verlag.

Bhattacharya, D., and L. Medlin. 1998. Algal phylogeny and the origin of land plants. *Plant Physiol.* 116: 9–15.

Bignell, D. E. 1989. Relative assimilations of ^{14}C-labelled microbial tissues and ^{14}C-labelled plant fibre ingested with leaf litter by millipede *Glomeris marginata* under experimental conditions. *Soil Biol. Biochem.* 21: 819–828.

Blackmore, S., and S. H. Barnes 1987. Embryophyte spore walls: Origin, homologies and development. *Cladistics* 3: 185–195.

Blokker, P., S. Schouten, J. S. Damste, H. van den Ende, and J. W. de Leeuw. 1997. Chemical structure of aliphatic biopolymers in cell walls of *Tetraedron minimum, Pediastrum boryanum,* and *Scenedesmus communis. Phycologia* 36: 9 (abstr.).

Bockelie, J. F. 1994. Plant roots in core. In S. K. Donovan, ed. *The Paleobiology of Trace Fossils*, pp. 177–199. New York: John Wiley.

Bonamo, P. M., and H. P. Banks. 1967. *Tetraxylopteris schmidtii:* Its fertile parts and its relationships within the Aneurophytales. *Am. J. Bot.* 54: 755–768.

Bonamo, P. M., H. P. Banks, and J. D. Grierson. 1988. *Leclercqia, Haskinsia,* and the role of leaves in delineation of Devonian lycopod genera. *Bot. Gaz.* 149: 222–239.

Bormann, B. T., D. Wang, F. H. Bormann, G. Benoit, R. April, and R. Snyder. 1998. Rapid plant induced weathering and soil development in an aggrading experimental ecosystem. *Biogeochemistry* 43: 129–155.

Boucot, A. J. 1990. Silurian and pre-Upper Devonian bio-events. In E. G. Kauffman and O. H. Walliser, eds. *Extinction Events in Earth History (Lecture Notes in Earth Sciences* 30), pp. 125–132. Berlin: Springer.

Boucot, A. J., L. M. Cumming, and H. Jaeger. 1967. Contributions to the age of the Gaspé Sandstone and Gaspé Limestone. *Geol. Surv. Can. Pap.* 67-25: 1–22.

Boucot, A. J., J. F. Dewey, D. L. Dinely, R. Fletcher, W. K. Fyson, J. G. Griffin, C. F. Hickox, W. S. McKerrow, and A. M. Ziegler. 1974. Geology of the Arisaig area, Antigonish County, Nova Scotia. *Geol. Soc. Am. Spec. Pap.* 139: 191.

Boudet, A.-M. 1998. A new view of lignification. *Trends Plant Sci.* 3: 67–71.

Boyer, J. S., and L. C. Matten. 1996. Anatomy of *Eospermatopteris eriana* from the upper Middle Devonian (=Givetian) of New York. *International Organisation of Paleobotany V*, Abstracts Volume, p. 13. Santa Barbara: Department of Geological Sciences.

Bradshaw, M. A. 1981. Paleoenvironmental interpretations and systematics of Devonian trace fossils from the Taylor Group (lower Beacon Supergroup), Antarctic. *New Zeal. J. Geol. Geophys.* 24: 615–652.

Brand, U. 1989. Global climatic changes during the Devonian-Mississippian: Stable isotope biogeochemistry of brachiopods. *Palaeogeog. Palaeoclim. Palaeoecol.* 75: 311–329.

Brattsten, L. B. 1979. Biochemical defense mechanisms in herbivores against plant allelochemicals. In G. S. Rosenthal and D. H. Janzen, eds. *Herbivores. Their Interaction with Secondary Plant Metabolism.* New York: Academic Press.

Brauckmann, K. 1977. Neue Arachniden-Funde (Scorpionida, Trigonotarbida) aus dem westdeutschen Unter-Devon. *Geol. Paláont.* 21: 73–85.

———. 1994. Zwei neue Arachniden-Funde (Trigonotarbida) aus dem Unter-Devon der Eifel. *Jber. Naturwiss. Ver. Wuppertal* 47: 168–173.

Brauer, D. F. 1981. Heterosporous, barinophytacean plants from the Upper Devonian of North America and a discussion of the possible affinities of the Barinophytaceae. *Rev. Palaeobot. Palynol.* 33: 347–362.

Braun, A. 1997. Vorkommen, Untersuchungsmethoden und Bedeutung tierischer Cuticulae in kohligen Sedimentgesteinen des Devons und Karbons. *Palaeontographica B* 245: 83–156 (plates 1–15).

Brewer, R. 1976. *Fabric and Mineral Analysis of Soils*, 2nd ed. New York: R. E. Krieger.

Bridge, J. S., and M. L. Droser. 1985. An unusual marginal-marine lithofacies from the Upper Devonian clastic wedge. In D. L. Woodrow and W. D. Sevon, eds. *The Catskill Delta* (*Geol. Soc. Am. Spec. Pap.* 201), pp. 143–161.

Bridge, J. S., and E. A. Gordon. 1985. Quantitative interpretation of ancient river systems in the Oneonta Formation, Catskill Magnafacies. In D. L. Woodrow and W. D. Sevon, eds. *The Catskill Delta* (*Geol. Soc. Am. Spec. Pap.* 201), pp. 163–182.

Bridge, J. S., D. H. Griffing, and C. L. Hotton. 1998. Early Devonian coastal-fluvial environments and plant paleoecology, Gaspé Bay, Québec, Canada. Abstract from Geological Society of London/British Sedimentological Research Group Meeting, London, December 7–8, p. 4.

Bridge, J. S., and R. C. Titus. 1986. Non-marine bivalves and associated burrows in the Catskill Magnafacies of New York State. *Palaeogeog. Palaeoclimatol. Palaeoecol.* 55: 65–77.

Bridge, J. S., and B. J. Willis. 1994. Marine transgressions and regressions recorded in Middle Devonian shore-zone deposits of the Catskill clastic wedge. *Geol. Soc. Am. Bull.* 106: 1440–1458.

Brisebois, D. 1981. Geologie de la région de Gaspé. Ministère de l'énergie et des réssources, direction générale des énergies conventionnelles, rapport intérimaire. *DPV*-824: 1–19.

Brook, A. J., and D. B. Williamson. 1988. The survival of desmids on the drying mud of a small lake. In F. E. Round, ed. *Algae and the Aquatic Environment*, pp. 185–196. Bristol: BioPress.

Brooks, J. 1971. Some chemical and geochemical studies on sporopollenin. In J. Brooks, P. R. Grant, M. D. Muir, P. van Gijzel, and G. Shaw, eds. *Sporopollenin*, pp. 351–407. London: Academic Press.

Brown, D. H. 1983. Mineral nutrition. In A. J. E. Smith, ed. *Bryophyte Ecology*, pp. 383–444. New York: Chapman and Hall.

Brown, D. H., and R. M. Brown. 1991. Mineral cycling and lichens: The physiological basis. *Lichenologist* 23: 293–307.

Brown, R. C., and B. E. Lemmon. 1990. Polar organizers mark division axis prior to preprophase band formation in mitosis of the hepatic *Reboulia hemisphaerica* (Bryophyta). *Protoplasma* 156: 74–81.

———. 1992. Polar organizers in monoplastidic mitosis of hepatics (Bryophyta). *Cell Motility and the Cytoskeleton* 22: 72–77.

———. 1997. The quadripolar microtubule system in lower land plants. *J. Plant Res.* 110: 93–106.

Brown, R. C., B. E. Lemmon, and L. E. Graham. 1994. Morphogenetic plastid migration and microtubule arrays in mitosis and cytokinesis in the green alga *Coleochaete orbicularis*. *Am. J. Bot.* 81: 127–133.

Browning, A. J., and B. E. S. Gunning. 1979. Structure and function of transfer cells in the sporophyte haustorium of *Funaria hygrometrica* Hedw. II. Kinetics of uptake of labelled sugars and localization of absorbed products by freeze-substitution and autoradiography. *J. Exper. Bot.* 30: 1247–1264.

Brüggl, G. 1992. Gut passage, respiratory rate and assimilation efficiency of three millipedes from a deciduous wood in the Alps. In E. Meyer, K. Thaler, and W. Schedl, eds. *Advances in Myriapodology*, pp. 319–325. Proceedings of the 8th International Congress of Myriapodology, Innsbruck, 1990. *Berichte des Naturwissenschaftlich-Medizinischen Vereins in Innsbruck*, Suppl. 10. Universitätsverlag Wagner, Innsbruck.

Buggisch, W. 1991. The global Frasnian-Famennian "Kellwasser Event". *Geol. Rundschau* 80: 49–72.

Bultynck, P., M. Coen-Aubert, L. Dejonghe, J. Godefroid, L. Hance, D. Lacroix, A. Preat, P. Stainier, P. Steemans, M. Streel, and F. Tourneur. 1991. Les formations du Dévonien moyen de la Belgique. *Mem. Servir Explication Cartes Geol. Min. Belg.* 30: 1–106.

Burchette, T. P. 1981. European Devonian reefs: A review of current concepts and models. In D. F. Toomey, ed. *European Fossil Reef Models*, pp. 85–142. Tulsa, OK: Soc. Econ. Paleontol. Mineral., Spec. Publ. 30.

Burgess, N. D. 1991. Silurian cryptospores and miospores from the type Llandovery area, south-west Wales. *Palaeontology* 34: 575–599.

Burgess, N. D., and D. Edwards. 1991. Classification of uppermost Ordovician to Lower Devonian tubular and filamentous macerals from the Anglo-Welsh Basin. *Bot. J. Linn. Soc.* 106: 41–66.

Burgess, N. D., and J. B. Richardson. 1991. Silurian cryptospores and miospores from the type Wenlock area, Shropshire, England. *Palaeontology* 34: 601–628.

———. 1997. Late Wenlock to early Pridoli cryptospores and miospores from South and Southwest Wales. *Palaeontographica* B236: 1–44.

Burne, R. V., J. Bauld, and P. De Deckker. 1980. Saline lake charophytes and their geological significance. *J. Sediment. Petrol.* 50: 281–293.

Buurman, P. 1975. Possibilities of paleopedology. *Sedimentology* 22: 289–298.

Cai, C.-Y., and L. Chen. 1996. On a Chinese Givetian lycopod, *Longostachys latisporophyllus* Zhu, Hu and Feng, emend.: Its morphology, anatomy and reconstruction. *Palaeontographica B* 238: 1–43.

Cai, C.-Y., Y.-W. Dou, and D. Edwards. 1993. New observations on a Pridoli plant assemblage from north Xinjiang, northwest China, with comments on its evolutionary and palaeogeographical significance. *Geol. Mag.* 130: 155–170.

Cai, C.-Y., and X. X. Li. 1982. Subdivision and correlation of continental Devonian strata in China. In Nanjing Institute of Geology and Paleontology, Academia Sinica, ed. *Stratigraphic Correlation Charts in China with Explanatory Text*, pp. 103–123. Beijing: Beijing Science Press.

Cai, C.-Y., and Y. Wang. 1995. Devonian floras. In X.-X. Li, ed. *Fossil floras of China through the Geological Ages*, pp. 28–77. Guangzhou, China: Guangdong Science and Technology Press.

Caldeira, K. 1992. Enhanced Cenozoic chemical weathering and the subduction of pelagic carbonate. *Nature* 357: 578–581.

Campbell, S. E. 1979. Soil stabilization by a prokaryotic desert crust: Implications for Precambrian land biota. *Origins of Life* 9: 335–348.

Cant, D. J., and R. G. Walker. 1976. Development of a braided-fluvial facies model for the Devonian Battery Point Sandstone, Quebec. *Can. J. Earth Sci.* 13: 102–119.

Capelle, D. G., and S. G. Driese. 1995. Changes in paleosol morphology along a muddy coastal margin-to-alluvial transect: Irish Valley and Sherman Creek members (Catskill Formation, Upper Devonian), central Pennsylvania (abst.). *Geol. Soc. Am.* (Abst. with Programs), vol. 27, no. 1, p. 33.

Caputo, M. V. 1985. Late Devonian glaciation in South America. *Palaeogeog. Palaeoclim. Palaeoecol.* 51: 291–317.

Caputo, M. V., and J. C. Crowell. 1985. Migration of glacial centers across Gondwana during Paleozoic Era. *Geol. Soc. Am. Bull.* 96: 1020–1036.

Carluccio, L. M., F. M. Hueber, and H. P. Banks. 1966. *Archaeopteris macilenta*, anatomy and morphology of its frond. *Am. J. Bot.* 53: 719–730.

Carroll, R. L. 1992. The primary radiation of terrestrial vertebrates. *Ann. Rev. Earth Planet. Sci.* 20: 45–84.

Caudill, M. R., S. G. Driese, and C. I. Mora. 1996. Preservation of a paleo-Vertisol and an estimate of Late Mississippian paleo-precipitation. *J. Sediment. Res.* 66: 58–70.

Cawley, J. L., R. C. Burruss, and H. D. Holland. 1969. Chemical weathering in Central Iceland: An analog of pre-Silurian weathering. *Science* 165: 391–392.

Cerling, T. E. 1984. The stable isotopic composition of modern soil carbonate and its relationship to climate. *Earth Planet. Sci. Lett.* 71P: 229–240.

———. 1991. Carbon dioxide in the atmosphere: Evidence from Cenozoic and Mesozoic paleosols. *Am. J. Sci.* 291: 377–400.

Chacko, P. M. 1966. A new variety of *Nitella terrestris* (var. *minor* var. nov.) from Kerala, S. India. *Current Science* 35: 630–631.

Chaloner, W. G. 1967. Spores and land-plant evolution. *Rev. Palaeobot. Palynol.* 1: 83–93.

———. 1989. Fossil charcoal as an indicator of palaeoat-mospheric oxygen level. *J. Geol. Soc. Lond.* 146: 171–174.

Chaloner, W. G., A. J. Hill, and W. S. Lacey. 1977. First Devonian platyspermic seed and its implications in gymnosperm evolution. *Nature* 265: 233–235.

Chaloner, W. G., A. C. Scott, and J. Stephenson. 1991. Fossil evidence for plant-arthropod interactions in the Palaeozoic and Mesozoic. *Phil. Trans. R. Soc. Lond.* B33: 177–186.

Chaloner, W. G., and A. Sheerin. 1979. Devonian macrofloras. In M. R. House, C. T. Scrutton, and M. G. Bassett, eds. *The Devonian System*. Palaeontol. Soc., Spec. Pap. Palaeontol. 23: 145–161.

Chapman, R. L., and M. A. Buchheim. 1991. Ribosomal RNA gene sequences: Analysis and significance in the phylogeny and taxonomy of green algae. *Crit. Rev. Plant Sci.* 10: 343–368.

Christiansen, K. 1964. Bionomics of Collembola. *Ann. Rev. Entomol.* 9: 147–178.

Chuvashov, B., and R. Riding. 1984. Principal floras of Paleozoic marine calcareous algae. *Palaeontology* 27: 487–500.

Clack, J. A. 1977. Devonian tetrapod trackways and trackmakers: A review of the fossils and footprints. *Palaeogeogr. Palaeoclimatol. Palaeoecol.* 130: 227–250.

Claeys, P., and J.-G. Casier. 1994. Microtektite-like impact glass associated with the Frasnian-Famennian boundary mass extinction. *Earth Planet. Sci. Lett.* 122: 303–315.

Claeys, P., J.-G. Casier, and S. V. Margolis. 1992. Microtektites and mass extinctions: Evidence for a Late Devonian asteroid impact. *Science* 257: 1102–1104.

Claoué-Long, J. C., P. J. Jones, J. Roberts, and S. Maxwell. 1992. The numerical age of the Devonian-Carboniferous boundary. *Geol. Mag.* 129: 281–291.

Cleal, C. J., and B. A. Thomas. 1995. *Palaeozoic Palaeobotany of Great Britain*, pp. 80–92. *Geological Conservation Review Series*. London: Chapman and Hall.

Cochran, M. F., and R. A. Berner. 1993. Enhancement of silicate weathering rates by vascular land plants: Quantifying the effect. *Chem. Geol.* 107: 213–215.

———. 1996. Promotion of chemical weathering by higher plants; Field observations on Hawaiian basalts. *Chem. Geol.* 132: 71–77.

Cocks, L. R. M., and C. R. Scotese. 1991. The global biogeography of the Silurian Period. In M. G. Bassett, P. D. Lane, and D. Edwards, eds. *The Murchison Symposium: Proceedings of an International Conference on the Silurian System.* Spec. Pap. Palaeontol. 44: 1–34.

Coleman, A. W. 1983. The roles of resting spores and akinetes in chlorophyte survival. In G. A. Fryxell, ed. *Survival Strategies of the Algae*, pp. 1–21. Cambridge: Cambridge University Press.

Colton, G. W. 1970. The Appalachian Basin - Its depositional sequences and their geologic relationships. In G. W. Fisher, F. J. Pettijohn, J. C. Reed, Jr., and K. N. Weaver, eds. *Studies of Appalachian Geology, Central and Southern*, pp. 5–47. New York: John Wiley.

Comeau, P. G., and J. P. Kimmins. 1989. Above- and below-ground biomass and production of lodgepole pine on sites with differing soil moisture regimes. *Can. J. For. Res.* 19: 447–454.

Cook, M. E., L. E. Graham, C. E. J. Botha, and C. A. Lavin. 1997. Comparative ultrastructure of plasmodesmata of *Chara* and selected bryophytes: Toward an elucidation of the evolutionary origin of plant plasmodesmata. *Am. J. Bot.* 84: 1169–1178.

Cook, M. E., L. E. Graham, and C. A. Lavin 1998. Cytokinesis and nodal anatomy in the charophycean green alga *Chara zeylanica. Protoplasma* 203: 65–74.

Cooper-Driver, G. A., and M. Bhattacharya. 1998. Role of phenolics in plant evolution. *Phytochemistry* 49: 1165–1174.

Copper, P. 1986. Frasnian/Famennian mass extinction and cold-water oceans. *Geology* 14: 835–839.

Corna, O. 1970. Plant remains in the Ordovician of the Bohemiam Massif. *Geolog. Carpathica* 21: 183–186.

Cornet, B., T. L. Phillips, and H. N. Andrews. 1976. The morphology and variation in *Rhacophyton ceratangium* from the Upper Devonian and its bearing on frond evolution. *Palaeontographica B* 158: 105–129.

Cotter, E. 1978. The evolution of fluvial style, with special reference to the central Appalachian Paleozoic. In A. D. Miall, ed. *Fluvial Sedimentology*, pp. 361–383. Calgary, Alberta: Canadian Society of Petroleum Geologists, Mem. 5.

Cotter, E., and S. G. Driese. 1998. Incised valley fills and other evidence of sea-level fluctuations affecting deposition of the Catskill Formation (Upper Devonian), Appalachian foreland basin of Pennsylvania. *J. Sed. Res.* 68: 347–361.

Coulombe, C. E., J. B. Dixon, and L. P. Wilding. 1996. Mineralogy and chemistry of Vertisols. In N. Ahmad and A. Mermut, eds. *Vertisols and Technologies for Their Management: Developments in Soil Science* 24: 115–200. New York: Elsevier.

Coulombe, C. E., L. P. Wilding, and J. B. Dixon. 1996. Overview of Vertisols: Characteristics and impacts on society. *Adv. Agron.* 57: 289–375.

Craig, H. 1965. The measurement of oxygen isotope paleotemperatures. In *Stable Isotopes in Oceanographic Studies and Paleotemperatures*, pp. 1–24. Pisa, Italy: Consiglio Nazionale delle Richerche, Laboratorio di Geologia Nucleare.

Crawford, C. S. 1992. Millipedes as model detritivores. *Berichte des Naturwissenschaftlich-Medizinischen Vereins in Innsbruck*, Suppl. 10: 277–288.

Crawford, C. S., K. Bercovitz, and M. R. Warburg. 1987. Regional environments, life-history patterns and habitat use of spirostreptid millipedes in arid regions. *Zool. J. Linn. Soc.* 89: 63–88.

Cross, A. T. 1983. Plants of Devonian-Mississippian black shales, Eastern Interior, U.S.A. *Am. Assoc. Petr. Geol. Bull.* 67: 444–445.

Crowe, J. H., and J. S. Clegg, eds. 1973. *Anhydrobiosis.* Stroudsburg, PA: Dowden, Hutchinson and Ross.

———. 1978. *Dry Biological Systems.* New York: Academic Press.

Crowe, J. H., F. A. Hoekstra and L. M. Crowe. 1992. Anhydrobiosis. *Ann. Rev. Physiol.* 54: 579–599.

Crowley, T. J., and S. K. Baum. 1991. Estimating Carboniferous sea-level fluctuations from Gondwanan ice extent. *Geology* 19: 975–977.

Cuerda, A. J., C. Cingolani, O. Arrondo, E. Morel, and D. Ganuza. 1987. Primer registro de plantas vasculares en la Formación Villavicencio, Precordillera de Mendoza, Argentina. *IV Congreso Lantinoamericano de Paleontología. Actas* 1: 179–183.

Cullmann, F., A. Klaus-Peter, and H. Becker. 1993. Bisbibenzyls and lignans from *Pellia epiphylla. Phytochemistry* 34: 831–834.

Cusick, F. 1954. Experimental and analytical studies of pteridophytes. XXV. Morphogenesis in *Selaginella Willdenovii* Baker. II. Angle-meristems and angle-shoot. *Ann. Bot., N. S.* 17: 171–181.

Daber, R. 1960. *Eogaspesiea gracilis* n. g., n. sp. *Geologie* 9: 418–425.

———. 1971. *Cooksonia*—one of the most ancient psilophytes—widely distributed, but rare. *Botanique (Nagpur)* 2: 35–40.

Daeschler, E. B., and N. Shubin. 1995. Tetrapod origins. *Paleobiol.* 21: 404–409.

Dannenhoffer, J. M., and P. M. Bonamo. 1989. *Rellimia thomsonii* from the Givetian of New York: Secondary growth in three orders of branching. *Am. J. Bot.* 76: 1312–1325.

Davis, J. S. 1972. Survival records in the algae, and the survival role of certain algal pigments, fat and mucilaginous substances. *The Biologist* 54: 52–93.

Dawson, J. W. 1859. On fossil plants from the Devonian rocks of Canada. *Q. J. Geol. Soc. Lond.* 15: 477–488.

———. 1871. The fossil plants of the Devonian and Upper Silurian Formations of Canada. *Geol. Surv. Can.*, pp. 1–92.

———. 1881. Notes on new Erian (Devonian) plants. *Q. J. Geol. Soc. Lond.* 37: 299–308.

De Freitas, T., J. C. Harrison, and R. Thorsteinsson. 1993. New field observations on the geology of Bathurst Island, Arctic Canada. Part A: Stratigraphy and sedimentology. *Geol. Surv. Canada, Cur. Res.*, Paper 93–1B; pp. 1–10.

Degens, E. T., and S. Epstein. 1962. Relationship between $^{18}O/^{16}O$ ratios in coexisting calcites and dolomites from recent and ancient sediments. *Am. Assoc. Petr. Geol. Bull.* 28: 23–44.

Delwiche, C. F., L. E. Graham, and N. Thomson. 1989. Lignin-like compounds and sporopollenin in *Coleochaete*, an algal model for land plant ancestry. *Science* 245: 399–401.

Demming-Adams, B., and W. W. Adams. 1996. The role of the xanthophyll cycle carotenoids in the protection of photosynthesis. *Trends Plant Sci.* 1: 21–26.

Dickson, J. A. D. 1965. Modified staining technique for carbonates in thin section. *Nature* 205: 587.

———. 1966. Carbonate identification and genesis as revealed by staining. *J. Sediment. Petrol.* 36: 491–505.

Diemer, J. A. 1992. Sedimentology and alluvial stratigraphy of the upper Catskill Formation, south-central Pennsylvania. *Northeastern Geol.* 14: 121–136.

DiMichele, W. A., and R. M. Bateman. 1996. Plant

palaeoecology and evolutionary inference: Two examples from the Paleozoic. *Rev. Palaeobot. Palynol.* 90: 223–247.

DiMichele, W. A., J. I. Davis, and R. G. Olmstead. 1989. Origins of heterospory and the seed habit: The role of heterochrony. *Taxon* 38: 1–11.

DiMichele, W. A., and R. Hook. 1992. Paleozoic terrestrial ecosystems. In A. K. Behrensmeyer, J. D. Damuth, W. A. DiMichele, R. Potts, H. Sues, S. L. Wing, eds. *Terrestrial Ecosystems through Time*, pp. 205–325. Chicago: University of Chicago Press.

DiMichele, W. A., and T. L. Phillips. 1996. Clades, ecological amplitudes, and ecomorphs: Phylogenetic effects and persistence of primitive plant communities in the Pennsylvanian wetland tropics. *Palaeogeog. Palaeoclim. Palaeoecol.* 127: 83–105.

DiMichele, W. A., and J. E. Skog. 1992. The Lycopsida: A symposium. *Ann. Mo. Bot. Gard.* 79: 447–449.

Dineley, D. L. 1963. The "red stratum" of the Silurian Arisaig Series, Nova Scotia, Canada. *J. Geol.* 71: 523–524.

———. 1984. *Aspects of a Stratigraphic System: The Devonian.* New York: Wiley.

Doran, J. B. 1980. A new species of *Psilophyton* from the Lower Devonian of northern New Brunswick, Canada. *Can. J. Bot.* 58: 2241–2262.

Doran, J. B., P. G. Gensel, and H. N. Andrews. 1978. New occurrences of trimerophytes from the Devonian of eastern Canada. *Can. J. Bot.* 56: 3052–3068.

Douglas, C. J. 1996. Phenylpropanoid metabolism and lignin biosynthesis: From weeds to trees. *Trends Plant Sci.* 1: 171–178.

Drever, J. I. 1994. The effect of land plants on weathering rates of silicate minerals. *Geochim. Cosmochim. Acta* 58: 2325–2332.

Drever, J. I., and J. Zobrist. 1992. Chemical weathering of silicate rocks as a function of elevation in the southern Swiss Alps. *Geochim. Cosmochim. Acta* 56: 3209–3216.

Driese, S. G., and J. L. Foreman. 1991. Traces and related chemical changes in a Late Ordovician paleosol *Glossifungites* ichnofacies, southern Appalachians, USA. *Ichnos* 1: 183–194.

———. 1992. Paleopedology and paleoclimatic implications of Late Ordovician vertic paleosols, Juniata Formation, southern Appalachians. *J. Sediment. Petrol.* 62: 71–83.

Driese, S. G., and C. I. Mora. 1993. Physico-chemical environment of pedogenic carbonate formation in Devonian vertic palaeosols, central Appalachians, USA. *Sedimentology* 40: 199–216.

Driese, S. G., C. I. Mora, E. Cotter, and J. L. Foreman. 1992. Paleopedology and stable isotope chemistry of Late Silurian vertic paleosols, Bloomsburg Formation, central Pennsylvania. *J. Sediment. Petrol.* 62: 825–841.

Driese, S. G., C. I. Mora, and J. M. Elick. 1997. Morphology and taphonomy of root traces and stump casts of the earliest trees (Middle to Late Devonian), Pennsylvania and New York, U.S.A. *Palaios* 12: 524–537.

Driese, S. G., K. Srinivasan. 1992. Paleopedology and paleoclimatic implications of Late Ordovician vertic paleosols, southern Appalachians. *J. Sediment. Petrol.* 62: 71–83.

Driese, S. G., K. Srinivasan, C. I. Mora, and F. W. Stapor. 1994. Paleoweathering of Mississippian Monteagle Limestone preceding development of a lower Chesterian transgressive systems tract and sequence boundary, middle Tennessee and northern Alabama. *Geol. Soc. Am. Bull.* 106: 866–878.

Dubinin, V. B. 1962. Class Acaromorpha: Mites or gnathosomic chelicerate arthropods. In B. B. Rodendorf, ed. *Fundamentals of Paleontology*, pp. 447–473. Moscow: Academy of Sciences of the USSR [in Russian].

Dudal, R., and H. Eswaran. 1988. Distribution, properties and classification of Vertisols. In L. P. Wilding and R. Puentes, eds. *Vertisols: Their Distribution, Properties, Classification and Management*, pp. 1–22. College Station, TX: Texas A&M University Printing Center.

Duffy, S. S. 1980. Sequestration of plant natural products by insects. *Ann. Rev. Entomol.* 25: 447–477.

Dufka, P. 1995. Upper Wenlock miospores and cryptospores derived from a Silurian volcanic island in the Prague Basin (Barrandian area, Bohemia). *J. Micropaleontol.* 14: 67–79.

Dulong, F. T., and C. B. Cecil. 1989. Stratigraphic variation in the bulk sample mineralogy of Pennsylvanian underclays from the central Appalachian basin. Carboniferous Geology of the United States, American Geophysical Union, *Field Trip Guidebook*, Washington, D.C., T143: 112–118.

Dunlop, J. A. 1994. Filtration mechanisms in the mouthparts of tetrapulmonate arachnids (Trigonotarbida, Araneae, Amblypygi, Uropygi, Schizomida). *Bull. Br. Arachnol. Soc.* 9: 267–273.

———. 1996a. A trigonotarbid arachnid from the Upper Silurian of Shopshire. *Palaeontology* 39: 605–614.

———. 1996b. A redescription of the trigonotarbid arachnid *Pocononia whitei* (Ewing 1930). *Paläontol. Zeit.* 70: 145–151.

———. 1996c. Systematics of the fossil arachnids. *Rev. Suisse Zool. vol. hors série I* Août: 173–184.

Dunn, P. A. 1988. *Dynamics of $\delta^{13}C$ and $\delta^{18}O$ Variation in the Devono-Carboniferous Succession of Belgium and Ireland.* Ann Arbor, MI: University of Michigan, unpubl. M.S. thesis.

Dustin, C., and G. Cooper-Driver. 1993. Changes in phenolic production in the hay-scented fern (*Dennstaedtia punctilobula*) in relation to resource availability. *Biochem. System. Ecol.* 20: 99–106.

Dyer, B. D., N. N. Lyalikova, D. Murray, M. Doyle, G. M. Kolesov, and W. E. Krumbein. 1989. Role for microorganisms in the formation of iridium anomalies. *Geology* 17: 1036–1039.

Edwards, D. 1972. A *Zosterophyllum* fructification from the Lower Old Red Sandstone of Scotland. *Rev. Palaeobot. Palynol. (Jubilee Volume for Professor S. Leclercq)* 14: 77–83.

———. 1975. Some observations on the fertile parts of *Zosterophyllum myretonianum* Penhallow from the

Lower Old Red Sandstone of Scotland. *Trans. R. Soc. Edinb.* 69: 251–265.

———. 1979a. The early history of vascular plants based on late Silurian and early Devonian floras of the British Isles. In A. L. Harris, C. H. Holland, and B. E. Leake, eds. *The Caledonides of the British Isles—Reviewed*, pp. 405–410. London: Geological Society.

———. 1979b. A late Silurian flora from the Lower Old Red Sandstone of south-west Dyfed. *Palaeontology* 22: 23–52.

———. 1980. Early land floras. In A. L. Panchen, ed. *The Terrestrial Environment and the Origin of Land Vertebrates*, pp. 55–85. New York: Academic Press.

———. 1982. Fragmentary non-vascular plant microfossils from the Late Silurian of Wales. *Bot. J. Linn. Soc.* 84: 223–256.

———. 1986. Dispersed cuticles of putative non-vascular plants from the Lower Devonian of Britain. *Bot. J. Linn. Soc.* 93: 259–275.

———. 1990. Constraints on Silurian and Early Devonian phytogeographic analysis based on megafossils. In W. S. McKerrow., and C. R. Scotese, eds. *Palaeozoic Palaeogeography and Biogeography*, Mem. no. 12, pp. 233–242. London: Geological Society.

———. 1993. Cells and tissues in the vegetative sporophytes of early land plants. *New Phytol.* 125: 225–247.

———. 1994. Towards an understanding of pattern and process in the growth of early vascular plants. In D. S. Ingram, ed. *Shape and Form in Plants and Fungi*, pp. 39–59. London: Linnean Society.

———. 1996. New insights into early land ecosystems: A glimpse of a Lilliputian world. *Rev. Palaeobot. Palynol.* 90: 159–174.

———. 1997. Charting diversity in early land plants: Some challenges for the next millenium. In K. Iwatsuki and P. H. Raven, eds. *Evolution and Diversification of Land Plants*, pp. 3–26. Tokyo: Springer-Verlag.

———. 1998. Climate signals in Palaeozoic land plants. *Phil. Trans. R. Soc. Lond.* B353: 141–157.

Edwards, D., G. D. Abbott, and J. A. Raven. 1996. Cuticles of early land plants: A palaeoecophysiological evaluation. In G. Kerstiens, ed. *Plant Cuticles—An integrated functional approach*, pp. 1–31. Oxford: BIOS Scientific.

Edwards, D., M. G. Bassett, and E. C. W. Rogerson. 1979. The earliest vascular land plants: Continuing the search for proof. *Lethaia* 12: 313–324.

Edwards, D., and J. L. Benedetto. 1985. Two new species of herbaceous lycopods from the Devonian of Venezuela with comments on their taphonomy. *Palaeontology* 28: 599–618.

Edwards, D., and C. M. Berry. 1991. Silurian and Devonian. In C. J. Cleal, ed. *Plant Fossils in Geological Investigations: The Palaeozoic*, pp. 117–153. Chichester: Ellis Horwood Series in Applied Geology.

Edwards, D., and M. S. Davies. 1990. Interpretations of early land plant radiations: "Facile adaptationist guesswork" or reasoned speculation? In P. D. Taylor and G. P. Larwood, eds. *Major Evolutionary Radiations*, pp. 351–376. The Systematics Association Spec. vol. 42. Oxford: Clarendon Press.

Edwards, D., K. L. Davies, and L. Axe. 1992. A vascular conducting strand in the early land plant *Cooksonia*. *Nature* 357: 683–685.

Edwards, D., K. L. Davies, J. B. Richardson, and L. Axe. 1995. The ultrastructure of spores of *Cooksonia pertoni*. *Palaeontology* 38: 153–168.

Edwards, D., K. L. Davies, J. B. Richardson, C. H. Wellman, and L. Axe. 1996. Ultrastructure of *Synorisporites downtonensis* and *Retusotriletes* cf. *coronadus* in spore masses from the Prídolí of the Welsh Borderland. *Palaeontology* 39: 783–800.

Edwards, D., J. G. Duckett, and J. B. Richardson. 1995. Hepatic characters in the earliest land plants. *Nature* 374: 635–636.

Edwards, D., G. Ewbank, and G. D. Abbott. 1997. Flash pyrolysis of the outer cortical tissues in Lower Devonian *Psilophyton dawsonii*. *Bot. J. Linn. Soc.* 124: 345–360.

Edwards, D., M. Fairon-Demaret, and C. M. Berry. 2000. Plant megafossils in Devonian stratigraphy: A progress report. *Cour. Forschungsinst. Senckenberg* vol. 220: 25–37.

Edwards, D., and U. Fanning. 1985. Evolution and environment in the late Silurian-early Devonian: The rise of the pteridophytes. *Phil. Trans. R. Soc. Lond.* B309: 147–165.

Edwards, D., U. Fanning, K. L. Davies, L. Axe, and J. B. Richardson. 1995. Exceptional preservation in Lower Devonian coalified fossils from the Welsh Borderland: A new genus based on reniform sporangia lacking thickened borders. *Bot. J. Linn. Soc.* 117: 233–254.

Edwards, D., U. Fanning, and J. B. Richardson. 1994. Lower Devonian coalified sporangia from Shropshire: *Salopella* Edwards and Richardson and *Tortilicaulis* Edwards. *Bot. J. Linn. Soc.* 116: 89–110.

Edwards, D., J. Feehan, and D. G. Smith. 1983. A late Wenlock flora from Co. Tipperary, Ireland. *Bot. J. Linn. Soc.* 86: 19–36.

Edwards, D., and P. Kenrick. 1986. A new zosterophyll from the Lower Devonian of Wales. *Bot. J. Linn. Soc.* 92: 269–283.

Edwards, D., H. Kerp, and H. Hass. 1998. Stomata in early land plants: An anatomical and ecophysiological approach. *J. Exp. Bot.* 49 (Spec. Issue, March): 255–278.

Edwards, D., and J. B. Richardson. 1974. Lower Devonian (Dittonian) plants from the Welsh Borderland. *Palaeontology* 17: 311–324.

———. 1996. Review of *in situ* spores in early land plants. In J. Jansonius and D. C. McGregor, eds. *Palynology: Principles and Applications*, vol. 1, *Principles*, pp. 391–407. Salt Lake City, UT: American Association of Stratigraphic Palynologists Foundation.

———. 2000. Progress in reconstructing vegetation on the Old Red Sandstone Continent: Two *Emphanisporites* producers from the Lochkovian of the Welsh Borderland. *Geol. Soc. Lond. Spec. Publ.* 180: 355–370.

Edwards, D., and R. Riding. 1993. Chlorophyta. In M. J. Benton, ed. *The Fossil Record 2*, pp. 16–18. London: Chapman and Hall.

Edwards, D., and E. C. W. Rogerson. 1979. New records of fertile Rhyniophytina from the late Silurian of Wales. *Geol. Mag.* 116: 93–98.

Edwards, D., and V. Rose. 1984. Cuticles of *Nematothallus*: A further enigma. *Bot. J. Linn. Soc.* 88: 35–54.

Edwards, D., and P. A. Selden. 1993. The development of early terrestrial ecosystems. *Bot. J. Scot.* 46: 337–366.

Edwards, D., P. A. Selden, J. B. Richardson, and L. Axe. 1995. Coprolites as evidence for plant-animal interaction in Siluro-Devonian terrestrial ecosystems. *Nature* 377: 329–331.

Edwards, D., and C. H. Wellman. 1996. Older plant macerals (excluding spores). In J. Jansonius and D. C. McGregor, eds. *Palynology: Principles and Applications*, vol. 1, *Principles*, pp. 383–387. Salt Lake City, UT: American Association of Stratigraphic Palynologists Foundation.

Edwards, D., C. H. Wellman, and L. Axe. 1998. The fossil record of early land plants and interrelationships between primitive embryophytes: Too little too late? In J. W. Bates, N. W. Ashton, and J. G. Duckett, eds. *Bryology for the 21st Century*, pp. 15–43. Leeds, U.K.: Maney Publishing and British Bryological Society.

———. 1999. Tetrads in sporangia and spore masses from the Upper Silurian and Lower Devonian of the Welsh Borderland. *Bot. J. Linn. Soc.* 130: 111–115.

Edwards, D. S. 1986. *Aglaophyton major*, a non-vascular plant from the Devonian Rhynie Chert. *Bot. J. Linn. Soc.* 93: 173–204.

Edwards, D. S., and A. G. Lyon. 1983. Algae from the Rhynie Chert. *Bot. J. Linn. Soc.* 86: 37–55.

Ekart, D. D., and T. E. Cerling. 1996. PCO_2 during deposition of the Late Jurassic Morrison Formation and other paleoclimatic/ecologic data as inferred by stable carbon and oxygen isotope analyses (abst.). *Geol. Soc. Am., Abstr. with Progr.* 28(7): A252.

Eldridge, J., D. Walsh, and C. R. Scotese. 1996. *Plate Tracker, Paleomap Project*. Arlington, TX: University of Texas at Arlington, Department of Geology.

Elick, J. M., S. G. Driese, and C. I. Mora. 1996. Evidence for deep Early to Middle Devonian root systems: Prével Member, Battery Point Formation, Gaspé, Québec (abst.). *Geol. Soc. Am., Abstr. with Progr.* 28(7): A105.

———. 1998a. Very large plant and root traces from the Early to Middle Devonian: Implications for early terrestrial ecosystems and atmospheric p(CO_2) estimations. *Geology* 26: 143–146.

———. 1998b. Drainage patterns, plants and soils in an evolving coastal to fluvial environment, Battery Point Formation, Gaspé, Québec. Abstract from Geological Society of London/British Sedimentological Research Group Meeting, London, December 7–8, p. 9.

Elix, J. A. 1996. Biochemistry and secondary metabolites. In T. H. Nash, ed. *Lichen Biology*, Cambridge, England: Cambridge University Press.

Elliott, T. 1986. Deltas. In H. G. Reading, ed. *Sedimentary Environments and Facies*, 2nd ed., pp. 113–154. Oxford: Blackwell Scientific.

El-Saadawy, W., and W. S. Lacey. 1979. Observations on *Nothia aphylla* Lyon ex Høeg. *Rev. Palaeobot. Palynol.* 27: 119–147.

Emiliani, C., and D. B. Ericson. 1991. The glacial/inter-glacial temperature range of the surface water of the oceans at low latitudes. In H. P. Taylor, Jr., J. R. O'Neil, and I. R. Kaplan, eds. *Stable Isotope Geochemistry*, Spec. Publ. 3, pp. 223–228. London: Geochemical Society.

Ettensohn, F. R., G. R. Dever, and J. S. Grow. 1988. A paleosol interpretation for profiles exhibiting exposure "crusts" from the Mississippian of the Appalachian Basin. *Geol. Soc. Am. Spec. Pap.* 216: 49–79.

Ettensohn, F. R., M. L. Miller, S. B. Dillman, T. D. Elam, K. L. Geller, D. R. Swager, G. Markowitz, R. D. Woock, and L. S. Barron. 1988. Characterization and implications of the Devonian-Mississippian black-shale sequence, eastern and central Kentucky, U.S.A.: Pycnoclines, transgression, regression, and tectonism. In N. J. McMillan, A. J. Embry, and D. J. Glass, eds. *Devonian of the World*, vol. 2: *Sedimentation*, pp. 323–345. Calgary, Alberta: Canadian Society of Petroleum Geologists, Mem. 14.

Ewbank, G., D. Edwards, and G. D. Abbott. 1996. Chemical characterization of Lower Devonian vascular plants. *Org. Geochem.* 25: 461–473.

Fairon-Demaret, M. 1969. *Dixopodoxylon goense* (gen. et sp. nov.), a new form genus from the Middle Devonian of Belgium. *Bull. Acad. R. Belg. Cl. Sci.* 55: 372–386.

———. 1981. Le genre *Leclercqia* Banks, H. P., Bonamo, P. M. et Grierson, J. D. 1972 dans le Devonien Moyen de Belgique. *Bull. Inst. R. Sci. Nat. Belg.* 31-XII: 1–10.

———. 1986. Some uppermost Devonian megafloras: A stratigraphical review. *Ann. Soc. Géol. Belg.* 109: 43–48.

———. 1996. *Dorinnotheca streelii* Fairon-Demaret, gen. et sp. nov., a new early seed plant from the upper Famennian of Belgium. *Rev. Palaeobot. Palynol.* 93: 217–233.

Fairon-Demaret, M., and H. P. Banks. 1978. Leaves of *Archaeosigillaria vanuxemii*, a Devonian lycopod from New York. *Am. J. Bot.* 65, 246–249.

Fairon-Demaret, M., and C.-S. Li. 1993. *Lorophyton goense* gen. et sp. nov. from the Lower Givetian of Belgium and a discussion of the Middle Devonian Cladoxylopsida. *Rev. Palaeobot. Palynol.* 77: 1–22.

Fairon-Demaret, M., and S. E. Scheckler. 1987. Typification and redescription of *Moresnetia zalesskyi* Stockmans 1948, a Late Devonian (Famennian) cupulate seed plant. *Bull. Inst. R. Sci. Nat. Belg. (Science de la Terre)* 57: 183–199.

Fang, W. 1989. Paleozoic paleomagnetism of the South China Block and the Shan Thai Block: The composite nature of southeast Asia. Dissertation, University of Michigan, Ann Arbor.

Fang, W., R. Van der Voo, and Q. Z. Liang. 1989. Devonian paleomagnetism of Yunnan province across the Shan Thai-South China suture. *Tectonics* 8: 939–952.

Fanning, U., D. Edwards, and J. B. Richardson. 1990. Further evidence for diversity in late Silurian land vegetation. *J. Geol. Soc. Lond.* 147: 725–728.

Fanning, U., D. Edwards, and J. B. Richardson. 1991. A new rhyniophytoid from the late Silurian of the Welsh Borderland. *Neues Jahrb. Geol. Paläontol. Abh.* 183: 37–47.

———. 1992. A diverse assemblage of early land plants

from the Lower Devonian of the Welsh Borderland. *Bot. J. Linn. Soc.* 109: 161–188.

Fanning, U., J. B. Richardson, and D. Edwards. 1988. Cryptic evolution in an early land plant. *Evol. Trends Plants* 2: 13–24.

———. 1991. A review of *in situ* spores in Silurian land plants. In S. Blackmore and S. H. Barnes, eds. *Pollen and spores, patterns of diversification*, pp. 25–47. The Systematics Association Spec. vol. 44. Oxford: Clarendon Press.

Farquhar, G., J. Ehleringer, and K. Hubick. 1989. Carbon isotope discrimination and Photosynthesis. *Ann. Rev. Plant Physiol. Plant Molec. Biol.* 40: 503–537.

Faulkner, H. 1990. Vegetation cover density variations and infiltration patterns on piped alkali sodic soils: Implications for the modeling of overland flow in semi-arid areas. In J. B. Thornes, ed. *Vegetation and Erosion*, pp. 317–346. New York: Wiley and Sons.

Feakes, C. R., and G. J. Retallack. 1988. Recognition and chemical characterization of fossil soils developed on alluvium: A Late Ordovician example. In J. Reinhardt and W. R. Sigleo, eds. *Paleosols and Weathering through Geologic Time: Principles and Applications*, pp. 35–48. Boulder, CO: Geol. Soc. Am. Spec. Pap. 216.

Feist, R., ed. 1990. *Guide Book of the Field Meeting, Montagne Noire 1990*. Montpellier, France: International Union of Geological Sciences Subcommission on Devonian Stratigraphy.

Feist, M., and R. Feist. 1997. Oldest record of a bisexual plant. *Nature* 385: 401.

Feist, M., and N. Grambast-Fessard. 1991. The genus concept in Charophyta: Evidence from Palaeozoic to Recent. In R. Riding, ed. *Calcareous Algae and Stromatolites*, pp. 189–203. New York: Springer-Verlag.

Fensome, R. A., F. J. R. Taylor, G. Norris, W. A. S. Sargeant, D. I. Wharton, and G. L. Williams. 1993. *A Classification of Living and Fossil Dinoflagellates*. American Museum of Natural History, Micropaleontology Publication # 7. Hanover, MA: Micropaleontology Press.

Ferguson, L. M. 1990. Insecta: Microcoryphia and Thysanura. In D. Dindal, ed. *Soil Biology Guide*, pp. 935–963. New York: Wiley-Interscience.

Fisher, R. F., and S. R. Long. 1992. *Rhizobium*-plant signal exchange. *Nature* 357: 655.

FitzPatrick, E. A. 1993. *Soil Microscopy and Micromorphology*. New York: John Wiley and Sons.

Fogel, R. 1985. Roots as primary producers in belowground ecosystems. In A. H. Fitter, D. Atkinson, D. J. Read, and M. B. Usher, eds. *Ecological Interactions in Soil: Plants, Microbes and Animals*, pp. 23–36. Oxford: Blackwell.

Fong, P., J. B. Zedler, and R. M. Donohoe. 1993. Nitrogen vs. phosphorus limitation of algal biomass in shallow coastal lagoons. *Limnol. Oceanogr.* 38: 906–923.

Foster, C. B., and G. E. Williams. 1991. Late Ordovician-early Silurian age for the Mallowa Salt of the Carribuddy Group, Canning Basin, Western Australia, based on occurrences of *Tetrahedraletes medinensis* Strother and Traverse 1979. *Austral. J. Earth Sci.* 38: 223–228.

Fraenkel, G. S. 1959. The raison d'etre of secondary plant substances. *Sciences* 129: 1466–1470.

Frakes, L. A., J. E. Francis, and J. I. Syktus. 1992. *Climate Modes of the Phanerozoic*. Cambridge: Cambridge University Press.

Frederick, S. E., P. J. Gruber, and N. E. Tolbert. 1973. The occurrence of glycolate dehydrogenase and glycolate oxidase in green plants. An evolutionary survey. *Plant Physiol.* 52: 318–323.

Galloway, D. J. 1993. Global environmental change: Lichens and chemistry. *Bibliotheca Lichenologica* 53: 87–95.

———. 1995. Lichens in Southern hemisphere temperate rainforest and their role in the maintenance of biodiversity. In D. Allsopp, R. R. Colwell, and D. L. Hawksworth, eds. *Microbial Diversity and Ecosystem Function*, pp. 125–135. Oxford, U.K.: CAB International.

Galtier, J., and B. Meyer-Berthaud. 1996. The early seed-plant *Tristichia tripos* (Unger) *comb. nov.* from the Lower Carboniferous of Saalfeld, Thuringia. *Rev. Palaeobot. Palynol.* 93: 299–315.

Galtier, J., and T. L. Phillips. 1996. Structure and evolutionary significance of Palaeozoic ferns. In J. M. Camus, M. Gibby, and R. J. Johns, eds. *Pteridology in Perspective*, pp. 417–433. Kew: Royal Botanic Gardens.

Gargas, A., P. T. DePriest, M. Grube, and A. Tehler. 1995. Multiple origins of lichen symbioses in fungi suggested by SSU rDNA phylogeny. *Science* 268: 1492–1495.

Garratt, M. J. 1978. New evidence for a Silurian (Ludlow) age for the earliest *Baragwanathia* flora. *Alcheringa* 2: 217–224.

Garratt, M. J., and R. B. Rickards. 1984. Graptolite biostratigraphy of early land plants from Victoria, Australia. *Proc. Yorkshire Geol. Soc.* 44: 377–384.

Garrels, R. M., and A. Lerman. 1984. Coupling of the sedimentary sulfur and carbon cycles—An improved model. *Am. J. Sci.* 284: 989–1007.

Geiger, H. 1990. Biflavonoids in bryophytes. In H. D. Zinsmeister and R. Mues, eds. *Bryophytes: Their Chemistry and Chemical Taxonomy. Proceedings of the Phytochemical Society of Europe.* 29: 161–170.

Geldsetzer, H. H. J., W. D. Goodfellow, and D. J. McLaren. 1993. The Frasnian-Famennian extinction event in a stable cratonic shelf setting: Trout River, Northwest Territories, Canada. *Palaeogeog. Palaeoclim. Paleoecol.* 104: 81–95.

Geldsetzer, H. H. J., W. D. Goodfellow, D. J. McLaren, and M. J. Orchard. 1987. Sulfur-isotope anomaly associated with the Frasnian-Famennian extinction, Medicine Lake, Alberta, Canada. *Geology* 15: 393–396.

Geng, B. Y. 1983. *Stachyophyton* gen. nov. discovered from the Lower Devonian of Yunnan and its significance. *Acta Bot. Sin.* 25: 574–579.

———. 1985. *Huia recurvata*—A new plant from the Lower Devonian of southeastern Yunnan, China. *Acta Bot. Sin.* 27: 419–426.

———. 1992a. Studies on Early Devonian flora of Sichuan. *Acta Phytotax. Sin.* 30: 197–211.

———. 1992b. *Amplectosporangium*—A new genus of

plant from the Lower Devonian of Sichuan, China. *Acta Bot. Sin.* 34: 450–455.

Gensel, P. G. 1976. *Renalia huebeni*, a new plant from the Lower Devonian of Gaspé. *Rev. Palaeobot. Palynol.* 22: 19–37.

———. 1979. Two *Psilophyton* species from the Lower Devonian of eastern Canada with a discussion of morphological variation within the genus. *Palaeontographica B* 168: 81–99.

———. 1980. Devonian *in situ* spores: A survey and discussion. *Rev. Palaeobot. Palynol.* 30: 101–132.

———. 1982a. On the contributions of Sir J. W. Dawson to the study of early land plants (Devonian) and current ideas concerning their nature, diversity and evolutionary relationships. *Proc. Third N. Am. Paleontol. Conv.*, vol. 1: 199–204.

———. 1982b. *Oricilla*, a new genus referable to the zosterophyllophytes from the late Early Devonian of northern New Brunswick. *Rev. Palaeobot. Palynol.* 37: 345–359.

———. 1984. A new Lower Devonian plant and the early evolution of leaves. *Nature* 309: 785–787.

———. 1986. Diversification of land plants in the Early and Middle Devonian. In T. W. Broadhead, ed. *Land Plants: Notes for a Short Course*, pp. 64–80. Knoxville, TN: University of Tennessee, Department of Geological Sciences, *Studies in Geology* no. 15.

———. 1992. Phylogenetic relationships of the zosterophylls and lycopsids: Evidence from morphology, paleoecology and cladistic methods of inference. *Ann. Mo. Bot. Gard.* 79: 450–473.

Gensel, P. G., and H. N. Andrews. 1984. *Plant Life in the Devonian.* New York: Praeger.

———. 1987. The evolution of early land plants. *Am. Sci.* 75: 478–489.

Gensel, P. G., H. N. Andrews, and W. H. Forbes. 1975. A new species of *Sawdonia* with notes on the origin of microphylls and lateral sporangia. *Bot. Gaz.* 136: 50–62.

Gensel, P. G., N. G. Johnson, and P. K. Strother. 1990. Early land plant debris (Hooker's "waifs and strays"?). *Palaios* 5: 520–547.

Gensel, P. G., A. E. Kasper, and H. N. Andrews. 1969. *Kaulangiophyton*, a new genus of plants from the Devonian of Maine. *Torrey Bot. Club Bull.* 96: 265–276.

Gensel, P. G., and A. R. White. 1983. The morphology and ultrastructure of spores of the Early Devonian trimerophyte *Psilophyton* (Dawson) Hueber and Banks. *Palynology* 7: 221–233.

Gerrienne, P. 1988. Early Devonian plant remains from Marchin (north of Dinant Synclinorium, Belgium). I. *Zosterophyllum deciduum* sp. nov. *Rev. Palaeobot. Palynol.* 55: 317–335.

———. 1992a. Les plantes emsiennes de Fooz-Wépion (bord nord du Synclinorium de Dinant, Belgique). II. *Urpicalis steemansii* gen. et sp. nov. *C. R. Acad. Sci. Paris*, Ser. 314: 851–857.

———. 1992b. The Emsian plants from Fooz-Wépion (Belgium). III. *Foozia minuta* gen. et sp. nov., a new taxon with probable cladoxylalean affinities. *Rev. Palaeobot. Palynol.* 74: 139–157.

———. 1995. Les fossiles végétaux du Dévonien inférieur de Marchin (bord nord du Synclinorium de Dinant, Belgique). III. *Psilophyton parvulum* nov. sp. *Geobios* 28: 131–144.

———. 1996a. A biostratigraphic method based on a quantification of fossil tracheophyte characters—Its application to the Lower Devonian Posongchong flora (Yunnan Province, China). *Palaeobotanist* 45: 194–200.

———. 1996b. Contribution à l'étude paléobotanique du Dévonien inférieur de Belgique: Les genres nouveaux *Ensivalia* et *Faironella*. *Acad. R. Belg. Cl. Sc.* Coll. 4, Ser. 3, 1: 1–94.

———. 1996c. Lower Devonian plant remains from Marchin (northern margin of Dinant Synclinorium, Belgium). IV. *Odonax borealis* gen. et sp. nov. *Rev. Palaeobot. Palynol.* 93: 89–106.

———. 1997. The fossil plants from the Lower Devonian of Marchin (northern margin of Dinant Synclinorium, Belgium). V. *Psilophyton genseliae* sp. nov., with hypotheses on the origin of Trimerophytina. *Rev. Palaeobot. Palynol.* 98: 303–324.

———. 1999. Lower Devonian plant mesofossils from the Parana Basin, Brazil: General introduction, description, age significance and correlation with floral succession from Laurussia and Gondwana. In M. A. C. Rodrigues and E. Periera, eds. *Ordovician/Devonian palynostratigraphy in western Gondwana: Update, problems and perspectives*, pp. 165–178. Faculdade de Geologia, Rio de Janeiro.

Gevers, T. W., and A. Twomey. 1982. Trace fossils and their environment in Devonian (Silurian?) Lower Beacon strata in the Asgard Ranges, Victoria Land, Antarctica. In C. Craddock, ed. *Antarctic Geoscience*, pp. 639–648. Madison, WI: Wisconsin Press.

Ghosh, P., S. K. Bhattacharya, and R. A. Jani. 1995. Palaeoclimate and palaeovegetation in central India during the Upper Cretaceous based on stable isotope composition of the paleosol carbonates. *Palaeogeogr. Palaeoclimatol. Palaeoecol.* 114: 285–296.

Gill, S., and K. Yemane. 1996. Implications of a Lower Pennsylvanian Ultisol for equatorial Pangean climates and early, oligotrophic, forest ecosystems. *Geology* 24: 905–908.

Gillespie, W. H., G. W. Rothwell, and S. E. Scheckler. 1981. The earliest seeds. *Nature* 293: 462–464.

Goldhammer, R. K., and R. D. Elmore. 1984. Paleosols capping regressive carbonate cycles in the Pennsylvanian Black Prince Limestone, Arizona. *J. Sediment. Petrol.* 54: 1124–1137.

Goldring, W. 1924. The Upper Devonian forest of seed ferns in Eastern New York. *N. Y. State Mus. Bull.* 251: 50–92.

———. 1926. New Upper Devonian plant material. *N. Y. State Mus. Bull.* 267: 85–87.

———. 1927. The oldest known petrified forest. *Scientific Monthly* 24: 514–529.

Goldstein, R. H. 1991. Stable isotope signatures associated with palaeosols, Pennsylvanian Holder Formation, New Mexico. *Sedimentology* 38: 67–77.

Golonka, J., M. I. Ross, and C. R. Scotese. 1994. Phanerozoic paleogeographic and paleoclimatic modeling maps: Pangea: Global environments and resources. *Can. Soc. Petrol. Geol. Mem.* 17: 1–47. eds.

Good, B. H., and R. L. Chapman. 1978. The ultrastructure of *Phycopeltis* (Chroolepidaceae: Chlorophyta). I. Sporopollenin in the cell walls. *Am. J. Bot.* 65: 27–33.

Goodarzi, F., T. Gentzis, and A. F. Embry. 1989. Organic petrology of two coal-bearing sequences from the Middle to Upper Devonian of Melville Island, Arctic Canada. In *Contributions to Canadian Coal Geoscience*, pp. 120–130. Geological Survey of Canada.

Goodarzi, F., T. Gentzis, and J. C. Harrison. 1994. Petrology and depositional environment of Upper Devonian coals from Eastern Melville Island, Arctic Canada. *Geol. Surv. Canada Bull.* 450: 203–213.

Goodarzi, F., and Q. Goodbody. 1990. Nature and depositional environment of Devonian coals from western Melville Island, Arctic Canada. *Int. J. Coal Geol.* 14: 175–196.

Goodfellow, W. D., and I. R. Jonasson. 1984. Ocean stagnation and ventilation defined by δ ^{34}S secular trends in pyrite and barite, Selwyn Basin, Yukon. *Geology* 12: 583–586.

Gordon, E. A. 1988. Body and trace fossils from the Middle-Upper Devonian Catskill Magnafacies, southeastern New York, U.S.A. In N. J. McMillan, A. F. Embry, and D. J. Glass, eds. *Devonian of the World*, vol. 2, pp. 139–155. Calgary, Alberta: Canadian Society of Petroleum Geologists, Mem. 14.

Gordon, E. A., and J. S. Bridge. 1987. Evolution of Catskill (Upper Devonian) river systems. *J. Sediment. Petrol.* 57: 234–249.

Gordon, M. S., and E. C. Olson. 1995. *Invasions of the Land.* New York: Columbia University Press.

Goudie, A. S. 1983. Chapter 4: Calcretes. In A. S. Goudie and K. Pye, eds. *Chemical Sediments and Geomorphology: Precipitates and Residua in the Near-surface Environment*, p. 93–131. New York: Academic Press.

Graham, J. B., R. Dudley, N. M. Aguilar, and C. Gans. 1995. Implications of the late Palaeozoic oxygen pulse for physiology and evolution. *Nature* 375: 117–120.

Graham, L. E. 1982. The occurrence, evolution and phylogenetic significance of parenchyma in *Coleochaete* Bréb. (Chlorophyta). *Am. J. Bot.* 69: 447–454.

——. 1984. *Coleochaete* and the origin of land plants. *Am. J. Bot.* 71: 603–608.

——. 1985. The origin of the life cycle of land plants. *Am. Sci.* 73: 178–186.

——. 1990a. Meiospore formation in charophycean algae. In S. Blackmore and R. B. Knox, eds. *Microspores: Evolution and Ontogeny*, pp. 43–54. London: Academic Press.

——. 1990b. Phylum Chlorophyta, Class Charophyceae, Orders Chlorokybales, Klebsormidiales, Coleochaetales. In L. Margulis, J. O. Corliss, M. Melkonian, and D. J. Chapman, eds. *Handbook of Protoctista*, pp. 636–640. Boston, MA: Jones and Bartlett.

——. 1993. *Origin of Land Plants.* New York: John Wiley and Sons.

——. 1996. Green algae to land plants: An evolutionary transition. *J. Plant Res.* 109: 241–251.

Graham, L. E., C. F. Delwiche, and B. D. Mishler. 1991. Phylogenetic connections between the "green algae" and the "bryophytes." *Adv. Bryol.* 4: 213–244.

Graham, L. E., J. M. Graham, W. A. Russin, and J. M. Chesnick. 1994. Occurrence and phylogenetic significance of glucose utilization by charophycean algae: Glucose enhancement of growth in *Coleochaete orbicularis. Am. J. Bot.* 81: 423–432.

Graham, L. E., and Y. Kaneko. 1991. Subcellular structures of relevance to the origin of land plants (embryophytes) from green algae. *Crit. Rev. Plant Sci.* 10: 323–342.

Graham, L. E., and L. W. Wilcox. 1983. The occurrence and phylogenetic significance of putative placental transfer cells in the green alga *Coleochaete. Am. J. Bot.* 70: 113–120.

Granoff, J. A., P. G. Gensel, and H. N. Andrews. 1976. A new species of *Pertica* from the Devonian of eastern Canada. *Palaeontographica B* 155: 119–128.

Grant, M. C. 1990. Phylum Chlorophyta, Class Charophyceae, Order Charales. In L. Margulis, J. O. Corliss, M. Melkonian, and D. J. Chapman, eds. *Handbook of Protoctists*, pp. 641–648. Boston, MA: Jones and Bartlett.

Gray, J. 1965. Extraction techniques. In B. Kummel and D. Raup, eds. *Handbook of Paleontological Techniques*, pp. 530–587. San Francisco: W. H. Freeman.

——. 1984. Ordovician-Silurian land plants: The interdependence of ecology and evolution. In M. G. Bassett and J. D. Lawson, eds. *Autecology of Silurian Organisms.* Spec. Pap. Palaeontol. no. 32, pp. 281–295. The Palaeontological Association.

——. 1985. The microfossil record of early land plants: Advances in understanding of early terrestrialization, 1970–1984. In W. G. Chaloner and J. D. Lawson, eds. *Evolution and Environment in the Late Silurian and Early Devonian. Phil. Trans. R. Soc. Lond.* B309: 167–195.

——. 1988. Land plant spores and the Ordovician-Silurian boundary. *Bull. Br. Mus. Nat. Hist. (Geol.)* 43: 351–358.

——. 1991. *Tetrahedraletes, Nodospora,* and the "cross" tetrad: An accretion of myth. In S. Blackmore and S. H Barnes, eds. *Pollen and Spores: Patterns of Diversification.* The Systematics Association, Spec. vol. 44, pp. 49–87. Oxford: Clarendon Press.

——. 1993. Major Paleozoic land plant evolutionary bio-events. *Palaeogeogr. Palaeoclimatol. Palaeoecol.* 104: 153–169.

Gray, J., and A. J. Boucot. 1972. Palynological evidence bearing on the Ordovician-Silurian paraconformity in Ohio. *Bull. Geol. Soc. Am.* 83: 1299–1314.

——. 1977. Early vascular land plants: Proof and conjecture. *Lethaia* 10: 145–174.

——. 1978. The advent of land plant life. *Geology* 6: 489–492.

——. 1994. Early Silurian nonmarine animal remains and the nature of the early continental ecosystem. *Acta Palaeontol. Pol.* 38: 303–328.

Gray, J., A. J. Boucot, Y. Grahn, and G. Himes. 1992. A new record of early Silurian land plant spores from the Paraná Basin, Paraguay (Malvinokaffric Realm). *Geol. Mag.* 129: 741–752.

Gray, J., G. K. Colbath, A. de Faria, A. J. Boucot, and D. M. Rohr. 1985. Silurian-age fossils from the Paleozoic Paraná Basin, southern Brazil. *Geology* 13: 521–525.

Gray, J., D. Massa, and A. J. Boucot. 1982. Caradocian land plant microfossils from Libya. *Geology* 10: 197–201.

Gray, J., and W. Shear. 1992. Early life on land. *Am. Sci.* 80: 444–456.

Gray, J., J. N. Theron, and A. J. Boucot. 1986. Age of the Cedarberg Formation, South Africa and early land plant evolution. *Geol. Mag.* 123: 445–454.

Greenslade, P. J. M. 1983. Adversity selection and the habitat template. *Am. Nat.* 122: 352–365.

———. 1988. Reply to R. A. Crowson's "Comments on Insecta of the Rhynie Chert" [*Entomol. Gener.* 1985, 11 (1/2): 97–98]. *Entomol. Gener.* 13: 115–117.

Greenslade, P., and P. E. S. Whalley. 1986. The systematic position of *Rhyniella praecursor* Hirts and Maulik (Collembola). In R. Dalla, ed. *The Earliest Known Hexapod*, pp. 319–323. Second International Seminar on Apterygota, University of Siena, Siena.

Gregor, C. B. 1985. The mass-age distribution of Phanerozoic sediments. In N. J. Snelling, ed. *The Chronology of the Geological Record*, pp. 284–289. *Geol. Soc. Lond. Mem.* 10. Oxford: Blackwell.

Grierson, J. D. 1976. *Leclercqia complexa* (Lycopsida, Middle Devonian): Its anatomy, and the interpretation of pyrite petrifactions. *Am. J. Bot.* 63: 1184–1202.

Grierson, J. D., and H. P. Banks. 1963. Lycopods of the Devonian of New York State. *Palaeontogr. Am.* 4: 220–295.

———. 1983. A new genus of lycopods from the Devonian of New York State. *Bot. J. Linn. Soc.* 86: 81–101.

Grierson, J. D., and P. M. Bonamo. 1979. *Leclercqia complexa*: Earliest ligulate lycopod (Middle Devonian). *Am. J. Bot.* 66: 474–476.

Griffing, D. H., J. S. Bridge, and C. L. Hotton. 2000. Coastal-fluvial paleoenvironments and plant paleoecology of the Early Devonian (Emsian), Gaspé Bay, Québec, Canada. In P. F. Friend and B. P. J. Williams, eds. *New Perspectives on the Old Red Sandstone*. Geol. Soc. London Spec. Publ. (in press).

Griffiths, R. P., J. E. Baham, and B. A. Caldwell. 1994. Soil solution chemistry of ectomycorrhizal mats in forest soil. *Soil Biol. Biochem.* 26: 331–337.

Grime, J. P. 1977. Evidence for the existence of three primary strategies in plants and its relevance to ecological and evolutionary theory. *Am. Nat.* 111: 1169–1194.

Grosser, J. R., and K. F. Prossl. 1991. First evidence of the Silurian in Colombia: Palynostratigraphic data from the Quetame Massif, Cordillera Oriental. *J. Soc. Am. Earth Sci.* 4: 231–238.

Guildford, W. J., D. M. Schneider, J. Labovitz, and S. J. Opella. 1988. High resolution solid state ^{13}C NMR spectroscopy of sporopollenins from different taxa. *Plant Physiol.* 86: 134–136.

Gunnison, D., and M. Alexander. 1975a. Resistance and susceptibility of algae to decomposition by natural microbial communities. *Limnol. and Oceanogr.* 20: 64–70.

———. 1975b. Basis for the resistance of several algae to microbial decomposition. *Appl. Microbiol.* 29: 729–738.

Gwiazda, R. H., and W. S. Broecker. 1994. The separate and combined effects of temperature, soil pCO_2, and organic acidity on silicate weathering in the soil environment: Formulation of a model and results. *Glob. Biogeochem. Cycles* 8: 141–155.

Hagstrom, J. 1997. Land-derived palynomorphs from the Silurian of Gotland, Sweden. *Geol. Foeren. St. Foehr.* 119: 301–316.

Halas, S., A. Balinski, M. Gruszczynski, A. Hoffman, K. Malkowski, and M. Narkiewicz. 1992. Stable isotope record at the Frasnian/Famennian boundary in southern Poland. *Neues Jb. Geol. Paläontol. Monatsh.* 3: 129–138.

Halle, T. G. 1927. Fossil plants from Southwestern China. *Palaeontol. Sin.* Ser. A 1(2): 1–26.

———. 1936. On *Drepanophycus, Protolepidodendron* and *Protopteridium*, with notes on the Palaeozoic flora of Yunnan. *Palaeontol. Sin.* Ser. A 1(4): 1–38.

Halperin, A., and J. S. Bridge. 1988. Marine to nonmarine transitional deposits in the Catskill clastic wedge, south-central New York. In N. J. McMillan, A. F. Embry, and D. J. Glass, eds. *Devonian of the World*, vol. 2, pp. 107–124. Calgary, Alberta: Canadian Society of Petroleum Geologists.

Hance, L., L. Dejonghe, M. Fairon-Demaret, and P. Steemans. 1996. La Formation de Pepinster dans le Synclinorium de Verviers, entre Pepinster et Eupen (Belgique)—Contexte structural et stratigraphique. *Ann. Soc. Geol. Belg.* 117: 75–93.

Hanson, D. T., S. Swanson, L. E. Graham, and T. D. Sharkey. 1999. Evolutionary significance of isoprene emission from mosses. *Am. J. Bot.* 86(5): 634–639.

Hao, S. G. 1988. A new Lower Devonian genus from Yunnan, with notes on the origin of leaf. *Acta Bot. Sin.* 30: 441–448.

———. 1989a. A new zosterophyll from the Lower Devonian (Siegenian) of Yunnan, China. *Rev. Palaeobot. Palynol.* 57: 155–171.

———. 1989b. *Gumuia*—A new genus of the Lower Devonian from Yunnan. *Acta Bot. Sin.* 31: 954–961.

Hao, S. G., and C. B. Beck. 1991a. *Catenalis digitata*, gen. et sp. nov., a plant from the Lower Devonian (Siegenian) of Yunnan, China. *Can. J. Bot.* 12: 233–242.

———. 1991b. *Yunia dichotoma*, a Lower Devonian plant from Yunnan, China. *Rev. Palaeobot. Palynol.* 68: 181–195.

———. 1993. Further observations on *Eophyllophyton bellum* from the Lower Devonian (Siegenian) of Yunnan, China. *Palaeontographica B* 230: 27–41.

Hao, S. G., and P. G. Gensel. 1995. A new genus and species, *Celatheca beckii*, from the Siegenian (Early Devonian) of southeastern Yunnan, China. *Int. J. Plant Sci.* 156(6): 896–909.

———. 1998. Some new plant finds from the Posongchong Formation of Yunnan, and consideration of a possible phytogeographic similarity between South China and Australia during the Early Devonian. *Science in China* Ser. D 41: 1–13.

Haq, B. U., J. Hardenbol, and P. R. Vail. 1987. Chronology of fluctuating sea levels since the Triassic. *Science* 235: 1156–1167.

Harland, W. B., R. L. Armstrong, A. V. Cox, L. E. Craig, A. G. Smith, and D. G. Smith. 1990. *A Geologic Time Scale 1989*. Cambridge: Cambridge University Press.

Harrison, M. J. 1997. The arbuscular mycorrhizal symbiosis, an underground association. *Trends Plant Sci.* 2: 54–60.

Hartland-Rowe, R. 1972. The limnology of temporary waters and the ecology of the Euphyllopoda. In R. B. Clark and R. J. Wooton, eds. *Essays in Hydrobiology*, pp. 15–31. Exeter, U.K.: University of Exeter Press.

Hass, H., and W. Remy. 1991. *Huvenia kleui* nov. gen., nov. spec.—ein vertreter der Rhyniaceae aus dem höheren Siegen des Rheinischen Schiefergebirges. *Argum. Palaeobot.* 8: 141–168.

Hass, H., and N. P. Rowe. 1999. Thin sections and wafering. In T. P. Jones and N. P. Rowe, eds. *Fossil Plants and Spores: Modern Techniques*, pp. 76–81. London: Geological Society of London.

Hass, H., T. N. Taylor, and W. Remy. 1994. Fungi from the Lower Devonian Rhynie Chert: Mycoparasitism. *Am. J. Bot.* 81: 29–37.

Hedderson, T. A., R. L. Chapman, and W. L. Rootes. 1996. Phylogenetic relationships of bryophytes inferred from nuclear-encoded rRNA gene sequences. *Plant Syst. Evol.* 20: 213–224.

Hemsley, A. R., P. J. Barrie, and A. C. Scott. 1995. ^{13}C solid-state n.m.r. spectroscopy of fossil sporopollenins. Variation in composition independent of diagenesis. *Fuel* 74: 1009–1012.

Hemsley, A. R., W. G. Chaloner, A. C. Scott, and C. J. Groombridge. 1992. Carbon-13 solid-state nuclear magnetic resonance of sporopollenins from modern and fossil plants. *Ann. Bot.* 69: 545–549.

Hemsley, A. R., A. C. Scott, P. J. Barrie, and W. G. Chaloner. 1996. Studies of fossil and modern spore wall macromolecules using ^{13}C solid state NMR. *Ann. Bot.* 78: 83–94.

Hernick, L. V. 1996. *The Gilboa Fossils*. Rensselaerville, NY: Givetian Press.

Heslop-Harrison, J. 1979. Pollen walls as adaptive systems. *Ann. Mo. Bot. Gard.* 66: 813–829.

Hiesel, R., B. Combrettes, and A. Brennicke. 1994. Evidence for RNA editing in mitochondria of all major groups of land plants except the Bryophyta. *Proc. Nat. Acad. Sci. USA* 91: 629–633.

Hill, S. A., S. E. Scheckler, and J. F. Basinger. 1997. *Ellesmeris sphenopteroides*, gen. et sp. nov., a new zygopterid fern from the Upper Devonian (Frasnian) of Ellesmere, N.W.T., Arctic Canada. *Am. J. Bot.* 84: 85–103.

Hilton, J., and D. Edwards. 1996. A new Late Devonian acupulate preovule from the Taff Gorge, South Wales. *Rev. Palaeobot. Palynol.* 93: 235–252.

Hinton, H. E. 1968. Reversible suspension of metabolism and the origin of life. *Proc. R. Soc. Lond.* B171: 43–57.

Hirst, S. 1923. On some arachnid remains from the Old Red Sandstone (Rhynie Chert bed, Aberdeenshire). *Ann. Mag. Nat. Hist.* Ser. 9, 12: 455–474.

Hladil, J., and J. Kalvoda. 1993. Devonian boundary intervals of Bohemia and Moravia (Fieldtrip 3). In M. Narkiewicz, ed. *Excursion Guidebook for Global Boundary Events symposium*, pp. 29–50. Kielce, Poland.

Hochachka, P. W., and M. Guppy. 1987. *Metabolic Arrest and the Control of Biological Time*. Cambridge, MA: Harvard University Press.

Hodge, P. 1994. *Meteorite Craters and Impact Structures of the Earth*. Cambridge: Cambridge University Press.

Høeg, O. A. 1935. Further contributions to the Middle Devonian flora of Western Norway. *Norsk Geol. Tidsskr.* 15: 1–18.

———. 1942. The Downtonian and Devonian flora of Spitzbergen. *Norges Svalbard- og Ishavs-Unders. Skr.* 83: 1–228.

———. 1967. Psilophyta. In E. Boureau, ed. *Traité de Paléobotanique*, vol. II., pp 191–433. Paris: Masson.

Holland, H. D. 1978. *The Chemistry of the Atmosphere and Oceans*. New York: Wiley.

———. 1984. *The Chemical Evolution of the Atmosphere and Oceans*. Princeton: Princeton University Press.

Holser, W. T., J. B. Maynard, and K. M. Cruikshank. 1989. Modelling the natural cycle of sulphur through Phanerozoic time. In P. Brimblecombe and A. Y. Lein, eds. *Evolution of the Global Biogeochemical Sulphur Cycle*, pp. 21–56. Chichester: Wiley.

Hopkin, S. P., and H. J. Read. 1992. *The Biology of Millipedes*. Oxford: Oxford University Press.

Hopping, C. A. 1956. On a specimen of "*Psilophyton robustius*" Dawson, from the Lower Devonian of Canada. *Proc. R. Soc. Edinb.* B66: 10–28.

Horodyski, R. J., and L. P. Knauth. 1994. Life on land in the Precambrian. *Nature* 263: 494–498.

Horsfield, P. 1978. Evidence for xylem feeding by *Philaenus spumonus* (L.) (Homoptera; Cercopidae). *Exp. Appl. Entomol.* 24: 95–99.

Hoshaw, R. W., R. M. McCourt, and J.-C. Wang. 1990. Phylum Conjugaphyta. In L. Margulis, J. O. Corliss, M. Melkonian, and D. J. Chapman, eds. *Handbook of Protoctista*, pp. 119–131. Boston, MA: Jones and Bartlett.

Hoskins, D. M. 1961. Stratigraphy and paleontology of the Bloomsburg Formation of Pennsylvania and adjacent states. *PA Geol. Surv.*, 4th ser., *General Geological Report* 36.

Hou, H.-F., Q. Ji, and J.-X. Wang. 1988. Preliminary report on Frasnian-Famennian events in south China. In N. J. McMillan, A. J. Embry, and D. J. Glass, eds. *Devonian of the World*, vol. 3: *Paleontology, Paleoecology and Biostratigraphy*, pp. 63–69. Calgary, Alberta: Canadian Society of Petroleum Geologists, Mem. 14.

House, M. R. 1985. Correlation of mid-Palaeozoic ammonoid evolutionary events with global sedimentary perturbations. *Nature* 313: 17–22.

House, M. R., J. B. Richardson, W. G. Chaloner, J. R. L.

Allen, C. H. Holland, and T. S. Westoll. 1977. A correlation of Devonian Rocks in the British Isles. *Geol. Soc. Lond., Spec. Rep.* 7: 1–110.

Hsü, J. 1946. Plant fragments from Devonian beds in Central Yunnan, China. Iyenger Comemoration Volume, *J. Indian Bot. Soc.*, pp. 339–360.

———. 1966. On plant-remains from the Devonian of Yunnan and their significance in the identification of the stratigraphical sequence of this region. *Acta Bot. Sin.* 14(1): 50–69 (in Chinese with English abstract).

Hsü, K. J., and J. A. McKenzie. 1985. A "Strangelove" ocean in the earliest Tertiary. In E. T. Sundquist and W. S. Broecker, eds. *The Carbon Cycle and Atmospheric CO_2: Natural Variations Archean to Present*, pp. 487–492. Washington, DC: Am. Geophys. Union, Geophys. Monogr. 32.

Hueber, F. M. 1967. *Psilophyton*: The genus and the concept. In D. S. Oswald, ed. *International Symposium on the Devonian System*, pp. 815–822. Calgary, Alberta: Canadian Society of Petroleum Geologists.

———. 1971. *Sawdonia ornata*: A new name for *Psilophyton* princeps var. *ornatum. Taxon* 20: 641–642.

———. 1972. Early Devonian land plants from Bathurst Island, District of Franklin. *Geol. Surv. Canada, Paper* 71-28: 1–17.

———. 1983a. A new species of *Baragwanathia* from the Sextant Formation (Emsian), Northern Ontario, Canada. *Bot. J. Linn. Soc.* 86: 57–79.

———. 1983b. *Taeniocrada dubia* Kr. and W.: Its conducting strand of helically strengthened tubes. *Bot. Soc. Am. Misc. Publ.* 162: 58–59.

———. 1992. Thoughts on early lycopsids and zosterophylls. *Ann. Mo. Bot. Gard.* 79: 474–499.

———. 1996. A solution to the enigma of *Prototaxites. The Palaeontological Society Spec. Publ.* 8, p. 183. Sixth North American Paleontological Convention Abstracts.

Hueber, F. M., and H. P. Banks. 1967. *Psilophyton princeps*: The search for organic connection. *Taxon* 16: 81–85.

———. 1979. *Serrulacaulis furcatus* gen. et sp. nov., a new zosterophyll from the lower Upper Devonian of New York State. *Rev. Palaeobot. Palynol.* 28: 169–189.

Hueber, F. M., and J. D. Grierson. 1961. On the occurrence of *Psilophyton princeps* in the early Upper Devonian of New York. *Am. J. Bot.* 48: 473–479.

Hurley, N. F., and R. Van der Voo. 1990. Magnetostratigraphy, Late Devonian iridium anomaly, and impact hypotheses. *Geology* 18: 291–294.

Ishchenko, T. A. 1965. Devonskaya flora bolshozo donbassa. *Kiev: Tr. Inst. Geol. Nauk. Acad. Nauk. Ukraine SSR*, pp. 88, 30 plates.

———. 1975. *The Late Silurian Flora of Podolia*. Institute of Geological Science, Academy of Science of the Ukrainian SSR, Kiev.

Ishchenko, T. A., and A. A. Ishchenko. 1960. Novaja nakhodka kharofitov v Verkhnem silure Podolii. In Y. V. Teslenko, ed. *Sistematika i Evolutsiya Drevnikh Rasteni Ukrainy*, pp. 21–32, Sbornik Nauchnykh Trudov, Akad. Nauk Ukrainskoy SSR, Naukova Dumka, Kiev.

Iyengar, M. O. P. 1960. *Nitella terrestris* sp. nov., a terrestrial charophyte from South India. *Bull. Bot. Soc. Bengal* 12: 85–90.

Jablonski, D. 1991. Extinctions: A paleontological perspective. *Science* 253: 754–757.

James, N. P. 1983. Reefs. In P. A. Scholle, D. G. Bebout, and C. H. Moore, eds. *Carbonate Depositional Environments*, pp. 345–462. Tulsa, OK: American Association of Petroleum Geologists, Mem. 33.

Jaminski, J., T. J. Algeo, J. B. Maynard, and J. C. Hower. 1998. Microstratigraphic analysis of compositional variation in black shales: The Cleveland Member of the Ohio Shale (Upper Devonian), central Appalachian Basin, USA. In J. Schieber, W. Zimmerle, and P. S. Sethi, eds. *Shales and Mudstones*, vol. 1: *Basin Studies, Sedimentology, and Paleontology*, pp. 217–242. Stuttgart: Schweizerbart'sche.

Janvier, P., A. Blieck, P. Gerrienne, and T.-D. Thanh. 1987. Faune et flore de la Formation de Sika (Dévonien inférieur) dans la presquîle de Dô Son (Viêt Nam). *Bull. Mus. Nat. Hist. Nat. Paris* 4e, série 9, section C: 291–301.

Janvier, P., T.-D. Thanh, and P. Gerrienne 1989. Les placodermes, arthropodes et lycophytes des grès Dévoniens de Dô Son (Haïphong, Viêtnam). *Geobios* 22: 625–639.

Jarvik, E. 1996. The Devonian tetrapod *Ichthyostega. Fossils and Strata* 40: 1–213.

Jennings, J. R., E. E. Karrfalt, and G. W. Rothwell. 1983. Structure and affinities of *Protostigmaria eggertiana. Am. J. Bot.* 70: 963–974.

Jeram, A. J. 1994a. Scorpions from the Viseán of East Kirkton, West Lothian, Scotland, with a revision of the infraorder Mesoscorpionina. *Trans. R. Soc. Edinb. (Earth Sciences)* 84: 283–288.

———. 1994b. Carboniferous *Orthosterni* and their relationship to living scorpions. *Palaeontology* 37: 513–550.

Jeram, A. J., P. A. Selden, and D. Edwards. 1990. Land animals in the Silurian: Arachnids and myriapods from Shropshire, England. *Science* 250: 658–661.

Joachimski, M. M., and W. Buggisch. 1993. Anoxic events in the late Frasnian—Causes of the Frasnian-Famennian faunal crisis? *Geology* 21: 675–678.

Joachimski, M. M., W. Buggisch, and T. Anders. 1994. Mikrofazies, Conodontenstratigraphie und Isotopengeochemie des Frasne-Famenne-Grenzprofils Wolayer Gletscher (Karnische Alpen). *Abhandlungen der Geologischen Bundesanstalt (Austria)* 50: 183–195.

Johnson, E. W., D. E. G. Briggs, R. J. Suthren, J. L. Wright, and S. P. Tunnicliffe. 1994. Non-marine arthropod traces from the subaereal Ordovician Borrowdale Volcanic Group, English Lake District. *Geol. Mag.* 131: 395–406.

Johnson, J. G. 1974. Extinction of perched faunas. *Geology* 2: 479–482.

Johnson, J. G., G. Klapper, and C. A. Sandberg. 1985. Devonian eustatic fluctuations in Euramerica. *Geol. Soc. Am. Bull.* 96: 567–587.

Johnson, N. G. 1985. Early Silurian palynomorphs from the Tuscarora Formation in central Pennsylvania and their paleobotanical and geological significance. *Rev. Palaeobot. Palynol.* 45: 307–360.

Johnson, N. G., and P. G. Gensel. 1992. A reinterpretation of the Early Devonian land plant, *Bitelaria* Istchenko and Istchenko, 1979, based on new material from New Brunswick, Canada. *Rev. Palaeobot. Palynol.* 74: 109–138.

Johnsson, M. J. 1993. The system controlling the composition of clastic sediments. In M. J. Johnsson and A. Basu, eds. *Processes Controlling the Composition of Clastic Sediments*, pp. 1–19. Boulder, CO: Geological Society of America Spec. Pap. 284.

Jones, T. P., and W. G. Chaloner. 1991. Fossil charcoal, its recognition and palaeoatmospheric significance. *Palaeogeog. Palaeoclim. Palaeoecol.* 97: 39–50.

Kalvoda, J. 1986. Upper Frasnian and Lower Tournaisian events and evolution of calcareous foraminifera: Close links to climatic changes. In O. H. Walliser, ed. *Global Bio-Events*, pp. 225–236. Berlin: Springer, Lecture Notes in Earth Sciences 8.

———. 1990. Late Devonian-Early Carboniferous paleobiogeography of benthic foraminifera and climatic oscillations. In E. G. Kauffman and O. H. Walliser, eds. *Extinction Events in Earth History*, pp. 183–187. Berlin: Springer, Lecture Notes in Earth Sciences 30.

Kasper, A. E., and H. N. Andrews. 1972. *Pertica*, a new genus of Devonian plants from northern Maine. *Am. J. Bot.* 59: 897–911.

Kasper, A. E., P. G. Gensel, W. H. Forbes, and H. N. Andrews. 1988. Plant Paleontology in the state of Maine—A review. *Maine Geol. Surv.* 1: 109–128.

Kato, M., and R. Imaichi. 1994. Morphological diversity and evolution of vegetative organs in pteridophytes. In K. Iwatsuki and P. H. Raven, eds. *Evolution and Diversification of Land Plants*, pp. 27–43. Tokyo: Springer-Verlag.

Keller, C. K., and B. D. Wood. 1993. Possibility of chemical weathering before the advent of vascular land plants. *Nature* 364: 223–225.

Kenrick, P. 1988. Studies on Lower Devonian plants from South Wales. Unpublished Ph.D. thesis, University of Wales (Cardiff).

———. 1994. Alternation of generations in land plants: New phylogenetic and palaeobotanical evidence. *Biol. Rev.* 69: 293–330.

Kenrick, P., and P. R. Crane. 1991. Water-conducting cells in early fossil land plants: Implications for the early evolution of tracheophytes. *Bot. Gaz.* 152: 335–356.

———. 1997a. *The Origin and Early Diversification of Land Plants: A Cladistic Study.* Washington, DC: Smithsonian Institution Press.

———. 1997b. The origin and early evolution of plants on land. *Nature* 389: 33–39.

Kenrick, P., and D. Edwards. 1988. The anatomy of Lower Devonian *Gosslingia breconensis* Heard based on pyritized axes, with some comments on the permineralization process. *Bot. J. Linn. Soc.* 97: 95–123.

Kenrick, P., D. Edwards, and R. C. Dales. 1991. Novel

ultrastructure in water-conducting cells of the Lower Devonian plant *Sennicaulis hippocrepiformis. Palaeontology* 34: 751–766.

Kent, D. V. 1985. Paleocontinental setting for the Catskill Delta. In D. L. Woodrow and W. D. Sevon, eds. *The Catskill Delta.* Geological Society of America Spec. Pap. 201: 9–13.

Kent, D. V., and J. V. Miller. 1988. New perspectives from paleomagnetism. Paleozoic drift and Appalachian tectonics. *Lamont-Doherty Geological Observatory of Columbia University, Yearbook*, p. 12–17.

Kepferle, R. C., and J. D. Pollock. 1983. Correlations of the lower part of the Devonian oil shale in Kentucky. In *Proceedings of the 1983 Eastern Oil Shale Symposium*, pp. 35–40. Lexington, KY: Institute for Mining and Minerals Research.

Kepferle, R. C., J. D. Pollock, and L. S. Barron. 1982. Stratigraphy of the Devonian and Mississippian oil-bearing shales of central Kentucky. In *Proceedings of the 1982 Eastern Oil Shale Symposium*, pp. 137–149. Lexington, KY: Institute for Mining and Minerals Research.

Kethley, J. B., R. A. Norton, P. M. Bonamo, and W. A. Shear. 1989. A terrestrial alicorhagiid mite (Acari: Acariformes) from the Devonian of New York. *Micropaleontology* 35: 367–373.

Kevan, P. G., W. G. Chaloner, and D. O. B. Savile. 1975. Interrelationships of early terrestrial arthropods and plants. *Palaeontology* 18: 391–417.

Keyes, M. R., and C. C. Grier. 1981. Above- and below-ground net production in 40-year-old Douglas-fir stands on low and high productivity sites. *Can. J. For. Res.* 11: 599–605.

Kheirallah, A. M. 1979. Behavioural preference of *Julus scandinavius* (Myriapoda) to different species of leaf litter. *Oikos* 33: 466–471.

Kidston, R., and W. H. Lang. 1917. On Old Red Sandstone plants showing structure, from the Rhynie Chert Bed, Aberdeenshire. Part I. *Rhynia gwynne-vaughani*, Kidston and Lang. *Trans. R. Soc. Edinb.* 24: 761–784.

———. 1920a. On Old Red Sandstone plants showing structure, from the Rhynie Chert bed, Aberdeenshire. Part II. Additional notes on *Rhynia Gwynne-Vaughani*, Kidston and Lang; with descriptions of *Rhynia major*, n. sp., and *Hornia Lignieri*, n. g., n. sp. *Trans. R. Soc. Edinb.* 52: 603–627.

———. 1920b. On Old Red Sandstone plants showing structure, from the Rhynie Chert Bed, Aberdeenshire. Part III. *Asteroxylon Mackiei*, Kidston and Lang. *Trans. R. Soc. Edinb.* 52: 643–680.

———. 1921a. On Old Red Sandstone plants showing structure, from the Rhynie Chert Bed, Aberdeenshire. Part IV. Restorations of the vascular cryptogams, and discussion of their bearing on the general morphology of the Pteridophyta and the origin and organization of land-plants. *Trans. R. Soc. Edinb.* 52: 831–854.

———. 1921b. On Old Red Sandstone plants showing

structure, from the Rhynie Chert bed, Aberdeenshire. Part V. The Thallophyta occurring in the peat-bed; the succession of the plants throughout a vertical section of the bed, and the conditions of accumulation and preservation of the deposit. *Trans. R. Soc. Edinb.* 52: 855–902.

Killops, S. D., and V. J. Killops. 1993. *An Introduction to Organic Geochemistry.* New York: Longmans.

Kjellesvig-Waering, E. 1986. A restudy of the fossil Scorpionida of the world. *Palaeontographica* 55: 1–287.

Klappa, C. F. 1979a. Calcified filaments in Quaternary calcretes: Organo-mineral interactions in the subaerial vadose environment. *J. Sediment. Petrol.* 49: 955–968.

———. 1979b. Lichen stromatolites: Criterion for subaerial exposure and a mechanism for the formation of laminar calcretes (caliche). *J. Sediment. Petrol.* 49: 387–400.

———. 1980. Rhizoliths in terrestrial carbonates: Classification, recognition, genesis and significance. *Sedimentology* 27: 613–629.

Klepper, B. 1987. Origin, branching and distribution of root systems. In P. J. Gregory, J. V. Lake, and D. A. Rose, eds. *Root Development and Function,* pp. 103–124. Cambridge: Cambridge University Press.

Klitzsch, E., A. Lejal-Nicol, and D. Massa. 1973. Le Siluro-Dévonien à psilophytes et lycophytes du bassin de Mourzouk (Libye). *Compt. Rend. Acad. Sci., Paris,* Série D, 277: 2465–2467.

Kluessendorf, J. J., D. G. Mikulic, and M. R. Carman. 1988. Distribution and depositional environments of the westernmost Devonian rocks in the Michigan Basin. In N. J. McMillan, A. J. Embry, and D. J. Glass, eds. *Devonian of the World,* vol. 1: *Regional Syntheses,* pp. 251–263. Calgary, Alberta: Canadian Society of Petroleum Geologists, Mem. 14.

Knoll, A. H., K. J. Niklas, P. G. Gensel, and B. H. Tiffney. 1984. Character diversification and patterns of evolution in early vascular plants. *Paleobiology* 10: 34–47.

Knoll, A. H., K. J. Niklas, and B. H. Tiffney. 1979. Phanerozoic land plant diversity in North America. *Science* 206: 1400–1402.

Knoll, M. A., and W. C. James. 1987. Effect of the advent and diversification of vascular land plants on mineral weathering through geologic time. *Geology* 15: 1099–1102.

Koes, R. E., F. Quattricchio, and J. N. M. Mol. 1994. The flavonoid biosynthetic pathway in plants: Function and evolution. *Bioessays* 16: 123–124.

Köhler, H.-R., G. Alberti, and V. Storch. 1991. The influence of the mandibles of Diplopoda on the food—A dependence of fine structure and assimilation efficiency. *Pedobiology* 35: 108–116.

Kolattukudy, P. E. 1984. Biochemistry and function of cutin and suberin. *Can. J. Bot.* 62: 2918–2933.

Kotyk, M. E. 1997. Late Silurian and Early Devonian land plants from Bathurst Island, Canadian Arctic Archipelago. *Am. J. Bot. Suppl.* 84: 135–136.

———. 1998. Late Silurian and Early Devonian Fossil

Plants of Bathurst Island, Arctic Canada. M.Sc. thesis, University of Saskatchewan, Saskatoon, Canada.

Kotyk, M. E., and J. Basinger. 2000. The early Devonian (Pragian) zosterophyll *Bathurstia denticulata* Hueber. *Can. J. Bot.* 78(2): 193–207.

Krantz, G. W., and E. E. Lindquist. 1979. Evolution of phytophagous mites (Acari). *Ann. Rev. Entomol.* 24: 121–158.

Krassilov, V. A., and R. M. Schuster. 1984. Paleozoic and Mesozoic fossils. In R. M. Schuster, ed. *New Manual of Bryology,* vol. 2. The Hattori Botanical Laboratory, pp. 1172–1193.

Kraus, O., and M. Kraus. 1994. Phylogenetic system of the Tracheata (Mandibulata): on "Myriapoda"-Insecta interrelationships, phylogenetic age and primary ecological niches. *Verh. Naturwiss. Ver. Hamburg* 34: 5–31.

Kräusel, R., and H. Weyland. 1926. Beiträge zur Kenntnis der Devonflora. II. *Abh. Senckenb. Naturforsch. Ges.* 49: 115–155.

———. 1933. Die Flora des böhmischen Mitteldevons (Stufe Hh1 Barrande = h Kettner-Kodym). *Palaeontographica B* 78: 1–46.

———. 1949. Pflanzenreste aus dem Devon. XIV. *Gilboaphyton* und die Protolepidophytales. *Senckenbergiana* 30: 129–152.

Kräusel, R., and H. Weyland. 1961. Über *Psilophyton robustius* Dawson. *Palaeontographica B* 108: 11–21.

Krebs, W. 1974. Devonian carbonate complexes of central Europe. In L. F. Laporte, ed. *Reefs in Time and Space,* pp. 155–208. Tulsa, OK: Soc. Econ. Paleontol. Mineral., Spec. Publ. 18.

Kroken, S. B., L. E. Graham, and M. E. Cook. 1996. Occurrence and evolutionary significance of resistant cell walls in charophytes and bryophytes. *Am. J. Bot.* 83: 1241–1254.

Kubitski, K. 1987. Phenylpropanoid metabolism in relation to land plant origin and diversification. *J. Plant Physiol.* 131: 17–24.

Kump, L. R. 1991. Interpreting carbon-isotope excursions: Strangelove oceans. *Geology* 19: 299–302.

Kump, L. R., and R. M. Garrels. 1986. Modeling atmospheric O_2 in the global sedimentary redox cycle. *Am. J. Sci.* 286: 337–360.

Kurmann, M. H., and J. A. Doyle. 1994. *Ultrastructure of Fossil Spores and Pollen: Its Bearing on Relationships among Fossil and Living Groups.* Kew: Royal Botanic Gardens.

Labandeira, C. C., B. S. Beall, and F. M. Hueber. 1988. Early insect diversification: Evidence from a Lower Devonian bristletail from Quebec. *Science* 242: 913–916.

Labandeira, C. C., and T. L. Phillips. 1996. Insect fluid-feeding on Upper Pennsylvanian tree ferns (Palaeodictyoptra, Marattiales) and the early history of the piercing-and-sucking functional feeding group. *Ann. Entomol. Soc. Am.* 89: 157–183.

Lacombe, E. 1997. Cinnamoyl-CoA reductase, the first committed enzyme of lignin branch biosynthetic pathway: Cloning, expression and phylogenetic relationships. *Plant J.* 11: 429–441.

Lakova, I., P. M. Gocev, and S. Yanev. 1992. Palynostratigraphy and geological setting of the Lower Palaeozoic allochthon of the Dervent Heights, S.E. Bulgaria. *Geol. Balcanica* 22: 71–88.

Lang, W. H. 1927. Contributions to the study of the Old Red Sandstone flora of Scotland. VI. On *Zosterophyllum myretonianum*, Penh., and some other plant-remains from the Carmyllie Beds of the Lower Old Red Sandstone. *Trans. R. Soc. Edinb.* 55: 443–455.

———. 1937. On the plant-remains from the Downtonian of England and Wales. *Phil. Trans. R. Soc. Lond.* B227: 245–291.

Lange, O. L., T. G. A. Green, and H. Ziegler. 1988. Water status related photosynthesis and carbon isotope discrimination in species of the lichen genus *Pseudocyphellaria* with green or blue-green photobionts and in photosymbiodemes. *Oecologia* 75: 494–501.

Lawrence, D. A. 1986. Sedimentology of the Lower Devonian Battery Point Formation, eastern Gaspé Peninsula. Ph.D. thesis, University of Bristol, Bristol, U.K.

Lawrence, D. A., and B. R. Rust. 1988. The Devonian clastic wedge of eastern Gaspé and the Acadian orogeny. In N. J. McMillan, A. F. Embry, and D. J. Glass, eds. *Devonian of the World*, vol. II, pp. 53–64. Calgary, Alberta: Canadian Society of Petroleum Geologists.

Lawrence, D. A., and B. P. J. Williams. 1987. Evolution of drainage systems in response to Acadian deformation: The Devonian Battery Point Formation, Eastern Canada. *Society of Economic Paleontologists and Mineralogists Spec. Publ.* 39: 287–300.

Leclercq, S. 1942. Quelques plantes fossiles recueillies dans le Dévonien inférieur des environs de Nonceveux (Bordure orientale du bassin de Dinant). *Bull. Soc. Géol. Belg.* 65: 193–211.

Leclercq, S., and H. N. Andrews. 1960. *Calamophyton bicephalum*, a new species from the Middle Devonian of Belgium. *Ann. Mo. Bot. Gard.* 47: 1–23.

Leclercq, S., and H. P. Banks. 1962. *Pseudosporochnus nodosus* sp. nov., a Middle Devonian plant with Cladoxylalean affinities. *Palaeontographica B* 110: 1–34.

Leclercq, S., and P. M. Bonamo. 1971. A study of the fructification of *Milleria* (*Protopteridium*) *thomsonii* Lang from the Middle Devonian of Belgium. *Palaeontographica B* 136: 83–114.

———. 1973. *Rellimia thomsonii*, a new name for *Milleria* (*Protopteridium*) *thomsonii* Lang 1926 emend. Leclercq and Bonamo 1971. *Taxon* 22: 435–437.

Legrand, R. 1967. Ronquières. Documents géologiques. *Mém. Expl. Cartes Géol. Min. Belg.* 6: 1–60.

Le Herisse, A., C. Rubinstein, and P. Steemans. 1996. Lower Devonian palynomorphs from the Talacasto Formation, Cerro del Fuerte section, San Juan Precordillera, Argentina. In O. Fatka and T. Servais, eds. *Acritarcha in Praha*, pp. 497–515. *Acta Univ. Carol. Geol.* 40: 497–515.

Lejal-Nicol, A. 1975. Sur une nouvelle flore à Lycophytes du Dévonien inférieur de la Libye. *Palaeontographica B* 151: 52–96.

Lejal-Nicol, A., and D. Massa. 1980. Sur des végétaux du Dévonien inférieur de Libye. *Rev. Palaeobot. Palynol.* 29: 221–239.

LeJeune, V. 1986. *Sedimentation Alluviale et Paléosols Associés de la Formation D'Évieux (Synclinorium de Dinant).* Liege, Belgium: Université de Liège, Faculté des Sciences, Institute de Minéralogie.

Lemoigne, Y., and A. Yurina. 1983. *Xenocladia medullosina* Ch. A. Arnold (1940) 1952 du Dévonien Moyen du Kazakhstan (URSS). *Geobios* 16: 513–547.

Leopold, A. C., ed. 1986. *Membranes, Metabolism, and Dry Organisms.* Ithaca, NY: Comstock.

Leventhal, J. S. 1987. Carbon and sulfur relationships in Devonian shales from the Appalachian Basin as an indicator of environment of deposition. *Am. J. Sci.* 287: 33–49.

Lewis, L. A., B. D. Mishler, and R. Vilgalys. 1997. Phylogenetic relationships of the liverworts (Hepaticae), a basal embryophyte lineage, inferred from nucleotide sequence data of the chloroplast gene rbcL. *Molec. Phylogenet. Evol.* 7: 377–393.

Lewis, N. G., and L. B. Davin. 1994. Evolution of lignan and neolignan biochemical pathways. In D. Nes, ed. *Evolution of Natural Products*, 22, pp. 202–246. American Chemical Society.

Li, C.-S. 1990. *Minarodendron cathaysiense* (gen. et comb. nov.), a lycopod from the late Middle Devonian of Yunnan, China. *Palaeontographica B* 220: 97–117.

———. 1992. *Hsüa robusta*, an Early Devonian plant from Yunnan Province, China, and its bearing on some structures of early land plants. *Rev. Palaeobot. Palynol.* 71: 121–147.

Li, C.-S., and D. Edwards. 1992. A new genus and species of early land plants with novel strobilar construction from the Lower Devonian Posongchong Formation, Yunnan province, China. *Palaeontology* 35: 257–272.

———. 1995. A re-investigation of Halle's *Drepanophycus spinaeformis* Göpp. from the Lower Devonian of Yunnan Province, southern China. *Bot. J. Linn. Soc.* 118: 163–192.

———. 1996. *Demersatheca* Li and Edwards, gen. nov., a new genus of early land plants from the Lower Devonian, Yunnan province, China. *Rev. Palaeobot. Palynol.* 93: 77–88.

———. 1997. A new microphyllous plant from the Lower Devonian of Yunnan Province, China. *Am. J. Bot.* 84: 1441–1448.

Li, C.-S., and J. Hsü. 1987. Studies on a new Devonian plant *Protopteridophyton devonicum* assigned to primitive fern from China. *Palaeontographica B* 207: 111–131.

Li, C.-S., F. M. Hueber, and C. L. Hotton. 2000. A neotype for *Drepanophycus spinaeformis* Goeppert 1852. *Can. J. Bot.* (in press).

Li, X.-X., and C.-Y. Cai. 1977. Early Devonian *Zosterophyllum* remains from southwest China. *Acta Palaeontol. Sin.* 16 (1): 12–34.

———. 1978. A type-section of Lower Devonian strata in southwest China with brief notes on the succession and correlation of its plant assemblages. *Acta Geol. Sin.* 52: 1–12 (in Chinese with English abstract).

———. 1979. Devonian floras of China. *Acta Stratigr. Sin.* 3: 90–95.

Li, X.-X., and X. Y. Wu. 1996. Late Paleozoic phytogeographic provinces in China and its adjacent regions. *Rev. Palaeobot. Palynol.* 90: 41–62.

Liao, W. H., H. K. Xu, Y. Wheng, Y. P. Ruan, C. Y. Cai, D. C. Mu, and L. C. Lu. 1978. The subdivision and correlation of the Devonian stratigraphy of S.W. China. In Institute of Geology and Mineral Resource, Chinese Academy of Geological Sciences, eds. *Symposium, Devonian System of South China*, pp. 193–213. Peking, P.P.: Geological Press.

Ligrone, R., and R. Gambardella. 1988. The sporophyte-gametophyte junction in bryophytes. *Adv. Bryol.* 3: 225–274.

Lindholm, R. C. 1987. *A Practical Approach to Sedimentology.* London: Allen and Unwin.

Lindroth, R. L. 1988. Adaptations of mammalian herbivores to plant chemical defenses. In K. C. Spencer, ed. *Chemical Mediation of Coevolution.* Washington, DC: American Institute of Biological Sciences.

Little, C. 1990. *The Terrestrial Invasion.* Oxford: Oxford University Press.

Liu, C., and Q. Z. Lian. 1984. Magnetostratigraphic research on the boundary line between Yulounsi Formation and Cuifengshan Group. *Sci. Bull.* 4: 232–234.

Lohmann, K. C., and J. C. G. Walker. 1989. The $\delta^{18}O$ record of Phanerozoic abiotic marine calcite cements. *Geophys. Res. Lett.* 16: 319–322.

Longton, R. E. 1988. *The Biology of Polar Bryophytes and Lichens.* Cambridge: Cambridge University Press.

———. 1992. The role of bryophytes and lichens in terrestrial ecosystems. In J. W. Bates and A. M. Farmer, eds. *Bryophytes and Lichens in a Changing Environment*, pp. 32–76. New York: Cambridge University Press.

Lu, P. Z., and Jernstedt, J. A. 1996. Rhizophore and root development in *Selaginella martensii*: Meristem transitions and identity. *Int. J. Plant Sci.* 157: 180–194.

Lynn, D. G., and M. Chang. 1990. Phenolic signals in cohabitation: Implications for plant development. *Ann. Rev. Plant Physiol.* 41: 497–526.

Lyon, A. G. 1964. Probable fertile region of *Asteroxylon mackiei* K. and L. *Nature* 203: 1082–1083.

Lyon, A. G., and D. Edwards. 1991. The first zosterophyll from the Lower Devonian Rhynie Chert, Aberdeenshire. *Trans. R. Soc. Edinb. (Earth Sciences)* 82: 323–332.

Lyons, T. W., R. A. Berner, and R. F. Anderson. 1993. Evidence for large pre-industrial perturbations of the Black Sea chemocline. *Nature* 365: 538–540.

Machel, H. G., and E. A. Burton. 1991. Factors governing cathodoluminescence in calcite and dolomite and their implications for studies of carbonate diagenesis. In C. E. Barker and O. C. Kopp, eds. *Luminescence Microscopy and Spectroscopy: Qualitative and Quantitative Applications.* Soc. Econ. Paleontol. and Mineral., Short Course Notes 25: 37–57.

Mack, G. H., and W. C. James. 1992. *Paleosols for Sedimentologists.* Boulder, CO: Geological Society of America, Short Course Notes.

Mägdefrau, K. 1952. *Vegetationsbilder der vorzeit.* Jena, Germany: Gustav Fischer.

Máguas, C., H. Griffiths, and M. S. J. Broadmeadow. 1995. Gas exchange and carbon iostope discrimination in lichens: Evidence for interactions between CO_2-concentrating mechanisms and diffusion limitation. *Planta* 196: 95–102.

Mamay, S. H., and R. M. Bateman. 1991. *Archaeocalamites lazarii* sp. nov.: The range of the Archaeocalamitaceae extended from the lowermost Pennsylvanian to the mid-Lower Permian. *Am. J. Bot.* 78: 489–496.

Mamet, B., A. Roux, M. Lapointe, and L. Gauthier. 1992. Algues ordoviciennes et Siluriennes de l'île d'Anticosti (Quebec, Canada). *Rev. Micropaleontol.* 35: 211–248.

Manhart, J. R. 1994. Phylogenetic analysis of green plant rbcL sequences. *Molec. Phylogenet. Evol.* 3: 114–127.

Manhart, J. R., and J. D. Palmer. 1990. The gain of 2 chloroplast tRNA introns marks the green algal ancestors of land plants. *Nature* 345: 268–270.

Marcelle, H. 1951. *Callixylon velinense* nov. sp., un bois à structure conservée du Dévonien de Belgique. *Bull. Acad. R. Belg. Cl. Sci.* Ser. 5, 37: 908–919.

Marchant, H. J., and J. D. Pickett-Heaps. 1973. Mitosis and cytokinesis in *Coleochaete scutata. J. Phycol.* 9: 461–471.

Markham, K. R. 1990. Bryophyte flavonoids, their structures, distribution, and evolutionary significance. In H. D. Zinsmeister and R. Mues, eds. *Bryophytes: Their Chemistry and Chemical Taxonomy, Proc. Phytochem. Soc. Eur.* 29: 143–170. New York: Oxford University Press.

Markham, K. R., A. Franke, D. R. Given, and P. Brownsey. 1990. Historical ozone level trends from herbarium specimen flavonoids. *Bull. Liaison-Groupe Polyphenols* 15: 230–235.

Marshall, J. E. A. 1991. Palynology of the Stonehaven Group, Scotland: Evidence for a Mid-Silurian age and its geological implications. *Geol. Mag.* 128: 283–286.

Martin, J. S., and J. B. Kirkham. 1989. Dynamic role of microvilli in peritrophic membrane formation. *Tissue Cell* 21: 627–638.

Matten, L. C. 1975. Additions to the Givetian Cairo flora from Eastern New York. *Torrey Bot. Club Bull.* 102: 45–52.

———. 1992. Studies on Devonian plants from New York State: *Stenokoleos holmesii* n. sp. from the Cairo flora (Givetian) with an alternative model for lyginopterid seed fern evolution. *Cour. Forschungsinst. Senckenberg* 147: 75–85.

———. 1996. Upper Givetian flora from Gilboa and Cairo, New York: Floristics and paleoenvironments. Abstracts–I.O.P.C. V–1996, p. 66. University of Californa–Santa Barbara: Department of Geological Sciences.

Matten, L. C., and H. P. Banks. 1969. *Stenokoleos bifidus* sp. n. in the Upper Devonian of New York State. *Am. J. Bot.* 56: 880–891.

Mattox, K. R., and K. D. Stewart. 1984. Classification of

the green algae: A concept based on comparative cytology. In D. E. G. Irving and D. M. John, eds. *Systematics of the Green Algae*, pp. 29–72. London: Academic Press.

Maynard, J. B. 1980. Sulfur isotopes of iron sulfides in Devonian–Mississippian shales of the Appalachian Basin: Control by rate of sedimentation. *Am. J. Sci.* 280: 772–786.

———. 1981. Carbon isotopes as indicators of dispersal patterns in Devonian-Mississippian shales of the Appalachian Basin. *Geology* 9: 262–265.

Maynard, J. B., R. Valloni, and H.-S. Yu. 1982. Composition of modern deep-sea sands from arc-related basins. In J. K. Leggett, ed. *Trench and Fore-arc Sedimentation*, pp. 551–561. London: Geological Society of London Spec. Publ. 10.

McBrayer, J. F. 1973. Exploitation of deciduous leaf litter by *Apheloria montana* (Diplopoda: Eurydesmidae). *Pedobiology* 13: 90–98.

McClure, H. A. 1988. The Ordovician-Silurian boundary in Saudi Arabia. *Bull. Br. Mus. (Nat. Hist.) Geol. Ser.* 43: 155–163.

McColl, J. G., and A. A. Pohlman. 1986. Soluble organic acids and their chelating influence on Al and other metal dissolution from forest soils. *Water Air Soil Pollut.* 31: 917–927.

McCourt, R. M. 1995. Green algal phylogeny. *Trends Ecol. Evol.* 10: 159–163.

McCourt, R. M., K. G. Karol, S. Kaplan, and R. W. Hoshaw. 1995. Using *rbcL* sequences to test hypotheses of chloroplast and thallus evolution in conjugating green algae (Zygnematales, Charophyceae). *J. Phycol.* 31: 989–995.

McCourt, R. M., K. G. Karol, M. Guerlesquin, and M. Feist. 1996. Phylogeny of extant genera of the family Characeae (Charales, Charophyceae) based on *rbcL* sequences and morphology. *Am. J. Bot.* 83: 125–131.

McCrea, J. M. 1950. The isotopic chemistry of carbonates and a paleotemperature scale. *J. Chem. Physics* 18: 849–857.

McElwain, J. C., and W. G. Chaloner. 1995. Stomatal density and index of fossil plants track atmospheric carbon dioxide in the Palaeozoic. *Ann. Bot.* 76: 389–395.

McGhee, G. R., Jr. 1996. *The Late Devonian Mass Extinction*. New York: Columbia University Press.

McGhee, G. R., Jr., C. J. Orth, L. R. Quintana, J. S. Gilmore, and E. J. Olsen. 1986. Late Devonian "Kellwasser Event" mass-extinction horizon in Germany: No geochemical evidence for a large-body impact. *Geology* 14: 776–779.

McGregor, D. C. 1973. Lower and Middle Devonian spores of eastern Gaspé, Canada. I. Systematics. *Palaeontographica B* 142: 1–77.

———. 1977. Lower and Middle Devonian spores of eastern Gaspé, Canada. II. Biostratigraphy. *Palaeontographica B* 163: 111–142.

McGregor, D. C., and G. Playford. 1992. Canadian and Australian Devonian spores: Zonation and correlation. *Geol. Soc. Can. Bull.* 438.

McLaren, D. J. 1982. Frasnian-Famennian extinctions. In L. T. Silver and P. H. Schultz, eds. *Geological Implications of Impacts of Large Asteroids and Comets on the Earth*, pp. 477–484. Boulder, CO: Geological Society of America, Spec. Pap. 190.

McLean, R. J., and G. F. Pessoney. 1971. Formation and resistance of akinetes of *Zygnema*. In B. C. Parker and R. M. Brown, Jr., eds. *Contributions in Phycology*, pp. 145–152. Lawrence, KS: Allen Press.

Meckel, L. D. 1970. Paleozoic alluvial deposition in the central Appalachians. In G. W. Fisher, F. J. Pettijohn, J. C. Reed, Jr., and K. N. Weaver, eds. *Studies of Appalachian Geology, Central and Southern*, pp. 49–67. New York: John Wiley and Sons.

Mehdy, M. 1994. Active oxygen species in plant defense against pathogens. *Plant Physiol.* 105: 467–472.

Melkonian, M., and B. Surek. 1995. Phylogeny of the Chlorophyta: Congruence between ultrastructural and molecular evidence. *Bull. Soc. Zool. Fr.* 120: 191–208.

Meyen, S. V. 1987. *Fundamentals of Palaeobotany*. London: Chapman and Hall.

Meyer, F. J. 1962. Das trophische Parenchym. A. Assimilationsgewebe. In W. Zimmermann and P. G. Ozenda, eds. *Handbuch der Pflanzenanatomie*, Bd. IV, Teil 7A. Berlin: Borntraeger.

Meyer-Berthaud, B., S. E. Scheckler, and J.-L. Bousquet. 2000. The development of *Archaeopteris*: New evolutionary characters from the structural analysis of an Early Famennian trunk from south-east Morocco. *Am. J. Bot.* 87: 456–468.

Meyer-Berthaud, B., S. E. Scheckler, and J. Wendt. 1999. *Archaeopteris* is the earliest known modern tree. *Nature* 398: 700–701.

Miller, H. A. 1982. Bryophyte evolution and geography. *Biol. J. Linn. Soc.* 18: 145–196.

Miller, M. A., and L. E. Eames. 1982. Palynomorphs from the Silurian Medina Group (Lower Llandovery) of the Niagara Gorge, Lewiston, New York, U.S.A. *Palynology* 6: 221–254.

Miller, M. F., and D. L. Woodrow. 1991. Shoreline deposits of the Catskill Deltaic Complex, Schoharie Valley, New York. In E. Landing and C. E. Brett, eds. *Dynamic Stratigraphy and Depositional Environments of the Hamilton Group (Middle Devonian) in New York State, Part II*, pp. 153–177. Albany: New York State Museum.

Mishler, B. D., L. A. Lewis, M. A. Buchheim, K. A. Renzaglia, D. J. Garbary, C. F. Delwiche, F. W. Zechman, T. S. Kantz, and R. L. Chapman. 1994. Phylogenetic relationships of the "green algae" and "bryophytes." *Ann. Mo. Bot. Gard.* 81: 451–483.

Moldowan, J. M., and N. M. Talyzina. 1998. Biogeochemical evidence for dinoflagellate ancestors in the Early Cambrian. *Science* 281: 1168–1169.

Moore, P. D. 1984. Clue to past climate in river sediment. *Nature* 308: 316.

Mora, C. I., and S. G. Driese. 1999. Palaeoclimatic significance and stable carbon isotopes of Palaeozoic red bed paleosols, Appalachian Basin, USA and Canada. In M. Thiry and R. Simon-Coinçon, eds. *Palaeoweathering, Palaeosurfaces and Related Continental Deposits*.

International Association of Sedimentologists Spec. Publ. 27: 61–84.

Mora, C. I., S. G. Driese, and L. A. Colarusso. 1996. Middle to late Paleozoic atmospheric CO_2 levels from soil carbonate and organic matter. *Science* 271: 1105–1107.

Mora, C. I., S. G. Driese, and P. G. Seager. 1991. Carbon dioxide in the Paleozoic atmosphere: Evidence from carbon-isotope compositions of pedogenic carbonate. *Geology* 19: 1017–1020.

Mora, C. I., D. E. Fastovsky, and S. G. Driese. 1993. Geochemistry and stable isotopes of paleosols: A short course manual for the annual meeting of the Geological Society of America convention. *Studies in Geology* no. 23. Knoxville, TN: University of Tennessee, Department of Geological Science.

Moreau-Benoit, A. 1994. Les spores des gres de Landevennec, dans la Coupe de Lanveoc, Lochkovien du Massif Armoricain, France. *Rev. Micropaleontol.* 37: 75–93.

Morel, E., D. Edwards, and M. Iñigez Rodriguez. 1995. The first record of *Cooksonia* from South America in Silurian rocks of Bolivia. *Geol. Mag.* 132: 449–452.

Morrow, D. W., and H. H. J. Geldsetzer. 1988. Devonian of the eastern Canadian Cordillera. In N. J. McMillan, A. J. Embry, and D. J. Glass, eds. *Devonian of the World*, vol. 1: *Regional Syntheses*, pp. 85–121. Calgary, Alberta: Canadian Society of Petroleum Geologists, Mem. 14.

Mosbrugger, V. 1990. The tree habit in land plants. *Lecture Notes in Earth Sciences* 28. Berlin: Springer.

Moulton, K. S., and R. A. Berner. 1998. Quantification of the effect of plants on weathering: Studies in Iceland. *Geology* 26: 895–898.

Muchez, P., C. Peeters, E. Keppens, and W. A. Viaene. 1993. Stable isotopic composition of paleosols in the Lower Viséan of eastern Belgium: Evidence of evaporation and soil-gas CO_2. *Chem. Geol.* 106: 389–396.

Mussa, D., L. Borghi, S. Bergamaschi, G. Schubert, E. Pereira, and M. A. C. Rodriguez. 1996. Estudo Preliminar da Tafoflora de Formação Furnas, Bacia do Paraná, Brasil. *Annal Acad. Braz. Ci.* 68: 65–89.

Mustafa, H. 1978a. Beiträge zur Devonflora II. *Argum. Palaeobot.* 5: 31–56.

———. 1978b. Beiträge zur Devonflora III. *Argum. Palaeobot.* 5: 91–132.

Mykura, W. 1983. Old Red Sandstone. In G. Y. Craig, ed. *Geology of Scotland*, 2nd ed, pp. 205–251. New York: Wiley.

Nathorst, A. G. 1902. Zur Oberdevonischen flora der Bären-insel. *Kongl. Svenska Vetenskaps-Akademiens Handlingar* 36: 1–60.

Nesbitt, H. W., and G. M. Young. 1982. Early Proterozoic climates and plate motions inferred from major element chemistry of lutites. *Nature* 299: 715–717.

Nesbitt, H. W., G. M. Young, S. M. McLennan, and R. R. Keays. 1996. Effects of chemical weathering and sorting on the petrogenesis of siliciclastic sediments, with implications for provenance studies. *J. Geol.* 104: 525–542.

Newman, E. I., and R. E. Andrews. 1973. Uptake of phosphorus and potassium in relation to root growth and root density. *Plant Soil* 38: 49–69.

Nicholas, J., and P. Bildgen. 1979. Relations between the location of karst bauxites in the northern hemisphere, the global tectonics and the climatic variations during geological time. *Palaeogeog. Palaeoclim. Palaeoecol.* 29: 205–239.

Nicholson, R. L., and R. Hammerschmidt. 1992. Phenolic compounds and their role in disease resistance. *Ann. Rev. Phytopathol.* 30: 369–389.

Nicoll, R. S., and P. E. Playford. 1993. Upper Devonian iridium anomalies, conodont zonation and the Frasnian-Famennian boundary in the Canning Basin, Western Australia. *Palaeogeog. Palaeoclim. Palaeoecol.* 104: 105–113.

Nie, S. Y., D. B. Rowley, and A. M. Ziegler. 1990. Constraints on the location of the Asia microcontinents in Palaeo-Tethys during the Late Palaeozoic. In W. S. McKerrow and C. R. Scotese, eds. *Palaeozoic Palaeogeography and Biogeography*, pp. 397–410. Geological Society Mem. no. 12.

Niklas, K. J. 1982. Chemical diversification and evolution of plants as inferred from paleobiochemical studies. In M. H. Nitecki, ed. *Biochemical Aspects of Evolutionary Biology*, pp. 29–91. Chicago: University of Chicago Press.

———. 1985. The evolution of tracheid diameter in early vascular plants and its implications on the hydraulic conductance of the primary xylem strand. *Evolution* 39: 1110–1122.

———. 1997. *The Evolutionary Biology of Plants*. Chicago: University of Chicago Press.

Niklas, K. J., and V. Kerchner. 1984. Mechanical and photosynthetic constraints on the evolution of plant shape. *Paleobiology* 10: 79–101.

Niklas, K. J., B. H. Tiffney, and A. H. Knoll. 1980. Apparent changes in the diversity of fossil plants. In M. K. Hecht, W. C. Steere, and B. Wallace, eds. *Evolutionary Biology*, vol. 12, pp. 1–89. New York: Plenum Press.

———. 1983. Patterns in vascular land plant diversification. *Nature* 303: 614–616.

———. 1985. Patterns in vascular land plant diversification: An analysis at the species level. In J. W. Valentine, ed. *Phanerozoic Diversity Patterns: Profiles in Macroevolution*, pp. 97–128. Princeton, NJ: Princeton University Press.

Norton, R. A., P. M. Bonamo, J. D. Grierson, and W. A. Shear. 1988. Oribatid mite fossils from a terrestrial Devonian deposit near Gilboa, New York. *J. Paleontol.* 62: 259–269.

Obrhel, J. 1959. Neue Pflanzenfunde in den Sbrsko-Schichten (Mitteldevon). *Vest. Ustred Ust. Geol. Svazek.* 34: 384–389.

———. 1961. Die Flora der Srbsko-Schichten (Givet) des mittelböhmischen Devons. *Sbornik Ustred Ust. Geol. Svazek.* 26: 7–46.

———. 1962. Die Flora der Pridoli-Schichten (Budnany-Stufe) des mittelböhmishcen Silurs. *Geologie* 11: 83–97.

———. 1968. Die Silur- und Devonflora des Barrandiums. *Palaeontol. Abh. Berlin* 2: 635–793.

Ohyama, K., Y. Ogura, K. Oda, K. Yamato, E. Ohta, Y. Nakamura, M. Takemura, N. Nozato, K. Akashi, T. Kanegae, and Y. Yamada. 1991. Evolution of organellar genomes. In S. Osawa and T. Honjo, eds. *Evolution of Life: Fossils, Molecules, and Culture,* pp. 187–198. Tokyo: Springer-Verlag.

Oliver, M. J. 1991. Influence of protoplasmic water loss on the control of protein synthesis in the desiccation-tolerant moss *Tortula ruralis. Plant Physiol.* 97: 1501–1511.

Oostendorp, D. 1987. The bryophytes of the Paleozoic and the Mesozoic. *Bryophytorum Bibliotheca* 34. Berlin-Stuttgart: J. Cramer.

Orth, C. J., M. Attrep, and L. R. Quintana. 1990. Iridium abundance patterns across bio-event horizons in the fossil record. In V. L. Sharpton and P. D. Ward, eds. *Global Catastrophes in Earth History: An Interdisciplinary Conference on Impacts, Volcanism, and Mass Mortality,* pp. 45–59. Boulder, CO: Geological Society of America, Spec. Pap. 247.

Palm, C. A., and P. A. Sanchez. 1991. Nitrogen release from the leaves of some tropical legumes as affected by lignin and polyphenol contents. *Soil Biol. Biochem.* 23: 83–88.

Pedder, A. E. H. 1982. The rugose coral record across the Frasnian/Famennian boundary. In L. T. Silver and P. H. Schultz, eds. *Geological Implications of Impacts of Large Asteroids and Comets on the Earth,* pp. 485–489. Boulder, CO: Geological Society of America, Spec. Pap. 190.

Peters, N. K., and D. P. S. Verma. 1990. Phenolic compounds as regulators of gene expression in plant-microbe interactions. *Molec. Plant-Microbe Interact.* 3: 4–8.

Petrosyan, N. M. 1968. Stratigraphic importance of the Devonian flora of the USSR. In D. H. Oswald, ed. *International Symposium on the Devonian System,* Calgary 1967, vol. 2, pp. 579–586. Calgary, Alberta: Canadian Society of Petroleum Geologists.

Pfefferkorn, H. W., and K. Fuchs. 1991. A field classification of fossil plant substrate interactions. *Neues Jahrb. Geol. Palaeontol. Abh.* 183: 17–36.

Phillips, T. L. 1979. Reproduction of heterosporous arborescent lycopods in the Mississippian-Pennsylvanian of Euramerica. *Rev. Palaeobot. Palynol.* 27: 239–289.

Phillips, T. L., M. J. Avcin, and J. M. Schopf. 1975. Gametophytes and young sporophyte development in *Lepidocarpon. Abstr. Bot. Soc. Am.* Corvallis, OR, p. 23. Lawrence, KS: Allen Press.

Pickett-Heaps, J. D. 1967a. Ultrastructure and differentiation in *Chara* sp. I. Vegetative cells. *Austral. J. Biol. Sci.* 20: 539–551.

———. 1967b. Ultrastructure and differentiation in *Chara* sp. II. Mitosis. *Austral. J. Biol. Sci.* 20: 883–894.

Pickett-Heaps, J. D., and H. J. Marchant. 1972. The phylogeny of the green algae: A new proposal. *Cytobios* 6: 255–264.

Piearce, T. G. 1989. Acceptability of pteridophyte litters to *Lumbricus terrestris* and *Oniscus asellus,* and implications for the nature of ancient soils. *Pedobiology* 33: 91–100.

Pirozynski, K. A. 1981. Interactions between fungi and plants through the ages. *Can. J. Bot.* 59: 1824–1827.

Playford, P. E., A. E. Cockbain, E. C. Druce, and J. L. Wray. 1976. Devonian stromatolites from the Canning Basin, Western Australia. In M. R. Walter, ed. *Stromatolites,* pp. 543–563 (*Developments in Sedimentology* 20). Amsterdam: Elsevier.

Playford, P. E., D. J. McLaren, C. J. Orth, J. S. Gilmore, and W. D. Goodfellow. 1984. Iridium anomaly in the Upper Devonian of the Canning Basin, Western Australia. *Science* 226: 437–439.

Plumstead, E. P. 1967. A general review of the Devonian fossil plants found in the Cape System of South Africa. *Palaeontol. Africana* 10: 1–83.

Polis, G. A. 1990. *The Biology of Scorpions.* Stanford, CA: Stanford University Press.

Pollard, J. E., and E. F. Walker. 1984. Reassessment of sediments and trace fossils from Old Red Sandstone (Lower Devonian) of Dunure, Scotland, described by John Smith (1909). *Geobios* 17: 567–581.

Popp, B. N., T. F. Anderson, and P. A. Sandberg. 1986. Brachiopods as indicators of original isotopic compositions in some Paleozoic limestones. *Geol. Soc. Am. Bull.* 97: 1262–1269.

Popp, J. H., and M. Barcellos-Popp. 1986. Análise estratigráfia da seqüência deposicional devoniana da Bacia do Paraná (Brasil). *Revista Brasiliera de Geociências* 16: 187–194.

Pratt, L. M., T. L. Phillips, and J. M. Dennison. 1978. Evidence of non-vascular land plants from the early Silurian (Llandoverian) of Virginia, U.S.A. *Rev. Palaeobot. Palynol.* 25: 121–149.

Price, P. W. 1988. An overview of organismal interactions in ecosystems in evolutionary and ecological time. *Agricul. Ecosyst. Environ.* 24: 369–377.

Proctor, M. C. F. 1982. Physiological ecology: Water relations, light and temperature responses, carbon balance. In A. J. E. Smith, ed. *Bryophyte Ecology,* pp. 333–381. London: Chapman and Hall.

———. 1984. Structure and ecological adaptation. In A. F. Dyer and J. G. Duckett, eds. *The Experimental Biology of Bryophytes,* pp. 9–37. London: Academic Press.

Proctor, V. W. 1962. Viability of *Chara* oospores taken from migratory water birds. *Ecology* 43: 528–529.

———. 1967. Storage and germination of *Chara* oospores. *J. Phycol.* 3: 90–92.

Qiu, Y.-L., Y. Cho, J. C. Cox, and J. D. Palmer. 1998. The gain of three mitochondrial introns identifies liverworts as the earliest land plants. *Nature* 394: 671–674.

Quilhot, W., M. E. Hidalgo, E. Fernandez, W. Pena, and E. Flores. 1992. Posible rol biologico de metabolitos secundarios en liquenes antarticos. *Serie Cientifica Insitituto Antarctico Chileno* 42: 53–59.

Racki, G. 1990. Frasnian/Famennian event in the Holy Cross Mts., central Poland: Stratigraphic and ecologic aspects. In E. G. Kauffman and O. H. Walliser, eds. *Extinction Events in Earth History*, pp. 169–181 (*Lecture Notes in Earth Sciences* 30). Berlin: Springer,.

Racki, G., M. Szulczewski, S. Skompski, and J. Malec. 1993. Frasnian/Famennian and Devonian/Carboniferous boundary events in carbonate sequences of the Holy Cross Mountains (Fieldtrip 1). In M. Narkiewicz, ed. *Excursion Guidebook for Global Boundary Events Symposium*, pp. 5–18. Kielce, Poland.

Ragan, M. A., T. J. Parsons, T. Sawa, and N. A. Straus. 1994. 18S ribosomal DNA sequences indicate a monophyletic origin of Charophyceae. *J. Phycol.* 30: 490–500.

Rahmanian, V. D. 1979. Stratigraphy and sedimentology of the Upper Devonian Catskill and uppermost Trimmers Rock Formations in central Pennsylvania. Unpublished Ph.D. thesis, Pennsylvania State University, University Park, PA.

Rasmussen, S., C. Wolff, and H. Rudolph. 1996. 4'-0-b-*D*-glucosyl-*cis*-*p*-coumaric acid—A natural constituent of *Sphagnum fallax* cultivated in bioreactors. *Phytochemistry* 42: 81–87.

Rateev, M. A., and P. P. Timofeev. 1979. Palygorskite and sepiolite associated with arid facies in Carboniferous rocks of the Russian platform. In E. S. Belt and R. W. Macqueen, eds. *Neuvième Congrès International de Stratigraphie et de Géologie du Carbonifere*, vol. 3, pp. 669–673. Carbondale, IL: Southern Illinois University Press.

Raubeson, L. A., and R. K. Jansen. 1992. Chloroplast DNA evidence on the ancient evolutionary split in vascular land plants. *Science* 255: 1697–1699.

Raup, D. M., and G. E. Boyajian. 1988. Patterns of generic extinction in the fossil record. *Paleobiology* 14: 109–125.

Raven, J. A. 1986. Evolution of plant life forms. In T. J. Givnish, ed. *On the Economy of Plant Form and Function*, pp. 421–492. Cambridge: Cambridge University Press.

———. 1993. The evolution of vascular plants in relation to quantitative functioning of dead water-conducting cells and stomata. *Biol. Rev.* 68: 337–363.

Raymond, A. 1987. Paleogeographic distribution of Early Devonian plant traits. *Palaios* 2: 113–132.

Raymond, A., and C. Metz. 1995. Laurussian land-plant diversity during the Silurian and Devonian: Mass extinction, sampling bias, or both? *Paleobiology* 21: 74–91.

Rayner, R. J. 1983. New observations on *Sawdonia ornata* from Scotland. *Trans. R. Soc. Edinb.* (*Earth Sciences*) 74: 79–93.

Rayner, R. J. 1984. New finds of *Drepanophycus spinaeformis* Göppert from the Lower Devonian of Scotland. *Trans. R. Soc. Edinb.* (*Earth Sciences*) 75: 353–363.

Reitz, E., and T. Heuse. 1994. Palynofazies im Oberordovizium des Saxsothuringikums. *Neues Jahrb. Geol. Palaentol. Monatshefte* 6: 348–360.

Remy, W., P. G. Gensel, and H. Hass. 1993. The gametophyte generation of some Early Devonian land plants. *Int. J. Plant Sci.* 154: 35–58.

Remy, W., and H. Hass. 1986. *Gothanophyton zimmermanni* nov. gen., nov. spec., eine Pflanze mit komplexem Stelär-Körper aus dem Emsian. *Argum. Palaeobot.* 7: 9–69.

———. 1991a. *Kidstonophyton discoides* nov. gen., nov. spec., ein Gametophyt aus dem Chert von Rhynie (Unterdevon, Schottland). *Argum. Palaeobot.* 8: 29–45.

———. 1991b. Gametophyten und Sporophyten im Unterdevon—Fakten und Spekulationen. *Argum. Palaeobot.* 8: 193–223.

———. 1996. New information on gametophytes and sporophytes of *Aglaophyton major* and inferences about possible environmental adaptations. *Rev. Palaeobot. Palynol.* 90: 175–193.

Remy, W., H. Hass, and S. Schultka. 1992. *Sciadophyton* Steinmann emend. Kräusel et Weyland (1930)—Der einzige Vertreter eines unterdevonischen Bauplanes? *Cour. Forschungsinst. Senckenberg* 147: 87–91.

Remy, W., D. Remy, and H. Hass. 1997. Organisation, Wuchsformen und Lebensstratigien früher Landpflanzen des Unterdevons. *Bot. Jahrb. Syst.* 119: 509–562.

Remy, W., S. Schultka, and H. Hass. 1986. *Anisophyton gothani*, nov. gen., nov. spec. und hinweise zur stratigraphie der südlichen Wilbringhäuser Scholle. *Argum. Palaeobot.* 7: 79–107.

Remy, W., T. N. Taylor, H. Hass, and H. Kerp. 1994. Four hundred-million-year-old vesicular arbuscular mycorrhizae. *Proc. Natl. Acad. Sci. USA* 91: 11841–11843.

Renault, S., J. L. Bonnemain, L. Faye, and J. P. Guadillere. 1992. Physiological aspects of sugar exchange between the gametophyte and the sporophyte of *Polytrichum formosum*. *Plant Physiol.* 100: 1815–1822.

Restropo, J. J., and J. Toussaint. 1988. Terranes and continental accretion in the Colombian Andes. *Episodes* 11: 189–193.

Retallack, G. J. 1985. Fossil soils as grounds for interpreting the advent of large plants and animals on land. *Phil. Trans. R. Soc. Lond.* B309: 105–142.

———. 1986. The fossil record of soils. In V. P. Wright, ed. *Paleosols: Their Recognition and Interpretation*. pp. 1–57. Oxford: Blackwell Scientific (also Princeton, NJ: Princeton University Press).

———. 1988. Field recognition of paleosols. In J. Reinhardt and W. R. Sigleo, eds. *Paleosols and Weathering through Geologic Time*. Geological Society of America Spec. Pap. 216: 1–20.

———. 1990. *Soils of the Past: An Introduction to Paleopedology*. Boston: Unwin-Hyman.

———. 1992. Paleozoic paleosols. In I. P. Martini and W. Chesworth, eds. *Weathering, Soils and Paleosols*, pp. 543–564. Amsterdam: Elsevier.

———. 1993. Late Ordovician paleosols of the Juniata Formation near Potters Mills, PA. In S. G. Driese, ed. *Paleosols, Paleoclimate and Paleoatmospheric CO$_2$: Paleozoic Paleosols of Central Pennsylvania*. Knoxville, TN: University of Tennessee, Department of Geological Sciences, *Studies in Geology* no. 22: 33–50.

———. 1997. Early forest soils and their role in Devonian global change. *Science* 276: 583–585.

Retallack, G., and C. R. Feakes. 1987. Trace fossil evidence for Late Ordovician animals on land. *Science* 235: 61–63.

Retallack, G. J., and J. Germán-Heins. 1994. Evidence from paleosols for the geological antiquity of rain forest. *Science* 265: 499–502.

Rice, C. M., W. A. Ashcroft, D. J. Batten, A. J. Boyce, J. B. D. Caulfield, A. E. Fallick, M. J. Hole, E. Jones, M. J. Pearson, G. Rogers, J. M. Saxton, F. M. Sturat, N. H. Trewin, and G. Turner. 1995. A Devonian auriferous hot springs system, Rhynie, Scotland. *J. Geol. Soc. Lond.* 152: 229–250.

Richardson, J. B. 1967. Some British Lower Devonian spore assemblages and their stratigraphic significance. *Rev. Palaeobot. Palynol.* 1: 111–129.

Richardson, J. B. 1985. Lower Palaeozoic sporomorphs: Their stratigraphical distribution and possible affinities. *Phil. Trans. R. Soc. Lond.* B309: 201–203.

———. 1988. Late Ordovician and Early Silurian cryptospores and miospores from northeast Libya. In A. El-Arnauti, B. Owens, and B. Thusu, eds. *Subsurface Palynostratigraphy of Northeast Libya*, pp. 89–109. Benghazi, Libya: Garyounis University Publications.

———. 1992. Origin and evolution of the earliest land plants. In J. W. Schopf, ed. *Major Events in the History of Life*, pp. 95–118. Boston, MA: Jones and Bartlett.

———. 1996a. Lower and middle Palaeozoic records of terrestrial palynomorphs. In J. Jansonius and D. C. McGregor, eds. *Palynology: Principles and Applications*, vol. 2: *Applications*, pp. 555–574. Salt Lake City, UT: American Association of Stratigraphic Palynologists Foundation.

———. 1996b. Abnormal spores and possible interspecific hybridization as a factor in the evolution of Early Devonian land plants. *Rev. Palaeobot. Palynol.* 93: 333–340.

———. 1996c. Taxonomy and classification of some new Early Devonian cryptospores from England. *Spec. Pap. Palaeontol.* 55: 7–40.

Richardson, J. B., and S. Ahmed. 1988. Miospores, zonation and correlation of Upper Devonian sequences from western New York State and Pennsylvania. In N. J. McMillan, A. J. Embry, and D. J. Glass, eds. *Devonian of the World*, vol. 3: *Paleontology, Paleoecology and Biostratigraphy*, pp. 541–558. Calgary, Alberta: Canadian Society of Petroleum Geologists, Mem. 14.

Richardson, J. B, P. M. Bonamo, and D. C. McGregor. 1993. The spores of *Leclercqia* and the dispersed spore morphon *Acinosporites lindlarensis* Riegel: A case of gradualistic evolution. *Bull. Nat. Hist. Mus. Lond. (Geol).* 49: 121–155.

Richardson, J. B., J. H. Ford, and F. Parker. 1984. Miospores, correlation and age of some Scottish Lower Old Red Sandstone sediments from the Strathmore region (Fife and Angus). *J. Micropalaeontol.* 3: 109–124.

Richardson, J. B., and D. C. McGregor. 1986. Silurian and Devonian spore zones of the Old Red Sandstone continent and adjacent regions. *Geol. Surv. Can. Bull.* 364: 1–79.

Richardson, J. B., and S. M. Rasul. 1990. Palynofacies in a Late Silurian regressive sequence in the Welsh Borderland and Wales. *J. Geol. Soc.* 147: 675–686.

Rickard, L. V. 1969. Stratigraphy of the Upper Silurian Salina Group, New York, Pennsylvania, Ohio, Ontario. *New York State Museum Science Service, Map and Chart Series* no. 12.

Rigby, J. K. 1979. Patterns in Devonian sponge distribution. In M. R. House, C. T. Scrutton, and M. G. Bassett, eds. *The Devonian System: Palaeontol. Soc., Spec. Pap. Paleontol.* 23: 225–228.

Ritchie, A. 1963. Palaeontological studies on Scottish Silurian fish beds. Unpublished doctoral thesis, University of Edinburgh.

Robinson, J. M. 1990. Lignin, land plants, and fungi: Biological evolution affecting Phanerozoic oxygen balance. *Geology* 15: 607–609.

Robl, T. L., and L. S. Barron. 1988. The geochemistry of Devonian black shales in central Kentucky and its relationship to inter-basinal correlation and depositional environment. In N. J. McMillan, A. J. Embry, and D. J. Glass, eds. *Devonian of the World*, vol. 2, *Sedimentation*, pp. 377–392. Calgary, Alberta: Canadian Society of Petroleum Geologists, Mem. 14.

Rogerson, E. C. W., D. Edwards, K. L. Davies, and J. B. Richardson. 1993. Identification of *in situ* spores in a Silurian *Cooksonia* from the Welsh Borderland. In M. E. Collinson and A. C. Scott, eds. *Studies in Palaeobotany and Palynology in Honour of Professor W. G. Chaloner, F.R.S. Spec. Pap. Palaeontol.* 49: 17–30.

Rolfe, W. D. I. 1980. Early invertebrate terrestrial fauna. In A. L. Panchen, ed. *The Terrestrial Environment and the Origin of Land Vertebrates*. London: Academic Press.

———. 1985. Early terrestrial arthropods: A fragmentary record. *Phil. Trans. R. Soc. Lond.* B309: 207–218.

Rolfe, W. D. I., E. N. K. Clarkson, and A. L. Panchen. 1994. Volcanism and early terrestrial biotas. *Trans. R. Soc. Edinb.* 84: parts 3, 4.

Rolfe, W. D. I., and J. K. Ingham. 1967. Limb structure, affinity and diet of the Carboniferous "centipede" *Arthropleura*. *Scott. J. Geol.* 3: 118–124.

Rosenthal, G. A., and D. H. Janzen. 1979. *Herbivores: Their Interaction with Secondary Plant Metabolites*. New York: Academic Press.

Rostek, F., G. Ruhland, F. C. Bassinot, P. J. Müller, L. D. Labeyrie, Y. Lancelot, and E. Bard. 1993. Reconstructing sea surface temperature and salinity using $\delta^{18}O$ and alkenone records. *Nature* 364: 319–321.

Rothwell, G. W. 1995. The fossil history of branching: Implications for the phylogeny of land plants. In P. C. Hoch and A. G. Stephenson, eds. *Experimental and Molecular Approaches to Plant Biosystematics*, pp. 71–86. St. Louis, MO: Missouri Botanical Garden.

Rothwell, G. W., and S. E. Scheckler. 1988. Biology of ancestral gymnosperms. In C. B. Beck, ed. *Origin and Evolution of Gymnosperms*, pp. 85–134. New York: Columbia University Press.

Rothwell, G. W., S. E. Scheckler, and W. H. Gillespie. 1989. *Elkinsia* gen. nov., a Late Devonian gymnosperm with cupulate ovules. *Bot. Gaz.* 150: 170–189.

Rothwell, G. W., and R. Serbet. 1994. Lignophyte phylogeny and the evolution of spermatophytes: A numerical cladistic analysis. *Syst. Bot.* 19: 443–482.

Rothwell, G. W., and D. C. Wight. 1989. *Pullaritheca longii* gen. nov. and *Kerryia mattenii* gen. et sp. nov., Lower Carboniferous cupules with ovules of the *Hydrasperma tenuis*-type. *Rev. Palaeobot. Palynol.* 60: 295–309.

Rozema, J., J. Van de Staaij, L. O. Bjorn, and M. Caldwell. 1997. UV-B as an environmental factor in plant stress: Stress and regulation. *Trends Ecol. Evol.* 12: 22–28.

Rubinstein, C. 1992. Esporas del Silurico Superior (Formacion Los Espejos) de la Precordillera Sanjuanina-Argentina. *Ser. Correl. Geol.* 9: 93–106.

———. 1993a. Primer registro de miosporas y acritarcos del Devonico Inferior, en el "Grupo Villavicencio," Precordillera de Mendoza, Argentina. *Ameghiniana* 30: 219–220.

———. 1993b. Investigaciones palinológicas en el Paleozoic Inferior de Argentina. *Zentral. Geol. Palaeontol.* 1: 217–230.

Russell, D. J., and J. F. Barker. 1983. Stratigraphy and geochemistry of the Kettle Point Formation, Ontario. In *Proceedings of the 1983 Eastern Oil Shale Symposium*, pp. 169–179. Lexington, KY: Institute for Mining and Minerals Research.

Russell, P. L. 1990. *Oil Shales of the World, Their Origin, Occurrence and Exploitation.* Oxford: Pergamon.

Russell, R. S. 1977. *Plant Root Systems: Their Function and Interaction with the Soil.* New York: McGraw-Hill.

Rust, B. R. 1984. Proximal braidplain deposits in the Middle Devonian Malbaie Formation of eastern Gaspé, Canada. *Sedimentology* 31: 675–695.

Sainz-Jimenez, C., and J. W. de Leeuw. 1986. Lignin pyrolysis products: Their structure and significance as biomarkers. *Org. Geochem.* 10: 869–876.

Sandberg, C. A., F. G. Poole, and J. G. Johnson. 1988. Upper Devonian of western United States. In McMillan, N. J., A. J. Embry, and D. J. Glass, eds. *Devonian of the World*, vol. 1: *Regional Syntheses*, pp. 183–220. Calgary, Alberta: Canadian Society of Petroleum Geologists, Mem. 14.

Sandberg, C. A., W. Ziegler, R. Dreesen, and J. L. Butler. 1988. Late Frasnian mass extinction: Conodont event stratigraphy, global changes, and possible causes. *Cour. Forschungsinst. Senckenberg* 102: 263–307.

Sarjeant, W. A. S. 1975. Plant trace fossils. In R. W. Frey, ed. *The Study of Trace Fossils*, pp. 163–179. New York: Springer-Verlag.

Schawaller, W., W. A. Shear, and P. M. Bonamo. 1991. The first Paleozoic pseudoscorpions. *Am. Mus. Novitates* 3009: 1–17.

Scheckler, S. E. 1974. Systematic characters of Devonian ferns. *Ann. Mo. Bot. Gard.* 61: 462–473.

———. 1975. A fertile axis of *Triloboxylon ashlandicum*, a progymnosperm from the Upper Devonian of New York. *Am. J. Bot.* 62: 923–934.

———. 1978. Ontogeny of progymnosperms. II. Shoots of Upper Devonian Archaeopteridales. *Can. J. Bot.* 56: 3136–3170.

———. 1984. Persistence of the Devonian plant group Barinophytaceae into the basal Carboniferous of Virginia, USA. In P. K. Sutherland and W. L. Manger, eds. *Ninth Internat. Congr. Stratigr. Geol. Carbon. 1979 at*

Urbana, IL. vol. 2: *Biostratigraphy*, pp. 223–228. Carbondale, IL: Southern Illinois University Press.

———. 1986a. Floras of the Devonian-Mississippian transition. In T. W. Broadhead, ed. *Land Plants*, pp. 81–96. Knoxville, TN: University of Tennessee, Department of Geological Sciences, *Studies in Geology* no. 15.

———. 1986b. Geology, floristics, and paleoecology of Late Devonian coal swamps from Appalachian Laurentia (U.S.A.). *Ann. Soc. Géol. Belg.* 109: 209–222.

———. 1986c. Old Red Continent facies in the late Devonian and early Carboniferous of Appalachian North America. *Ann. Soc. Géol. Belg.* 109: 223–236.

———. 1995. Progymnosperms have gymnospermous roots. The evolution of plant architecture. *Programme and Abstracts* no. 31. The Linnean Society of London and The Royal Botanic Gardens, Kew.

Scheckler, S. E., and H. P. Banks. 1971a. Anatomy and relationships of some Devonian progymnosperms from New York. *Am. J. Bot.* 58: 737–751.

———. 1971b. *Proteokalon*, a new genus of progymnosperms from the Devonian of New York State and its bearing on phylogenetic trends in the group. *Am. J. Bot.* 58: 874–884.

———. 1974. Periderm in some Devonian Plants. In Y. S. Murty, B. M. Johri, N. Y. Mohan Ram, and T. M. Varghese, eds. *Advances in Plant Morphology*, pp. 58–64. Meerut, India: Meerut City University Press.

Schindler, E. 1990. The Late Frasnian (Upper Devonian) Kellwasser Crisis. In E. G. Kauffman and O. H. Walliser, eds. *Extinction Events in Earth History*, pp. 151–159 (*Lecture Notes in Earth Sciences* 30). Berlin: Springer.

———. 1993. Event-stratigraphic markers within the Kellwasser Crisis near the Frasnian/Famennian boundary (Upper Devonian) in Germany. *Palaeogeog. Palaeoclim. Palaeoecol.* 104: 115–125.

Schlesinger, W. H. 1991. *Biogeochemistry: An Analysis of Global Change.* San Diego, CA: Academic Press.

Schlüter, U. 1982. The anal glands of *Rhapidostreptus virgator* (Diplopoda, Spirostrepsidae). I. Appearance during the intermoult cycle. *Zoomorphology* 100: 65–73.

———. 1983. The anal glands of *Rhapidostreptus virgator* (Diplopoda, Spirostrepsidae). II. Appearance during a moult. *Zoomorphology* 102: 79–86.

Schönlaub, H. P., M. Attrep, K. Boeckelmann, et al. 1992. The Devonian/Carboniferous boundary in the Carnic Alps (Austria): A multidisciplinary approach. *Jahrbuch der Geologischen Bundes-Anstalt (Austria)* 135: 57–98.

Schopf, J. M. 1969. Early Palaeozoic palynomorphs. In R. H. Tschudy and R. A. Scott, eds. *Aspects of Palynology*, pp. 163–192. New York: Wiley Interscience.

Schopf, J. M., E. Mencher, A. J. Boucot, and H. N. Andrews. 1966. Erect plants in the early Silurian of Maine. *U.S. Geol. Surv. Prof. Pap.* 550-D, D69-D75.

Schopf, J. M., and J. F. Schwietering. 1970. *The Foerstia Zone of the Ohio and Chattanooga Shales* (Prof. Pap. 1294-H). Washington, DC: U.S. Geological Survey.

Schuchman, P. G. 1969. *Pseudosporochnus* in the Middle

Devonian of New York State. M.S. thesis, Cornell University, Ithaca, NY.

Schultka, S. 1991. *Trigonotarbus stoermeri*, n. sp. ein Spinnentier aus de Bensberger Schichten (Ems/Unter-Devon) des Rheinischen Schiefergebirges. *Neues Jahrb. Geol. Paláont. Abh.* 183: 375–390.

Schultka, S., and H. Hass. 1997. *Stockmansella remyi* sp. nov. from the Eifelian: New aspects in the Rhyniaceae (sensu Hass et Remy, 1991). *Rev. Palaeobot. Palynol.* 97: 381–393.

Schumm, S. A. 1968. Speculations concerning paleohydrologic controls of terrestrial sedimentation. *Geol. Soc. Am. Bull.* 79: 1573–1588.

———. 1977. *The Fluvial System.* New York: Wiley.

Schuster, R., and I. J. Schuster. 1977. Ernährungs—und fortpglanzungsbiologiische Studien an der Milbenfamilie Nanorchestidae (Acari, Trombidiformes). *Zool. Anz.* 199: 89–94.

Schuster, R. M. 1966. *The Hepaticae and Anthocerotae of North America East of the 100th Meridian.* New York: Columbia University Press.

Schwartzman, D. W., and T. Volk. 1989. Biotic enhancement of weathering and the habitability of earth. *Nature* 340: 457–460.

Schweitzer, H.-J. 1965. Über *Bergeria mimerensis* und *Protolepidodendropsis pulchra* aus dem Devon Westspitzbergen. *Palaeontographica B* 115: 117–138.

———. 1966. Die Mitteldevon-Flora von Lindlar (Rheinland). 1. Lycopodiinae. *Palaeontographica B* 118: 93–112.

———. 1968. Pflanzenreste aus dem Devon Nord-Westspitzbergens. *Palaeontographica B* 123: 43–75.

———. 1969. Die Oberdevon-flora der Bäreninsel 2. Lycopodiinae. *Palaeontographica B (Paläophytology)* 126: 101–137.

———. 1972. Die Mitteldevonflora von Lindlar (Rheinland). 2. Filicineae—*Hyenia elegans* Kraüsel & Weyland. *Palaeontographica B* 137: 154–175.

———. 1973. Die Mitteldevon-Flora von Lindlar (Rheinland) 4. Filicinae—*Calamophyton primaevum* Kräusel & Weyland. *Palaeontographica B* 140: 117–150.

———. 1974. Zur mitteldevonischen Flora von Lindlar (Rheinland). *Bonner Paläobotanische Mitteilungen* December 1: 1–9.

———. 1980a. Die Gattungen *Taeniocrada* White und *Sciadophyton* Steinmann im Unterdevon des Rheinlandes. *Bonner Palaeobot. Mitteil.* no. 5.

———. 1980b. Über *Drepanophycus spinaeformis* Göppert. *Bonn. Paläobot. Mitteil.* 7: 1–29.

———. 1983. Die Unterdevonflora des Rheinlandes. 1. Teil. *Palaeontographica B* 189: 1–138.

———. 1992. Vorläufiger Bericht über die während der goewissenschaftlichen Spitzbergen-Expedition 1990 (SPE 90) erzielten paläobotanischen Ergebnisse. *Stutt. Geogr. Stud.* 117: 55–72.

Schweitzer, H.-J., and P. Giesen. 1980. Über *Taeniophyton inopinatum*, *Protolycopodites devonicus* und *Cladoxylon scoparium* aus dem Mitteldevon von Wuppertal. *Palaeontographica B* 173: 1–25.

Schweitzer, H.-J., and C.-S. Li. 1996. *Chamaedendron* nov.

gen., eine multisporangiate lycophyte aus dem Frasnium südchinas. *Palaeontographica B* 238: 45–69.

Schweitzer, H.-J., and L. C. Matten. 1982. *Aneurophyton germanicum* and *Protopteridium thomsonii* from the Middle Devonian of Germany. *Palaeontographica B* 184: 65–106.

Scotese, C. R., R. K. Bambach, C. Barton, R. Van der Voo, and A. M. Ziegler. 1979. Paleozoic base maps. *J. Geol.* 87: 217–277.

Scotese, C. R., and J. Golonka. 1992. *Paleomap-Paleogeographic Atlas.* Arlington, TX: University of Texas at Arlington, Department of Geology, *Paleomap Progress Report*, vol. 20.

Scotese, C. R., and W. S. McKerrow. 1990. Revised world maps and introduction. In W. S. McKerrow and C. R. Scotese, eds. *Palaeozoic Palaeogeography and Biogeography.* Geological Society of London Mem. 12(1): 1–21.

Scott, A. C. 1991. Evidence for plant-arthropod interactions in the fossil record. *Geol. Today* 7: 58–61.

———. 1992. Trace fossils of plant-arthropod interactions. In C. G. Maples and R. R. West, eds. *Trace fossils, Short Courses in Paleontology.* Knoxville, TN: University of Tennessee, Paleontological Society.

Scott, A. C., J. Stephenson, and W. Chaloner. 1992. Interaction and co-evolution of plants and arthropods during the Palaeozoic and Mesozoic. *Phil. Trans. R. Soc. Lond.* B335: 129–165.

Scott, A. C., and T. N. Taylor. 1983. Plant/animal interactions during the Upper Carboniferous. *Bot. Rev.* 49: 259–307.

Scott, R. J. 1994. Pollen exine—The sporopollenin enigma and the physics of pattern. In R. J. Scott and M. A. Stead, eds. *Molecular and Cellular Aspects of Plant Reproduction.* Society for Experimental Biology Seminar Series, 55: 49–81. New York: Cambridge University Press.

Scriber, J. M., and F. Slansky. 1981. The nutritional ecology of immature insects. *Ann. Rev. Entomol.* 26: 183–211.

Scrutton, C. T. 1988. Patterns of extinction and survival in Palaeozoic corals. In G. P. Larwood, ed. *Extinction and Survival in the Fossil Record*, pp. 65–88. Oxford: Clarendon.

Seastedt, T. R., and D. A. Crossley. 1984. The influence of arthropods on ecosystems. *BioScience* 34: 157–161.

Selden, P. A. 1981. Functional morphology of the prosoma of *Baltoeurypterus tetragonophthalmus* (Fischer) (Chelicerata: Eurypterida) *Trans. R. Soc. Edinb.* (*Earth Sciences*) 72: 9–48.

Selden, P. A., and D. Edwards. 1989. Colonisation of the land. In K. C. Allen and D. E. G. Briggs, eds. *Evolution and the Fossil Record*, pp. 122–152. London: Belhaven.

Selden, P. A., and D. Edwards. 1990. Colonisation of the land. In K. Allen and D. Briggs, eds. *Evolution and the Fossil Record*, pp. 122–152. Washington, DC: Smithsonian Institution Press.

Selden, P. A., and A. J. Jeram 1989. Palaeophysiology of terrestrialisation in the Chelicerata. *Trans. R. Soc. Edinb.* (*Earth Sciences*) 80: 303–310.

Selden, P. A., W. A. Shear, and P. M. Bonamo. 1991. A

spider and other arachnids from the Devonian of New York, and reinterpretation of Devonian Araneae. *Palaeontology* 34: 241–281.

Senkevich, M. A. 1975. New Devonian psilophytes from Kazakhstan. *Esheg. Vses Paleontol. Obschestva* 21: 288–298 (in Russian).

———. 1986. Fossil plants in the Tokrau horizon of the Upper Silurian. In I. F. Nikitin and S. M. Bandaletoc, eds. *The Tokrau Horizon of the Upper Silurian Series: Balkhash Segment.* Nuka: Alma-Ata (in Russian).

Senkevitch, M. A., A. L. Yurina, and A. D. Arkhangelskaya. 1993. On fructifications, morphology and anatomy of Givetian lepidophytes in Kazakhstan (USSR). *Palaeontographica B* 230: 43–58.

Sepkoski, J. J., Jr. 1979. A kinetic model of Phanerozoic taxonomic diversity. II. Early Paleozoic families and multiple equilibria. *Paleobiology* 5: 222–251.

———. 1986. Phanerozoic overview of mass extinctions. In D. M. Raup and D. Jablonski, eds. *Patterns and Processes in the History of Life,* pp. 277–295. Berlin: Springer.

———. 1996. Patterns of Phanerozoic extinction: A perspective from global data bases. In O. H. Walliser, ed. *Global Events and Event Stratigraphy in the Phanerozoic,* pp. 35–51. Berlin: Springer.

Serlin, B. S., and H. P. Banks. 1978. Morphology and anatomy of *Aneurophyton,* a progymnosperm from the late Middle Devonian of New York. *Palaeontographica Americana* 8: 343–359.

Sevon, W. D. 1985. Non-marine facies of the Middle and Late Devonian Catskill coastal alluvial plain. In D. L. Woodrow and W. D. Sevon, eds. *The Catskill Delta.* Geological Society of America Spec. Pap. 201: 79–90.

Sevon, W. D., J. M. Dennison, F. R. Ettensohn, W. T. Sevon, W. T. Kirchgasser. 1988. Middle and Upper Devonian stratigraphy and paleogeography of the central and southern Appalachians and eastern Midcontinent, U.S.A. In N. J. McMillan, A. F. Embry, and D. J. Glass, eds. *Devonian of the World,* pp. 277–301. Calgary, Alberta: Canadian Society of Petroleum Geologists Mem. 14, vol. I, Regional Syntheses.

Sevon, W. D., and D. L. Woodrow. 1985. Middle and Upper Devonian stratigraphy within the Appalachian basin. In D. L. Woodrow and W. D. Sevon, eds. *The Catskill Delta.* Geological Society of America Spec. Pap. 201: 1–13.

Seward, A. C. 1932. Fossil plants from the Bokkeveld and Witteberg Series of Cape Colony. *Q. J. Geol. Soc. Lond.* 88: 358–369.

Shaikin, I. M. 1987. Calcareous algae: Charophyta. In V. N. Dubatolov, ed. *Fossil Calcareous Algae: Morphology, Classification and Methods of Study.* Trudy Instituta Geologii i Geofiziki (Novosibirsk), vol. 674: 140–160. Akad. Nauk. SSSR, Sibirskoye Otdeleniye Inst. Geol. and Geofizik. Novosibirsk, USSR. (in Russian).

Shaw, G. 1971. The chemistry of sporopollenin. In J. Brooks, R. R. Grant, M. D. Muir, P. van Gijzel, and G. Shaw, eds. *Sporopollenin,* pp. 305–348. London: Academic Press.

Shear, W. A. 1990. Silurian-Devonian terrestrial arthropods. In D. G. Mikulic and S. J. Culver, eds. *Arthropod Paleobiology: Short Courses in Paleontology.* Knoxville, TN: University of Tennessee, Paleontological Society.

———. 1991. The early development of terrestrial ecosystems. *Nature* 351: 283–289.

———. 1992. End of the "Uniramia" taxon. *Nature* 359: 477–478.

———. 1998. The fossil record and evolution of the Myriapoda. In R. A. Fortey and R. H. Thomas, eds. *Arthropod Relationships.* Syst. Assoc. Spec. vol. ser. 55: 211–219.

———. 2000. *Gigantocharinus szatmaryi,* a new trigonotarbid arachnid from the Late Devonian of North America (Arachnida, Trigonotarbida). *J. Paleontol.* 74: 25–31.

Shear, W. A., and P. M. Bonamo. 1988. *Devonobiomorpha,* a new order of centipeds (Chilopoda) from the Middle Devonian of Gilboa, New York, USA, and the phylogeny of chilopod orders. *Am. Mus. Novitates* 2927: 1–30.

Shear, W. A., P. M. Bonamo, J. D. Grierson, W. D. I. Rolfe, E. L. Smith, and R. A. Norton. 1984. Early land animals in North America. *Science* 224: 492–494.

Shear, W. A., P. G. Gensel, and A. J. Jeram. 1996. Fossils of large terrestrial arthropods from the Lower Devonian of Canada. *Nature* 384: 555–557.

Shear, W. A., A. J. Jeram, and P. A. Selden. 1998. Centiped legs (Arthropoda, Chilopoda, Scutigeromorpha) from the Silurian and Devonian of Britain and the Devonian of North America. *Am. Mus. Novitates* 3231: 1–16.

Shear, W. A., and J. Kukalová-Peck. 1990. The ecology of Paleozoic terrestrial arthropods: The fossil evidence. *Can. J. Zool.* 68: 1807–1834.

Shear, W. A., J. Palmer, J. Coddington, and P. M. Bonamo. 1989. A Devonian spinneret: Early evidence of spiders and silk use. *Science* 246: 479–481.

Shear, W. A., W. Schawaller, and P. M. Bonamo. 1989. Palaeozoic record of pseudoscorpions. *Nature* 341: 527–529.

Shear, W. A., and P. A. Selden. 1995. *Eoarthropleura* from the Silurian of Britain and the Devonian of North America. *Neues Jahrb. Geol. Paläontol. Abh.* 196: 347–375.

Shear, W. A., P. A. Selden, W. D. I. Rolfe, P. M. Bonamo, and J. D. Grierson. 1987. New terrestrial arachnids from the Devonian of Gilboa, New York (Arachnida, Trigonotarbida). *Am. Mus. Novitates* 2901: 1–74.

Sherwood-Pike, M. A., and J. Gray. 1985. Silurian fungal remains: Probable records of the Class Ascomycetes. *Lethaia* 18: 1–20.

Shirley, B. W. 1996. Flavonoid biosynthesis: "New" functions for an "old" pathway. *Trends Plant Sci.* 1: 377–382.

Shoemaker, E. M., R. F. Wolfe, and C. S. Shoemaker. 1990. Asteroid and comet flux in the neighborhood of Earth. In V. L. Sharpton and P. D. Ward, eds. *Global Catastrophes in Earth History,* pp. 155–170. Boulder, CO: Geological Society of America, Spec. Pap. 247.

Shukla, J., and Y. Mintz. 1982. Influence of land-surface evapotranspiration on the Earth's climate. *Science* 215: 1498–1500.

Shukla, J., C. Nobre, and P. Sellers. 1990. Amazon deforestation and climate change. *Science* 247: 1322–1325.

Simakov, K. V. 1993. Biochronological aspects of the Devonian-Carboniferous crisis in the regions of the former USSR. *Palaeogeog. Palaeoclim. Palaeoecol.* 104: 127–137.

Simpson, G. G. 1953. *The Major Features of Evolution.* New York: Simon and Schuster.

Singer, M. J., and D. N. Munns. 1991. *Soils: An Introduction*, 2nd ed. New York: Macmillan.

Skog, J. E., and H. P. Banks. 1973. *Ibyka amphikoma*, gen. et sp. n., a new protoarticulate precursor from the late Middle Devonian of New York State. *Am. J. Bot.* 60: 366–380.

Snigirevskaya, N. S. 1984a. On the field technique of fossil wood collection in connection with the problem of the reconstruction of archaeopterids. *Botanicheskii Zhurnal (Academy of Sciences, USSR)* 69: 705–710 (in Russian with English summary).

———. 1984b. Root systems of *Archaeopteris*, Upper Devonian, Donbass. *Ann. J. Paleontol. Soc. Acad. Sci. USSR* (Leningrad) 27: 28–41. (English translation by O. A. Puretskaya).

———. 1988a. The Late Devonian—The time of the appearance of forests as the natural phenomenon. In *Contributed Papers: The Formation and Evolution of the Continental Biotas: 31st Session of the All-Union Palaeontological Society*, 50: 115–124 (in Russian).

———. 1988b. The origin and evolution of continental biota. *Science* (Leningrad) 31: 115–124. (English translation by O. A. Puretskaya.)

———. 1995. Archaeopterids and their role in the land plant cover evolution. *Botanicheskii Zhournal (Academy of Sciences, USSR)* 80: 70–75 (in Russian.)

Solomon, S. T., and G. M. Walkden. 1985. The application of cathodoluminescence to interpreting the diagenesis of an ancient calcrete profile. *Sedimentology* 32: 877–896.

Somero, G. N. 1992. Adapting to water stress: Convergence on common solutions. In G. N. Somero, C. B. Osmond, and C. L. Bolis, eds. *Water and Life, Comparative Analysis of Water Relationships at the Organismic, Cellular, and Molecular Levels*, pp. 3–18. Berlin: Springer-Verlag.

Southworth, D. 1974. Solubility of pollen exines. *Am. J. Bot.* 61: 36–44.

———. 1990. Exine biochemistry. In S. Blackmore and R. B. Knox, eds. *Microspores: Evolution and Ontogeny*, pp. 193–212. London: Academic Press.

Speck, T., and D. Vogellehner. 1988. Biophysikalische Untersuchungen zur Mechanostabilität Verschiedener Stelentypen und zur Art des Festigungssystems Früher "Gefässlandpflanzen." *Palaeontographica B* 210: 91–126.

———. 1994. Devonische Landpflanzen mit und ohne Hypodermales Sterom—Eine Biomechanische Analyse mit Überlegungen zur Frühevolution des Leit- und Festigungssystems. *Palaeontographica B* 233: 157–227.

Spicer, R. A. 1989. The formation and interpretation of plant fossil assemblages. *Adv. Bot. Res.* 16: 96–191.

Stafford, H. A. 1991. Flavonoid evolution: An enzymatic approach. *Plant Physiol.* 96: 680–685.

Stallard, R. F. 1985. River chemistry, geology, geomorphology, and soils in the Amazon and Orinoco basins. In J. I. Drever, ed. *The Chemistry of Weathering*, pp. 293–316. Dordrecht: Reidel.

Stanley, S. M. 1988. Paleozoic mass extinctions: Shared patterns suggest global cooling as a common cause. *Am. J. Sci.* 288: 334–352.

Staplin, F. L. 1961. Reef-controlled distribution of Devonian microplankton in Alberta. *Palaeontology* 4: 392–424.

Starke, T., and J. P. Gogarten. 1993. A conserved intron in the V-ATPase A subunit genes of plants and algae. *FEBS Letters* 315: 252–258.

Stearn, C. W. 1987. Effect of the Frasnian-Famennian extinction event on the stromatoporoids. *Geology* 15: 677–679.

Steemans, P. 1989. Etude palynostratigraphique du Devonien Inférieur dans l'ouest de l'Europe. *Mem. Servir Explication Cartes Geol. Min. Belg.* 27: 453.

———. 1995. Silurian and Lower Emsian spores in Saudi Arabia, *Rev. Palaeobot. Palynol.* 89: 91–104.

Steemans, P., and P. Gerrienne. 1984. La micro- et macroflore du Gedinnian de la Gileppe, Synclinorium de la Vesdre, Belgique. *Ann. Soc. Géol. Belg.* 107: 51–71.

Steemans, P., A. Le Herisse, and N. Bozdogan. 1996. Ordovician and Silurian cryptospores and miospores from southeastern Turkey. *Rev. Palaeobot. Palynol.* 93: 35–76.

Stein, W. E. 1981. Reinvestigation of *Arachnoxylon kopfii* from the Middle Devonian of New York State, U.S.A. *Palaeontographica B* 147: 90–117.

———. 1982. *Iridopteris eriensis* from the Middle Devonian with systematics of apparently related taxa. *Bot. Gaz.* 143: 401–416.

———. 1993. Modeling the evolution of stelar architecture in vascular plants. *Int. J. Plant Sci.* 154: 229–263.

Stein, W. E., and C. B. Beck. 1983. *Triloboxylon arnoldii* from the Middle Devonian of western New York. *Contrib. Mus. Palaeontol. University Mich.* 26: 257–288.

Stein, W. E., G. E. Harmon, and F. M. Hueber. 1993. *Spongiophyton* from the Lower Devonian of North America reinterpreted as a lichen. *Am. J. Bot. Suppl.* 80: 93.

Stein, W. E., and F. M. Hueber. 1989. The anatomy of *Pseudosporochnus: P. hueberi* from the Devonian of New York. *Rev. Palaeobot. Palynol.* 60: 311–359.

Stein, W. E., D. C. Wight, and C. B. Beck. 1983. *Arachnoxylon* from the Middle Devonian of southwestern Virginia. *Can. J. Bot.* 61: 1283–1299.

———. 1984. Possible alternatives for the origin of Sphenopsida. *Syst. Bot.* 9: 102–118.

Stepanov, S. A. 1975. Phytostratigraphy of the key sections of the Devonian of the marginal regions of the Kuznetsk Basin. *Trans. Siber. Inst. Geol. Geophys. Min. Res.* 211: 1–150.

Stephenson, J., and A. C. Scott. 1992. The geological history of insect-related plant damage. *Terra Nova* 4: 542–552.

Stewart, W. N., and G. W. Rothwell. 1993. *Palaeobotany and the Evolution of Plants*, 2nd ed. New York: Cambridge University Press.

Stockmans, F. 1948. Végétaux de Dévonian supérieur de la Belgique. *Mem. Mus. R. Hist. Nat. Belg.* 110: 1–85.

———. 1968. Végétaux Mésodévoniens récoltés aux confins du Massif du Brabant (Belgique). *Mem. Mus. R. Hist. Nat. Belg.* 159: 1–49.

Størmer, L. 1970. Arthropods from the Lower Devonian (Lower Emsian) of Alken-an-der-Mosel, Germany. Part 1. Arachnida. *Senckenb. Lethaea* 51: 335–369.

———. 1972. Arthropods from the Lower Devonian (Lower Emsian) of Alken an der Mosel, Germany. Part 2: Xiphosura. *Senckenb. Lethaea* 53: 1–29.

———. 1973. Arthropods from the Lower Devonian (Lower Emsian) of Alken-an-der-Mosel, Germany. Part 3: Eurypterida, Hughmilleriidae. *Senckenb. Lethaea* 54: 119–205.

———. 1974. Arthropods from the Lower Devonian (Lower Emsian) of Alken-an-der-Mosel, Germany. Part 4: Eurypterida, Drepanopteridae, and other groups. *Senckenb. Lethaea* 54: 359–451.

———. 1976. Arthropods from the Lower Devonian (Lower Emsian) of Alken-an-der-Mosel, Germany. Part 5. Myriapoda and additional forms, with general remarks on fauna and problems regarding invasion of land by arthropods. *Senckenb. Lethaea* 57: 87–183.

Steel, M. 1964. Une association de spores du Givétian inférieur de la Vesdre, a Goé (Belgique). *Ann. Soc. Géol. Belg.* 87: 2–30.

———. 1986. Miospore contribution to the Upper Famennian-Strunian event stratigraphy. *Ann. Soc. Géol. Belg.* 109: 75–92.

———. 1992. Climatic impact on Famennian miospore distribution. In *Fifth International Conference on Global Bioevents (IGCP 216), Göttingen, Germany, Abstracts*, pp. 108–109.

Steel, M., M. Fairon-Demaret, P. Gerrienne, S. Loboziak, and P. Steemans. 1990. Lower and Middle Devonian miospore-based stratigraphy in Libya and its relation to the megafloras and faunas. *Rev. Palaeobot. Palynol.* 66: 229–242.

Steel, M., K. Higgs, S. Loboziak, W. Riegel, and P. Steemans. 1987. Spore stratigraphy and correlation with faunas and floras in the type marine Devonian of the Ardenne-Rhenish regions. *Rev. Palaeobot. Palynol.* 50: 211–229.

Steel, M., and S. E. Scheckler. 1990. Miospore lateral distribution in the Upper Famennian alluvial-lagoonal to tidal facies from eastern United States and Belgium. *Rev. Palaeobot. Palynol.* 64: 315–324.

Strother, P. K. 1988. New species of *Nematothallus* from the Silurian Bloomsburg Formation of Pennsylvania. *J. Paleontol.* 62: 967–982.

———. 1991. A classification schema for the cryptospores. *Palynology* 15: 219–236.

———. 1993. Clarification of the genus *Nematothallus* Lang. *J. Paleontol.* 67: 1090–1094.

———. 1997. Acritarchs. In J. Jansonius and D. C. McGregor, eds. *Palynology: Principles and Applications*, vol. 1, pp. 81–106. American Association of Stratigraphic Palynologists Foundation.

Strother, P. K., S. Al-Hajri, and A. Traverse. 1996. New evidence for land plants from the lower Middle Ordovician of Saudi Arabia. *Geology* 24: 55–58.

Strother, P. K., and A. Traverse. 1979. Plant microfossils from Llandoverian and Wenlockian rocks of Pennsylvania. *Palynology* 3: 1–21.

Stubblefield, S. P., and T. N. Taylor. 1988. Tansley Review no. 12. Recent advances in palaeomycology. *New Phytol.* 108: 3–25.

Stubblefield, S. P., T. N. Taylor, and C. B. Beck. 1985. Studies of Paleozoic fungi. IV. Wood-decaying fungi in *Callixylon newberryi* from the Upper Devonian. *Am. J. Bot.* 72: 1765–1774.

Sud, Y. C., W. C. Chao, and G. K. Walker. 1993. Dependence of rainfall on vegetation: Theoretical considerations, simulation experiments, observations, and inferences from simulated atmospheric soundings. *J. Arid Environ.* 25: 5–18.

Summerfield, M. A. 1991. *Global Geomorphology*. Singapore: Longman.

Surek, B., U. Beemelmanns, M. Melkonian, and D. Bhattacharya. 1994. Ribosomal RNA sequence comparisons demonstrate an evolutionary relationsip between Zygnematales and charophytes. *Plant Syst. Evol.* 191: 171–181.

Sussman, A. S. 1965a. Physiology of dormancy and germination in the propagules of cryptogamic plants. In W. Ruhland, ed. *Encyclopedia of Plant Physiology*, vol. XV, Part 2, pp. 931–1025. Berlin: Springer-Verlag.

———. 1965b. Longevity and resistance of the propagules of bryophytes and pteridophytes. In W. Ruhland, ed. *Encyclopedia of Plant Physiology*, vol. XV, Part 2, pp. 1086–1093. Berlin: Springer-Verlag.

Swain, T., and G. A. Cooper-Driver. 1981. Biochemical evolution of early land plants. In K. J. Niklas, ed. *Paleobotany, Paleoecology, and Evolution Part* 1, pp. 103–123. New York: Praeger.

Swift, M. J., O. W. Heal, and J. M. Anderson. 1979. *Decomposition in Terrestrial Ecosystems*. Oxford: Blackwell Scientific.

Sym, S. D., and R. N. Pienaar. 1993. The class Prasinophyceae. *Progr. Phycol. Res.* 9: 281–376.

Szabo, I. M., E. G. A. Nasser, B. Striganova, Y. R. Rakhmo, K. Jáger, and M. Heydrich. 1992. Interactions among millipedes (Diplopoda) and their intestinal bacteria. In E. Meyer, K. Thaler, and W. Schedl, eds. *Advances in Myriapodology*, pp. 289–296. Proceedings of the 8th International Congress of Myriapodology, Innsbruck, 1990. *Berichte des Naturwissenschaftlich-Medizinischen*

Vereins in Innsbruck, Suppl. 10. Universitätsverlag Wagner, Innsbruck.

Sze, H. C. 1941. A new occurrence of the oldest land plants from the Chaotung district, Yunnan. *Bull. Geol. Soc. China* 21: 107–110.

Sztein, A. E., J. D. Cohen, J. P. Slovin, and T. J. Cooke. 1995. Auxin metabolism in representative land plants. *Am. J. Bot.* 82: 1514–1521.

Tajovsky, K. 1992. Feeding biology of the millipede *Glomeris hexasticha* (Glomeridae, Diplopoda). In E. Meyer, K. Thaler, and W. Schedl, eds. *Advances in Myriapodology*, pp. 305–311. Proceedings of the 8th International Congress of Myriapodology, Innsbruck, 1990. *Berichte des Naturwissenschaftlich-Medizinischen Vereins in Innsbruck*, Suppl. 10. Universitätsverlag Wagner, Innsbruck.

Takeda, R., J. Hasegawa, and M. Shiozaki. 1990. The first isolation of lignans, megacerotonic acid and anthocerotonic acid from non-vascular plants, anthocerotae (hornworts). *Tetrahedron Letters* 31: 4159–4162.

Takiguchi, Y., R. Imaichi, and M. Kato. 1997. Cell division patterns in the apices of subterranean axes and aerial shoot of *Psilotum nudum* (Psilotaceae): Morphological and phylogenetic implications for the subterranean axis. *Am. J. Bot.* 84: 588–596.

Talent, J. A., R. Mawson, A. S. Andrew, P. J. Hamilton, and D. J. Whitford. 1993. Middle Paleozoic extinction events: Faunal and isotopic data. *Palaeogeog. Palaeoclim. Palaeoecol.* 104: 139–152.

Tappan, H. 1980. *The Paleobiology of Plant Protists.* San Francisco: W. H. Freeman.

Tasch, P. 1957. Flora and fauna of the Rhynie Chert: A palaeoecological reevaluation of the published evidence. *Univ. Wichita Bull.* 32: 3–24.

Taylor, R. M., and W. A. Foster. 1994. Spider nectarivory (abstract). *Am. Arachnol.* 50: 10.

Taylor, T. N. 1990. Fungal associations in the terrestrial paleoecosystem. *Trends Ecol. Evol.* 5: 21–25.

Taylor, T. N., H. Hass, W. Remy, and H. Kerp. 1995. The oldest fossil lichen. *Nature* 378: 244.

Taylor, T. N., and J. M. Osborn. 1996. The importance of fungi in shaping the ecosystem. *Rev. Palaeobot. Palynol.* 90: 249–262.

Taylor, T. N., W. Remy, and H. Hass. 1992a. Fungi from the Lower Devonian Rhynie Chert: Chytridiomycetes. *Am. J. Bot.* 79: 1233–1241.

———. 1992b. Parasitism in a 400-million-year-old green alga. *Nature* 357: 493–494.

———. 1994. *Allomyces* in the Devonian. *Nature* 367: 601.

Taylor, T. N., W. Remy, H. Hass, and H. Kerp. 1994. 400 Million year old vesicular arbuscular mycorrhizae. *Proc. Nat. Acad. Sci.* 91: 11841–11843.

———. 1995. Fossil arbuscular mycorrhizae from the early Devonian. *Mycologia* 87: 560–573.

Taylor, T. N., and S. E. Scheckler. 1996. Devonian spore ultrastructure: *Rhabdosporites*. *Rev. Palaeobot. Palynol.* 93: 147–158.

Taylor, T. N., and E. L. Taylor. 1993. *The Biology and Evolution of Fossil Plants.* Englewood Cliffs, NJ: Prentice Hall.

Taylor, W. A. 1995a. Spores in earliest land plants. *Nature* 373: 391–392.

———. 1995b. Ultrastructure of *Tetrahedraletes medinensis* (Strother and Traverse) Wellman and Richardson, from the Upper Ordovician of southern Ohio. *Rev. Palaeobot. Palynol.* 85: 183–187.

———. 1996. Ultrastructure of lower Paleozoic dyads from southern Ohio. *Rev. Palaeobot. Palynol.* 92: 269–279.

———. 1997. Ultrastructure of lower Paleozoic dyads from southern Ohio. II: *Dyadospora murusattenuata*, functional and evolutionary considerations. *Rev. Palaeobot. Palynol.* 97: 1–8.

Tegelaar, E. W., H. Kerp, H. Visscher, P. A. Schenck, and J. W. de Leeuw. 1991. Bias of the paleobotanical record as a consequence of variations in the chemical composition of higher vascular plant cuticles. *Paleobiology* 17: 133–144.

Telford, P. G. 1988. Devonian stratigraphy of the Moose River Basin, James Bay lowland, Ontario, Canada. In N. J. McMillan, A. J. Embry, and D. J. Glass, eds. *Devonian of the World*, vol. 1: *Regional Syntheses*, pp. 123–132. Calgary, Alberta: Canadian Society of Petroleum Geologists, Mem. 14.

Tesakov, A. S., and A. S. Alekseev. 1992. Myriapod-like arthropods from the Lower Devonian of Kazakhstan. *Paleont. J.* 26: 18–23.

Thomas, B. A., and R. A. Spicer. 1987 *The Evolution and Palaeobiology of Land Plants.* London: Croom Helm.

Thompson, A. M. 1970. Lithofacies and formation nomenclature in upper Ordovician stratigraphy, central Appalachians. *Geol. Soc. Am. Bull.* 53: 533–538.

Thompson, A. M., and W. D. Sevon. 1982. Comparative sedimentology of Paleozoic clastic wedges in the central Appalachians, U.S.A. Hamilton, Ontario: *Eleventh International Congress on Sedimentology, Excursion 19B, Guidebook.*

Thompson, J. B., and C. R. Newton. 1988. Late Devonian extinction and episodic climatic cooling or warming? In N. J. McMillan, A. J. Embry, and D. J. Glass, eds. *Devonian of the World*, vol. 3: *Paleontology, Paleoecology and Biostratigraphy*, pp. 29–34. Calgary, Alberta: Canadian Society of Petroleum Geologists, Mem. 14.

Thomson, K. S. 1991. Where did Tetrapods come from? *Am. Sci.* 79: 488–490.

Thorne, C. R. 1990. Effects of vegetation on riverbank erosion and stability. In J. B. Thornes, ed. *Vegetation and Erosion*, pp. 125–144. New York: Wiley and Sons.

Tiffney, B. H. 1981. Diversity and major events in the evolution of land plants. In K. J. Niklas, ed. *Paleobotany, Paleoecology and Evolution*, pp. 193–230. New York: Praeger.

Tiffney, B. H., and K. J. Niklas. 1985. Clonal growth in land plants: A paleobotanical perspective. In J. B. C. Jackson, L. W. Buss, and R. E. Cook, eds. *Population Biology and Evolution of Clonal Organisms*, pp. 35–66. New Haven, CT: Yale University Press.

Tims, J. D., and T. C. Chambers. 1984. Rhyniophytina and Trimerophytina from the early land flora of Victoria, Australia. *Palaeontology* 27 (2): 265–279.

Toro, M., O. Chamon, R. Salguero, and C. Vargas. 1997. Las plantas de la Formacion Kirusillas (Silurico) en la region de La Angostura Departamento de Cochabamba. *Mem. del XII Congreso Geol. de Bolivia—Tarija, Bolivia*, pp. 523–529.

Trägårdh, I. 1909. *Speleorchestes*, a new genus of saltatorial Trombidiidae, which lives in termites' and ants' nests. *Ark. Zool.* 6(2): 1–14.

Trant, C. A., and P. G. Gensel. 1985. Branching in *Psilophyton*: A new species from the Lower Devonian of New Brunswick, Canada. *Am. J. Bot.* 72: 1256–1273.

Trewin, N. H. 1986. Palaeoecology and sedimentology of the Achanarras fish bed of the Middle Old Red Sandstone, Scotland. *Trans. R. Soc. Edinb.* (*Earth Sciences*) 77: 21–46.

———. 1994. Depositional environment and preservation of biota in the Lower Devonian hot-springs of Rhynie, Aberdeenshire, Scotland. *Trans. R. Soc. Edinb.* (*Earth Sciences*) 84: 433–442.

———. 1996. The Rhynie cherts: An Early Devonian ecosystem preserved by hydrothermal activity. Evolution of hydrothermal ecosystems on Earth (and Mars?), pp. 131–149. *Ciba Foundation Symposium*, 202. Chichester: Wiley.

Trewin, N. H., and R. G. Davidson. 1996. An Early Devonian lake and its associated biota in the Midland Valley of Scotland. *Trans. R. Soc. Edinb.* (*Earth Sciences*) 86: 233–246.

Trewin, N. H., and K. J. McNamara. 1995. Arthropods invade the land: Trace fossils and palaeoenvironments of the Tumblagooda Sandstone (?Late Silurian) of Kalbarri, Western Australia. *Trans. R. Soc. Edinb.* (*Earth Sciences*) 85: 177–210.

Trewin, N. H., and C. M. Rice. 1992. Stratigraphy and sedimentology of the Devonian Rhynie Chert locality. *Scott. J. Geol.* 28: 37–47.

Trivett, M. L. 1993. An architectural analysis of *Archaeopteris*, a fossil tree with pseudomonopodial and opportunistic adventitious growth. *Bot. J. Linn. Soc.* 111: 301–329.

Turner, R. E., and N. N. Rabalais. 1994. Coastal eutrophication near the Mississippi river delta. *Nature* 368: 619–621.

Urey, H. C. 1952. *The Planets: Their Origin and Development*. New Haven, CT: Yale University Press.

Van Bergen, P. F., M. E. Collinson, D. G. Briggs, J. W. de Leeuw, A. C. Scott, R. P. Evershed, and P. Finch. 1995. Resistant macromolecules in the fossil record. *Acta Bot. Neerl.* 44: 319–342.

Van Bergen, P. F., A. C. Scott, P. J. Barrie, J. W. deLeeuw, and M. E. Collinson. 1994. The chemical composition of Upper Carboniferous pteridosperm cuticles. *Org. Geochem.* 21: 107–112.

Van der Voo, R. 1988. Paleozoic paleogeography of North America, Gondwana and intervening displaced terranes: Comparisons of paleo-magnetism with paleoclimatology and biogeographical patterns. *Geol. Soc. Am. Bull.* 100: 311–324.

Van der Voo, R., A. N. French, and R. B. French. 1979. A paleomagnetic pole position from the folded Upper Devonian Catskill redbeds, and its tectonic implications. *Geology* 7: 345–348.

Van Winkle-Swift, K. P., and W. L. Rickoll. 1997. The zygospore wall of *Chlamydomonas monoica* (Chlorophyceae): Morphogenesis and evidence for the presence of sporopollenin. *J. Phycol.* 33: 655–665.

Vanstone, S. D. 1991. Early Carboniferous (Mississippian) paleosols from southwest Britain: Influence of climatic change on soil development. *J. Sediment. Petrol.* 61: 445–457.

Vavrdova, M. 1982. Recycled acritarchs in the uppermost Ordovician of Bohemia. *Cas. Mineral. Geol.* 27: 337–345.

———. 1984. Some plant microfossils of possible terrestrial origin from the Ordovician of Central Bohemia. *Vest. Ustred. Ust. Geol.* 59: 165–170.

———. 1988. Further acritarchs and terrestrial plant remains from the Late Ordovician at Hlásná Treban (Czechoslovakia). *Cas. Mineral. Geol.* 33: 1–10.

———. 1989. New acritarchs and miospores from the late Ordovician of Hlásná Treban, Czechoslovakia. *Cas. Mineral. Geol.* 34: 403–420.

———. 1990. Coenobial acritarchs and other palynomorphs from the Arenig/Llanvirn boundary, Prague Basin. *Vest. Ustred. Ust. Geol.* 65: 237–242.

Veizer, J., P. Fritz, and B. Jones. 1986. Geochemistry of brachiopods: Oxygen and carbon isotopic records of Paleozoic oceans. *Geochim. Cosmochim. Acta* 50: 1679–1696.

Velbel, M. A. 1993. Temperature dependence of silicate weathering in nature: How strong a negative feedback on long term accumulation of atmospheric CO_2 and global greenhouse warming? *Geology* 21: 1059–1062.

Velde, B. 1985. *Clay Minerals: A Physico-Chemical Explanation of Their Occurrence* (*Developments in Sedimentology* 40). Amsterdam: Elsevier.

Viles, H., and A. Pentecost. 1994. Problems in assessing the weathering action of lichens with an example of epiliths on sandstone. In D. A. Robinson and R. B. G. Williams, eds. *Rock Weathering and Landform Evolution*, pp. 99–116. New York: Wiley.

Walker, R. G. 1971. Nondeltaic depositional environments in the Catskill clastic wedge (Upper Devonian) of central Pennsylvania. *Geol. Soc. Am. Bull.* 82: 1305–1326.

Walker, R. G., and J. C. Harms. 1971. The "Catskill Delta": A prograding muddy shoreline in central Pennsylvania. *J. Geol.* 79: 381–399.

———. 1975. Shorelines of weak tidal activity: Upper Devonian Catskill Formation, central Pennsylvania. In R. Ginsburg, ed. *Tidal Deposits: A Casebook of Recent Examples and Fossil Counterparts*, pp. 103–108. New York: Springer-Verlag.

Wallace, M. W., R. R. Keays, and V. A. Gostin. 1991. Stromatolitic iron oxides: Evidence that sea-level changes can cause sedimentary iridium anomalies. *Geology* 19: 551–554.

Walliser, O. H. 1996a. Patterns and causes of global events. In O. H. Walliser, ed. *Global Events and Event Stratigraphy in the Phanerozoic*, pp. 7–19. Berlin: Springer.

———. 1996b. Global events in the Devonian and Carboniferous. In O. H. Walliser, ed. *Global Events and Event Stratigraphy in the Phanerozoic*, pp. 225–250. Berlin: Springer.

Walton, J. 1964. On the morphology of *Zosterophyllum* and other early Devonian plants. *Phytomorphology* 15: 155–160.

Wang, K. 1992. Glassy microspherules (microtektites) from an Upper Devonian limestone. *Science* 256: 1547–1550.

Wang, K., M. Attrep, Jr., and C. J. Orth. 1993. Global iridium anomaly, mass extinction, and redox changes at the Devonian-Carboniferous boundary. *Geology* 21: 1071–1074.

Wang, K., and B. D. E. Chatterton. 1993. Microspherules in Devonian sediments: Origins, geological significance, and contamination problems. *Can. J. Earth Sci.* 30: 1660–1667.

Wang, K., H. H. J. Geldsetzer, and B. D. E. Chatterton. 1994. A Late Devonian extraterrestrial impact and extinction in eastern Gondwana: Geochemical, sedimentological, and faunal evidence. In B. O. Dressler, R. A. F. Grieve, and V. L. Sharpton, eds. *Large Meteorite Impacts and Planetary Evolution*, pp. 111–120. Boulder, CO: Geological Society of America, Spec. Pap. 293.

Wang, K., H. H. J. Geldsetzer, W. D. Goodfellow, and H. R. Krouse. 1996. Carbon and sulfur isotope anomalies across the Frasnian-Famennian extinction boundary, Alberta, Canada. *Geology* 24: 187–191.

Wang, K., C. J. Orth, M. Attrep, Jr., B. D. E. Chatterton, H. Hou, and H. H. J. Geldsetzer. 1991. Geochemical evidence for a catastrophic biotic event at the Frasnian/Famennian boundary in south China. *Geology* 19: 776–779.

Wang, Y. 1994. Lower Devonian miospores from Gumu in the Wenshan District, Southeastern Yunnan. *Acta Micropalaeontol. Sin.* 11: 319–332.

Wang, Y., and C.-Y. Cai. 1996. Further observation of *Stachyophyton yunnanense* Geng from the Posongchong Formation (Siegenian) of SE Yunnan, China. *Acta Palaeontol. Sin.* 35: 99–108.

Wang, Y., J. Li, and R. Wang. 1997. Latest Ordovician cryptospores from southern Xinjiang, China. *Rev. Palaeobot. Palynol.* 99: 61–74.

Wang, Z., and B.-Y. Geng. 1997. A new Middle Devonian plant: *Metacladophyton tetrapetalum* gen. et sp. nov. *Palaeontographica B* 243: 85–102.

Waterman, P. G., and S. Mole. 1994. *Analysis of Phenolic Plant Metabolites*. Oxford: Blackwell Scientific.

Waters, D. A., and R. L. Chapman. 1996. Molecular phylogenetics and the evolution of green algae and land plants. In B. R. Chaudhary and S. B. Agrawal, eds. *Cytology, Genetics and Molecular Biology of Algae*, pp. 337–349. Amsterdam: SPB Academic.

Watson, A. 1992. Desert soils. In I. P. Martini and W. Chesworth, eds. *Weathering, Soils and Paleosols*, pp. 225–260. Amsterdam: Elsevier.

Weaver, C. E. 1967. Potassium, illite and the ocean. *Geochim. Cosmochim. Acta* 31: 2181–2196.

———. 1989. *Clays, Muds, and Shales*. Amsterdam: Elsevier.

Weaver, C. E., and K. C. Beck. 1977. *Miocene of the S.E. United States: A model for Chemical Sedimentation in a Peri-Marine Environment*. Amsterdam: Elsevier.

Weis, A. E., and M. R. Berenbaum. 1989. Herbivorous insects and green plants. In W. G. Abrahamson, ed. *Plant-Animal Interactions*. New York: McGraw-Hill.

Wellman, C. H. 1993a. A Lower Devonian sporomorph assemblage from the Midland Valley of Scotland. *Trans. R. Soc. Edinb. (Earth Sciences)* 84: 117–136.

———. 1993b. A land plant microfossil assemblage of Mid Silurian age from the Stonehaven Group, Scotland. *J. Micropalaeontol.* 12: 47–66.

———. 1995. "Phytodebris" from Scottish Silurian and Lower Devonian continental deposits. *Rev. Palaeobot. Palynol.* 84: 255–279.

———. 1996. Cryptospores from the type area of the Caradoc Series in southern Britain. *Spec. Pap. Palaeontol.* 55: 103–136.

———. 1999. Sporangia containing *Scylaspora* from the Lower Devonian of the Welsh Borderland. *Palaeontology* 42: 67–81.

Wellman, C. H., D. Edwards, and L. Axe. 1998a. Permanent dyads in sporangia and spore masses from the Lower Devonian of the Welsh Borderland. *Bot. J. Linn. Soc.* 127: 117–147.

———. 1998b. Ultrastructure of laevigate hilate cryptospores in sporangia and spore masses from the Upper Silurian and Lower Devonian of the Welsh Borderland. *Phil. Trans. R. Soc. Lond.* 353: 1983–2004.

Wellman, C. H., K. Habgood, G. Jenkins, and J. B. Richardson. 2000. A new plant assemblage (microfossil and megafossil) from the Lower Old Red Sandstone of the Anglo-Welsh Basin: Its implications for the palaeoecology of early terrestrial ecosystems. *Rev. Palaeobot. Palynol.* 109: 16–96.

Wellman, C. H., and J. B. Richardson. 1993. Terrestrial plant microfossils from Silurian inliers of the Midland Valley of Scotland. *Palaeontology* 36: 155–193.

———. 1996. Sporomorph assemblages from the "Lower Old Red Sandstone" of Lorne, Scotland. *Spec. Pap. Palaeontol.* 55: 41–101.

Wellman, C. H., R. G. Thomas, D. Edwards, and P. Kenrick. 1998. The Cosheston Group (Lower Old Red Sandstone) in southwest Wales: Age, correlation and palaeobotanical significance. *Geol. Mag.* 135: 397–412.

Wendt, J., and Z. Belka. 1991. Age and depositional environment of Upper Devonian (early Frasnian to early Famennian) black shales and limestones (Kellwasser Facies) in the eastern Anti-Atlas, Morocco. *Facies* 25: 51–90.

Weygoldt, P. M. 1969. *The Biology of Pseudoscorpions*. Cambridge: Harvard University Press.

White, D. 1907. A remarkable fossil tree trunk from the Middle Devonian of New York. *New York State Mus. Bull.* 107: 327–340.

Whitford, W. G., and D. W. Freckman. 1988. The role of soil biota in soil processes in the Chihuahuan Desert. In E. E. Whitehead, C. F. Hutchinson, B. N. Timmerman, and R. G. Varady, eds. *Arid Lands Today and Tomorrow*, pp. 1063–1073. Boulder, CO: Westview Press.

Wicander, R., and G. D. Wood. 1997. The use of micro-phytoplankton and chitinozoans for interpreting transgressive/regressive cycles in the Rapid Member of the Cedar Valley Formation (Middle Devonian), Iowa. *Rev. Palaeobot. Palynol.* 98: 125–152.

Wiggins, G. B., R. J. Mackay, and I. M. Smith. 1980. Evolutionary and ecological strategies of animals in annual temporary pools. *Arch. Hydrobiol.* (suppl.) 58: 97–206.

Wight, D. C. 1987. Non-adaptive change in early land plant evolution. *Paleobiology* 13: 208–214.

Wilcox, L. W., P. A. Fuerst, and G. L. Floyd. 1993. Phylogenetic relationships of four charophycean green algae inferred from complete nuclear-encoded small subunit rRNA gene sequences. *Am. J. Bot.* 80: 1028–1033.

Wilde, P., and W. B. N. Berry. 1984. Destabilization of the oceanic density structure and its significance to marine "extinction" events. *Palaeogeog. Palaeoclim. Palaeoecol.* 48: 143–162.

Wilding, L. P., and D. Tessier. 1988. Genesis of Vertisols: Shrink-swell phenomena. In L. P. Wilding and R. Puentes, eds. *Vertisols: Their Distribution, Properties, Classification and Management*, pp. 55–81. College Station, TX: Texas A&M University Printing Center.

Willis, B. J., and J. S. Bridge. 1988. Evolution of Catskill river systems, New York State. In N. J. McMillan, A. F. Embry, and D. J. Glass, eds. *Devonian of the World*, vol. 2: *Sedimentation*, pp. 85–106. Calgary, Alberta: Canadian Society of Petroleum Geologists, Mem. 14.

Wilson, H. W., and W. A. Shear. 2000. Microdecemplicida, a new order of minute arthropleurids from the Devonian of New York State, USA. *Proc. R. Soc. Edinb.* (in press).

Witzke, B. J. 1990. Palaeoclimatic constraints for Palaeozoic palaeolatitudes of Laurentia and Euramerica. In W. S. McKerrow and C. R. Scotese, eds. *Palaeozoic Palaeogeography and Biogeography*, Geol. Soc. Mem. 12: 57–73.

Woodrow, D. L. 1985. Paleogeography, paleoclimate and sedimentary processes of the Late Devonian Catskill Delta. In D. L. Woodrow and W. D. Sevon, eds. *The Catskill Delta*. Geological Society of America Spec. Pap. 201.

Woodrow, D. L., J. M. Dennison, F. R. Ettensohn, W. T. Sevon, and W. T. Kirchgasser. 1988. Middle and Upper Devonian stratigraphy and paleogeography of the central and southern Appalachians and eastern Midcontinent, U.S.A. In N. J. McMillan, A. J. Embry, and D. J. Glass, eds. *Devonian of the World*, vol. 1: *Regional Syntheses*, pp. 277–301. Calgary, Alberta: Canadian Society of Petroleum Geologists, Mem. 14.

Woodrow, D. L., and F. W. Fletcher. 1968. Upper Devonian aestivation tube-casts. *Geol. Soc. Am. Spec. Pap.* 121: 383–384.

Woodrow, D. L., F. W. Fletcher, and W. F. Ahrnsbrak. 1973. Paleogeography and paleoclimate at the deposition sites of the Devonian Catskill and Old Red Facies. *Geol. Soc. Am. Bull.* 84: 3051–3064.

Wright, J. L., L. Quinn, D. E. G. Briggs, S. H. Williams. 1995. A subaerial arthropod trackway from the Upper Silurian Clam Bank Formation of Newfoundland. *Can. J. Earth Sci.* B32: 304–313.

Wright, V. P. 1987. The ecology of two Early Carboniferous paleosols. In J. Miller, A. E. Adams, and V. P. Wright, eds. *European Dinantian Environments*, pp. 345–358. New York: Wiley.

———. 1990. A micromorphological classification of fossil and recent calcic and petrocalcic microstructures. In L. A. Douglas, ed. *Soil Micromorphology: A Basic and Applied Science*, pp. 401–407. Amsterdam: Elsevier.

Wright, V. P., and D. R. Robinson. 1988. Early Carboniferous floodplain deposits from South Wales: A case study of the controls on paleosol development. *J. Geol. Soc. Lond.* 145: 847–857.

Wright, V. P., and M. E. Tucker. 1991. Calcretes. *International Association of Sedimentologists Reprint Series* 2: 1–22. Oxford, U.K.: Blackwell Scientific.

Xu, D.-Y., Z. Yan, Q.-W. Zhang, Z.-D. Shen, Y.-Y. Sun, and L.-F. Ye. 1986. Significance of a $\delta^{13}C$ anomaly near the Devonian/Carboniferous boundary at the Muhua section, South China. *Nature* 321: 854–855.

Yan, Z., H.-F. Hou, and L.-F. Ye. 1993. Carbon and oxygen isotope event markers near the Frasnian-Famennian boundary, Luoxiu section, South China. *Palaeogeog. Palaeoclim. Palaeoecol.* 104: 97–104.

Yancey, P. H., M. E. Clark, S. C. Hand, R. D. Bowlus, and G. N. Somero. 1982. Living with water stress: Evolution of osmolyte systems. *Science* 217: 1214–1222.

Yapp, C. J., and H. Poths. 1992. Ancient atmospheric CO_2 pressures inferred from natural goethites. *Nature* 355: 342–344.

Yapp, C. J., and H. Poths. 1996. Carbon isotopes in continental weathering environments and variations in ancient atmospheric CO_2 pressure. *Earth Planet. Sci. Lett.* 137: 71–82.

Yeakel, L. S. 1962. Tuscarora, Juniata, and Bald Eagle paleocurrents and paleogeography in the central Appalachians. *Geol. Soc. Am. Bull.* 73: 1515–1539.

Yurina, A. L. 1988. The Middle and Upper Devonian floras of northern Eurasia. *Tr. Palaeontol. Inst. Akad. Nauk. SSSR.* (in Russian).

Zanten, B. O. van. 1976. Preliminary report on germination experiments designed to estimate the survival changes of moss spores during aerial transoceanic long-range dispersal in the Southern Hemisphere, with particular reference to New Zealand. *J. Hattori Bot. Lab.* 41: 133–140.

———. 1978a Experimental studies on trans-oceanic long-range dispersal of moss spores in the Southern Hemisphere. *Bryophytorum Bibliotheca* 13: 715–733.

———. 1978b. Experimental studies on trans-ocean long-range dispersal of moss spores in the Southern Hemisphere. *J. Hattori Bot. Lab.* 44: 455–482.

Zanten, B. O. van, and T. Pocs. 1981. Distribution and dispersal of bryophytes. *Adv. Bryol.* 1: 479–560.

Ziegler, A. M., K. S. Hansen, M. E. Johnson, M. A. Kelly,

C. R. Scotese, and R. Van der Voo. 1977. Silurian continental distributions, paleogeography, climatology, and biogeography. In M. W. McElhinney, ed. The past distribution of continents. *Tectonophysics* 40: 13–51.

Ziegler, A. M., C. R. Scotese, W. S. McKerrow, M. E. Johnson, and R. K. Bambach. 1979. Paleozoic paleogeography. *Ann. Rev. Earth Planet. Sci.* 7: 473–502.

Ziegler, W., and C. A. Sandberg. 1984. *Palmatolepis*-based revision of upper part of standard Late Devonian conodont zonation. In D. L. Clark, ed. *Conodont Biofacies and Provincialism*, pp. 179–194. Boulder, CO: Geological Society of America, Spec. Pap. 196.

———. 1990. The Late Devonian standard conodont zonation. *Cour. Forschungsinst. Senckenberg* 121: 1–115.

Zimmer, C. 1995. Coming on to the land. *Discover* 22: 119–127.

INDEX

Abrahamson, W. G., 48
Acacus Formation (Libya), 10
Acanthostega, 40
Acari (mites), 34, 35, 38, 43, 47
acetate-shikimate cycle, 168
acid hydrolysis, 145, 147, 149, 154, 164
acritarchs, 6, 13, 147, 192; at Gaspé Bay, 185, 188, 189, 190, 192, 195, 200–201; and salinity, 200–201
actinomycetes, 50, 243
adaptations, 1, 2; of charophyceans, 146–49, 157; and desiccation tolerance, 144, 145–46, 150; and herbivory, 41, 50; pre-, 146–49, 150, 151, 157; and specialization, 149, 150, 157
Adoketophyton, 105, 113, 115, 117; *subverticillatum*, 109, 117
aestivation burrows, 242, 246
Africa, 121, 230
Aglaophyton, 42, 80, 82, 218; *major*, 52, 53, 79, 81, 86, 90, 164, 240
Aglosperma quadripartita, 219
Alekseev, A. S., 38
algae, 33, 39, 44, 81, 170, 228, 243; and atmospheric CO_2, 176; charophycean, 140, 141, 142–49, 150; chlorophycean, 148; desiccation resistance of, 148–49; and Devonian plant hypothesis, 233, 235; as soil cover, 202; trebouxiophycean, 148; and weathering, 220, 229. *See also* cyanobacteria
algaenans, 148, 149
Algeo, Thomas J., 213–36, 240; Devonian plant hypothesis of, 213–14, 233–35
Algites (Palaeonitella) crani. See Palaeonitella cranii
Alken-an-der-Mosel (Germany), 30, 34, 36, 37, 39
allelochemicals, 48–49, 50
Allt Dhu (Brecon Beacons, Wales), 4, 11
Almond, J. E., 32, 38
amblypygids, 35, 37–38
ammonoids, 227, 228
amphipods, 33
Amplectosporangium, 118; *jiangyouensis*, 118
Anapaulia, 89; *moodyi*, 128
Anderson, L., 34

Andrews, H. N., 92, 124
Aneurophytales, 130, 131
aneurophytes, 89, 136
Aneurophyton, 117, 130, 131, 134, 135
angiosperms, 149, 151, 162, 171
animals: co-occurrence with plants of, 40–41; early terrestrial, 1, 29–51; fossils of, 30, 32, 33, 34, 37, 38, 39, 41, 44; herbivorous, 30, 41, 46–51, 171, 240; interactions with terrestrial plants of, 2, 32, 40–50; mammals, 50; marine, 29, 33, 34; predatory, 159; terrestrial, 30, 33; traces of, 240–42, 247
Anisophyton, 86, 109
annelids, 29, 31, 44
annulatus-sextantii spore assemblage zone, 183
anoxia, marine, 2, 227–36; and Devonian plant hypothesis, 233–34, 235
Antarctica, 31, 168
Anthocerotales, 168
Antrim Shale (Michigan Basin), 228
"Aphantomartids," 37
Apheloria montana, 44
Apiculiretusispora, 199; *arenorugosa*, 195
Appalachian Basin, 226, 228, 231, 237–53
Arabidopsis, 102
arachnids, 34, 36, 37, 43, 47
Arachnoxylon, 128; *minor*, 130
Araneae (spiders), 33, 34, 35, 37, 39, 43
arborescence, 30, 215–17; archaeopterid, 35, 36; and Devonian plant hypothesis, 233, 234, 235; lycopsid, 125, 127, 135, 138, 168, 214, 215, 225; and roots, 217, 244–46, 247
Archaeocalamites, 132
archaeognaths, 35, 40, 44
archaeopterids, 35, 36, 215, 235
Archaeopteris, 30, 120, 130, 238, 246, 248; height of, 138, 245; in plant assemblages, 135, 136, 137; spread of, 215, 217–20, 234
Archaeosigillaria, 125
Archaeosperma arnoldii, 219
Archanodon, 241
Archidesmus, 36; *macnicoli*, 38
Argentina, 10, 11, 12, 13, 27, 230

Printed in the USA
CPSIA information can be obtained
at www.ICGtesting.com
JSHW051458221024
72172JS00012B/105